T0231562

Swarm Intelligence Algorithms

Swarm Intelligence Algorithms
A Tutorial

Edited by
Adam Slowik

CRC Press
Taylor & Francis Group
Boca Raton London New York

CRC Press is an imprint of the
Taylor & Francis Group, an **informa** business

First edition published 2020
by CRC Press
6000 Broken Sound Parkway NW, Suite 300, Boca Raton, FL 33487-2742

and by CRC Press
2 Park Square, Milton Park, Abingdon, Oxon, OX14 4RN

© 2021 Taylor & Francis Group, LLC

CRC Press is an imprint of Taylor & Francis Group, LLC

Library of Congress Cataloging-in-Publication Data

Names: Slowik, Adam, editor.
Title: Swarm intelligence algorithms. A tutorial / edited by Adam Slowik.
Description: First edition. | Boca Raton : Taylor and Francis, 2020. |
Includes bibliographical references and index.
Identifiers: LCCN 2020018736 (print) | LCCN 2020018737 (ebook) |
ISBN 9781138384491 (hardback) | ISBN 9780429422614 (ebook)
Subjects: LCSH: Swarm intelligence. | Algorithms. | Mathematical
optimization.
Classification: LCC Q337.3 .S9246 2020 (print) | LCC Q337.3 (ebook) |
DDC 006.3/824--dc23
LC record available at https://lccn.loc.gov/2020018736
LC ebook record available at https://lccn.loc.gov/2020018737

ISBN: 978-1-138-38449-1 (hbk)
ISBN: 978-0-429-42261-4 (ebk)

Visit the Taylor & Francis Web site at
http://www.taylorandfrancis.com

and the CRC Press Web site at
http://www.crcpress.com

*— To my beloved wife Justyna,
and my beloved son Michal —*

*— My beloved late parents
Bronislawa and Ryszard,
I will always remember you —*

Contents

4 Bat Algorithm 43
Xin-She Yang and Adam Slowik

5 Cat Swarm Optimization 55
Dorin Moldovan, Viorica Chifu, Ioan Salomie, and Adam Slowik

6 Chicken Swarm Optimization 71
Dorin Moldovan and Adam Slowik

Preface

Swarm Intelligence (SI) is one of the areas of artificial intelligence which is being developed dynamically. As a precursor of SI we can mention three algorithms such as Stochastic Diffusion Search (SDS), Ant Colony Optimization (ACO), and Particle Swarm Optimization (PSO). The SDS algorithm was published in 1989 by Bishop. The SDS is the first SI metaheuristic where an agent-based probabilistic global search optimization technique is introduced. The ACO was published in 1992 by Dorigo in his Ph.D. thesis. The main inspiration for ACO was the actions of a real ant colony. The original ACO algorithm is a probabilistic optimization technique useful in discrete optimization for finding the best paths in the graphs. The PSO was published in 1995 by Kennedy et al. The main inspiration for the PSO algorithm was the social behavior of such organisms as birds (a bird flock) or fish (a fish school). The original PSO is a global optimization technique for a continuous domain.

SI is a relatively new branch of artificial intelligence that is used to model the collective intelligent behavior of social swarms in nature. A simple behavior of particular agents and self-organizing interaction among them are observed in nature, e.g., fish shoals, bird flocking, ant colonies. These behaviors were the inspiration for developing "artificial colonies of agents" which are able to solve difficult optimization problems. The concept of Swarm Intelligence can be described as follows:

"The simple behavior of individuals which are relatively simple in structure + interactions between individuals of the swarm over time = very complex collective behavior"

Based on the general concept of Swarm Intelligence many SI algorithms have been developed up until now. The SI algorithms represent the subfamily of nature inspired global optimization techniques. Other subfamilies are physical algorithms (such as simulated annealing, harmony search, and so on), evolutionary algorithms (such as genetic algorithms, genetic programming, and others), and immune algorithms (such as clonal selection algorithms, negative selection algorithms, and so on). The main advantages of the SI algorithms over traditional optimization techniques are as follows: SI algorithms start with a population of potential solutions not from a single point, SI algorithms do not require the derivative objective function, the solutions can cooperate with each other to share knowledge.

The family of SI algorithms is still growing. In the table below you can see the list of 45 algorithms (sorted by the date of their development) selected from the whole family of swarm algorithms. 24 of them (which are marked by gray color) are presented in this book in detail - in tutorial form.

SI Algorithm (Year)	Biological inspiration
Stochastic Diffusion Search (1989)	Tandem calling mechanism employed by one species of ants
Ant Colony Optimization (1992)	Real ant colonies using pheromone as a means of chemical messenger
Particle Swarm Optimization (1995)	Social behavior of bird flocking or fish schooling
Bee System (2001)	Foraging behavior of bee colonies
Bacterial Foraging (2002)	Social foraging behavior of Escherichia coli
Fish-swarm Algorithm (2002)	Fish behaviors such as preying and swarming
Beehive (2004)	Communicative and evaluative methods and procedures of honey bees
Bacterial Colony Chemotaxis (2005)	Bacterium's reaction to chemoattractants
Bee Colony Optimization (2005)	Bee colonies in nature
Bee Swarm Optimization (2005)	Behavior of real bees in nature
Virtual Bees (2005)	Swarm of bees and interactions between them when they find nectar
Cat Swarm (2006)	Behaviors of cats and their skills such as tracing and seeking
Artificial Bee Colony (2007)	Natural foraging behavior of real honey bees
Fast Bacterial Swarming (2008)	Foraging mechanism of Escherichia coli and the swarming pattern of birds
Bumblebees (2009)	Collective behavior of social insects
Cuckoo Search (2009)	Brood parasitic behavior of some cuckoo species
FireFly Algorithm (2009)	Behavior of fireflies and their flashing light (process of bioluminescence)
Glowworm Swarm Optimization (2009)	Luciferin induced glow of a glowworm which is used to attract mates/prey
Artificial Fish School Algorithm (2010)	Fish behaviors such as preying, swarming, following
Bat Algorithm (2010)	Echolocation characteristics of microbats
Cockroach Swarm Optimization (2010)	Social behavior of cockroaches
Hunting Search (2010)	Group hunting of animals such as lions, wolves, and dolphins
Bacterial Colony Optimization (2012)	Five basic behaviors of Escherichia coli bacteria in their whole lifecycle
Blind-Naked Mole-Rats (2012)	Social behavior of the blind naked mole-rats colony
Krill Herd (2012)	Herding behavior of krill individuals
Lion's Algorithm (2012)	Lion's social behavior that aids to keep the mammal strong in the world
Wolf Search (2012)	Wolves search for food and survive by avoiding their enemies
Fruit Fly Optimization (2013)	Behavior of fruit flies
Social Spider Optimization (2013)	Cooperative behavior of social-spiders which interact with each other
Chicken Swarm Optimization (2014)	Behavior of chickens when they search for food
Dispersive Flies Optimisation (2014)	Swarming behavior of flies over food sources
Grey Wolf Optimizer (2014)	Mimics the social dominant structure of the grey wolves pack
Elephant Herding (2015)	Herding behavior of the elephant groups
Monarch Butterfly Optimization (2015)	Migration of monarch butterflies
Crow Search Algorithm (2016)	Intelligent behavior of crows
Dolphin Swarm Algorithm (2016)	Dolphin's echolocation, information exchanges, cooperation
Dynamic Virtual Bats Algorithm (2016)	Bat's ability to manipulate frequency/wavelength of the emitted sound waves
Whale Optimization Algorithm (2016)	Social behavior of humpback whales - the bubble-net hunting strategy
Swarm Dolphin Algorithm (2016)	Social behaviors of dolphins
Artificial Wolf Pack Algorithm (2016)	Social behaviors of the wolf pack in scouting, calling and besieging
Grasshopper Optimisation (2017)	Behavior of grasshopper swarms in nature
Spotted Hyena Optimizer (2017)	Social relationship between spotted hyenas and their collaborative behavior
Salp Swarm Algorithm (2017)	Swarming behaviour of salps when navigating and foraging in oceans
Emperor Penguin Optimizer (2018)	Mimics the huddling behavior of emperor penguins
Seagull Optimization Algorithm (2019)	Migration and attacking behaviors of a seagull in nature

In each chapter, the given algorithm is presented by the pseudo-code, by the source-code in Matlab, and by the source-code in C++ programming language. Also, at the end of each chapter, the step-by-step numerical example is shown. This example presents the particular steps of the described algorithm in very detailed numerical manner. As an editor, I believe that such a way of presenting algorithms will be very helpful in the rapid understanding of their operation, and will give the reader a comprehensive answer to the question: How does the given algorithm work? Also, I would like to note that in the second volume of this book – entitled: *Swarm Intelligence Algorithms:*

Modifications and Applications – the modifications and the real-world engineering applications of each algorithm are presented.

At the end of this short preface, I would very much like to thank all the contributors for their hard work in preparation of the chapters for this book. I also would like to wish all readers enjoyment in reading this book.

Adam Slowik
Department of Electronics and Computer Science
Koszalin University of Technology, Koszalin, Poland

MATLAB® is a registered trademark of The MathWorks, Inc. For product information, please contact:

The MathWorks, Inc.
3 Apple Hill Drive
Natick, MA, 01760-2098 USA
Tel: 508-647-7000
Fax: 508-647-7001
E-mail: info@mathworks.com
Web: www.mathworks.com

Editor

ADAM SLOWIK (IEEE Member 2007; IEEE Senior Member 2012) received the B.Sc. and M.Sc. degrees in computer engineering and electronics in 2001 and the Ph.D. degree with distinction in 2007 from the Department of Electronics and Computer Science, Koszalin University of Technology, Koszalin, Poland. He received the Dr. habil. degree in computer science (intelligent systems) in 2013 from the Department of Mechanical Engineering and Computer Science, Czestochowa University of Technology, Czestochowa, Poland. Since October 2013, he has been an Associate Professor in the Department of Electronics and Computer Science, Koszalin University of Technology. His research interests include soft computing, computational intelligence, and, particularly, bio-inspired optimization algorithms and their engineering applications. He is a reviewer for many international scientific journals. He is an author or coauthor of over 80 refereed articles in international journals, two books, and conference proceedings, including one invited talk. Dr. Slowik is an Associate Editor of the IEEE TRANSACTIONS ON INDUSTRIAL INFORMATICS. He is a member of the program committees of several important international conferences in the area of artificial intelligence and evolutionary computation. He was a recipient of one Best Paper Award (IEEE Conference on Human System Interaction - HSI 2008).

Contributors

Adam Slowik
Department of Electronics and Computer Science
Koszalin University of Technology, Koszalin, Poland
e-mail: aslowik@ie.tu.koszalin.pl, adam.slowik@tu.koszalin.pl

Pushpendra Singh
Department of Electrical Engineering
Govt. Women Engineering College, Ajmer, India
e-mail: pushpendragweca@gmail.com

Nand K. Meena
School of Engineering and Applied Science
Aston University, Birmingham, B4 7ET, United Kingdom
e-mail: nkmeena@ieee.org

Jin Yang
School of Engineering and Applied Science
Aston University, Birmingham, B4 7ET, United Kingdom
e-mail: j.yang8@aston.ac.uk

Bahriye Akay
Department of Computer Engineering
Erciyes University, 38039, Melikgazi, Kayseri, Turkey
e-mail: bahriye@erciyes.edu.tr

Dervis Karaboga
Department of Computer Engineering
Erciyes University, 38039, Melikgazi, Kayseri, Turkey
e-mail: karaboga@erciyes.edu.tr

Sonam Parashar
Department of Electrical Engineering
Malaviya National Institute of Technology, Jaipur, 302017, India
e-mail: sonam_ee@yahoo.com

Neeraj Kanwar
Department of Electrical Engineering
Manipal University Jaipur, Jaipur, India
e-mail: nk12.mnit@gmail.com

Xin-She Yang
School of Science and Technology
Middlesex University, London NW4 4BT, United Kingdom
e-mail: x.yang@mdx.ac.uk

Dorin Moldovan
Department of Computer Science
Technical University of Cluj-Napoca, Romania
e-mail: dorin.moldovan@cs.utcluj.ro

Viorica Chifu
Department of Computer Science
Technical University of Cluj-Napoca, Romania
e-mail: viorica.chifu@cs.utcluj.ro

Ioan Salomie
Department of Computer Science
Technical University of Cluj-Napoca, Romania
e-mail: ioan.salomie@cs.utcluj.ro

Joanna Kwiecien
Department of Automatics and Robotics
AGH University of Science and Technology, Krakow, Poland
e-mail: kwiecien@agh.edu.pl

Ali Osman Topal
Department of Computer Engineering
Epoka University, Tirana, Albania
e-mail: aotopal@epoka.edu.al

Mohammad Majid al-Rifaie
School of Computing and Mathematical Sciences, University of Greenwich
Old Royal Naval College, Park Row, London SE10 9LS, United Kingdom
e-mail: m.alrifaie@greenwich.ac.uk

Krishnanand Kaipa
Department of Mechanical and Aerospace Engineering
Old Dominion University, Norfolk, Virginia, United States
e-mail: kkaipa@odu.edu

Debasish Ghose
Department of Aerospace Engineering
Indian Institute of Science, Bangalore, India
e-mail: dghose@iisc.ac.in

Szymon Lukasik
Faculty of Physics and Applied Computer Science
AGH University of Science and Technology, Krakow, Poland
e-mail: slukasik@agh.edu.pl

Ahmed F. Ali
Department of Computer Science
Suez Canal University, Ismaillia, Egypt
e-mail: ahmed_fouad@ci.suez.edu.eg

Mohamed A. Tawhid
Department of Mathematics and Statistics
Thompson Rivers University, Kamloops, BC, Canada V2C 0C8
e-mail: mtawhid@tru.ca

Ferhat Erdal
Department of Civil Engineering
Akdeniz University, Turkey
e-mail: eferhat@akdeniz.edu.tr

Osman Tunca
Department of Civil Engineering
Karamanoglu Mehmetbey University, Turkey
e-mail: osmantunca@kmu.edu.tr

Ali R. Kashani
Department of Civil Engineering
University of Memphis, Memphis, TN 38152, United States
e-mail: kashani.alireza@ymail.com, akashani@memphis.edu

Charles V. Camp
Department of Civil Engineering
University of Memphis, Memphis, TN 38152, United States
e-mail: cvcamp@memphis.edu

Hamed Tohidi
Department of Civil Engineering
University of Memphis, Memphis, TN 38152, United States
e-mail: htohidi@memphis.edu

Essam H. Houssein
Faculty of Computers and Information
Minia University, Minya, Egypt
e-mail: essam.halim@mu.edu.eg

Ibrahim E. Mohamed
Faculty of Computers and Information
South Valley University, Luxor, Egypt
e-mail: ibrahim.elsayed.ibrahim@gmail.com

Aboul Ella Hassanien
Faculty of Computers and Information
Cairo University, Cairo, Egypt
e-mail: aboitcairo@gmail.com

J. Mark Bishop
Department of Computing
Goldsmiths, University of London, SE14 6NW, United Kingdom
e-mail: m.bishop@gold.ac.uk

Moein Armanfar
Department of Civil Engineering
Arak University, Arak, Iran
e-mail: m.armanfar.ce@gmail.com

1

Ant Colony Optimization

Pushpendra Singh
Department of Electrical Engineering
Govt. Women Engineering College, Ajmer, India

Nand K. Meena
School of Engineering and Applied Science
Aston University, Birmingham, United Kingdom

Jin Yang
School of Engineering and Applied Science
Aston University, Birmingham, United Kingdom

Adam Slowik
Department of Electronics and Computer Science
Koszalin University of Technology, Koszalin, Poland

CONTENTS

1.1 Introduction

Ants are known as social insects (as are bees and termites) because they live and work together in well-structured ant colonies or communities of the same species. The ants can construct small to large sized colonies where population may vary from a few dozens to millions of individuals from different castes of sterility. These colonies are described as superorganisms because of their

1

social behavior. Ants are almost blind insects but effectively search for the food source. It is very interesting to know how this creature, being a simple individual, understands the importance of togetherness in finding the shortest route between their nest and food source [3].

"Many entomologists prefer to view ant colonies and the societies of other social insects as more like superorganisms than communities of individualized organisms"

Robert L. O'Connell

Ant colony optimization (ACO) was introduced in the early nineties. The social and optimal behavior of the ant colony was implemented through artificial ants. In this artificially developed population, an ant is a simple, independent, and asynchronous agent that collaborate to find an optimal solution for complex real-life optimization problems. ACO is a population based meta-heuristic optimization technique biologically inspired from the foraging behavior of an insects horde, i.e., ant colony. It is broadly defined as the class of model based search (MBS) techniques [4]. The MBS methods are quite effective to find the optimal solution for combinatorial optimization problems. On the basis of probabilistic modes, the MBS methods are bifurcated into the following classes:

a) algorithms which employ a specified probabilistic model without reforming the model configuration during the run.

b) algorithms which reform the probabilistic model in alternating phases.

The ant colony based techniques fall into the first class. The ACO technique, during the run, updates the parameter values of the probabilistic model in such a manner that it creates a rich probability to produce high quality results over time. This chapter presents the basic fundamentals of ant behaviors and the ACO technique followed by its application.

1.2   Ants' behavior

The ants are social creatures which prefer to live in groups. Ants use sensing clues for seeking the shortest route from their nest to a food source. These insects are capable to adopt the changes in the surrounding environment. In these ant colonies, each ant executes very simple group actions without knowing what other ants are doing; however everybody knows that the outcome is highly social and structured [1, 2]. They can easily find the next shortest route of their trailing aim even if the present one gets corrupted or is obstructed by an obstacle. This phenomenon is presented in Fig. 1.1. From the figure, it can be visualized that, a) the ants are traveling in a straight line to follow the shortest route; b) an obstacle breaks their journey and splits the path; c) ants are again seeking the next possible shortest path; and d) finally, ants are able to ensure the shortest path.

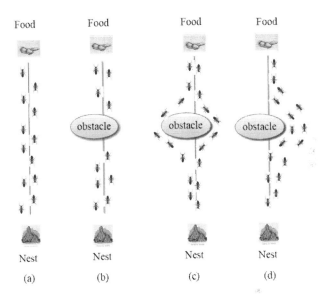

FIGURE 1.1
Traveling behavior of ant colony, a) ants following shortest path, b) an obstacle in ants' path, c) ants again seeking the shortest path, and d) ants finally locating the shortest path.

Ethnologists have found that this skill of ants is due to phenomena called *pheromone trails*. These help ants communicate with their fellows. This information also helps the ant colony to walk on the same path or to make a decision regarding a change of foraging path. Initially, ants randomly explore the area surrounding the nest to search for a food source. After seeking the food origin, ants return to their nest by carrying some food. While returning with food, they deposit an organic compound on the ground known as a *pheromone*. Usually, they deposit more pheromone when returning with food and produce less when searching for a food source. This deposited chemical on the ground guides their colleagues to reach a food source. Ants in the neighborhood presumably prefer to trace the path which has the highest pheromone concentration level. This continuous and indirect form of communication between the fellow ants by the pheromone trail helps the ant colony in seeking the shortest route between their nest and food source. As presented in Fig. 1.1, ants are traveling in a straight line as it clearly provides the shortest possible path between two points, i.e., nest and food source. The foraging and adjacent ants with food are using the same path. It has been assumed that all ants are traveling at normal speed; thus most of the ants will appear at the shortest path in an average time. Therefore, the pheromone concentration will increase at the shortest path as most of the ants follow this path. After some time, the difference in the amount of pheromone levels on these paths will be identified easily by new ants entering into the system. These new ants will prefer to travel on the shortest path which has a rich pheromone level, that is a straight line between nest and food source. In Fig. 1.1 b), the shortest path pheromone trail has been interrupted by an obstacle. The ants are unable to follow the shortest path in this situation therefore seeking a new path, as can be observed from Fig. 1.1 c). They randomly turn left and right seeking the path to reach their destination as they have no clue to follow. After some time, it has been observed from Fig. 1.1 d) that some ants found the shortest routes around the obstacle, and will rapidly accumulate the interrupted pheromone trail as compared to the ants who have taken a longer path. The pheromone concentration level will continuously increase as more ants follow the new shortest path. This increased pheromone concentration will help the new ants to follow the next searched shortest path as explained earlier. However, this chemical has an evaporation tendency. This constructive auto-catalytic process helps the ant colony seek the shortest path to their destination very soon.

1.3   Ant colony algorithm

The ACO technique has been inspired by the foraging behavior of an ant colony, discussed in previous sections. The ant's behavior is mathematical, modeled into some set of equations to find an optimal solution for combinatorial optimization problems. Since its induction, various real-life engineering optimization problems have been solved by using this method. Among these methods, *Ant System* (AS) is the most popular, productive, and oldest method. It was developed by M. Dorigo and is the most preferred operation used in ACO techniques [5, 6, 7, 8]. One of the important characteristics of this process is that at every iteration, the concentration of pheromone is updated for the ants who have participated in constructing the solution in iteration itself. The AS is described as follows: a) all ants are randomly initialized in the starting stage, and b) these m ants are set on the vertex of the build frame and a constant amount of pheromone trail intensity, i.e. $\tau(i,j) = 1 \ \forall \ (i,j) \in Allowed$, is alloted at all the edges, where, *Allowed* represents the set of feasible neighbors of the ant.

Each artificial ant is an agent with following characteristics [5]:

1) it prefers to walk on the highest pheromone probability frame,

2) the already visited edges are prohibited until the circuit is complete,

3) when journey ends, the substance known as the trail is updated at each visited edge.

At every solution building level or step, each ant incrementally adds its part of the substance to the partial building solution frame. Let us assume that kth ant of ith edge, at the tth building step, performs a random walk from ith edge to the next edge of the building j. At each edge, stochastic decisions are taken by ants to choose the next node or edge. These decisions are taken according to the transition probability of one edge to another with respect to the present edge of the ant. The transition probability of the kth ant located at the ith edge, to travel towards the jth edge, can be determined by using random proportional state transition law [5], defined as

$$P_k(i,j) = \begin{cases} \frac{[\tau(i,j)]^\alpha \cdot [\eta(i,j)^\beta]}{\sum_{c_{il} \in N(s^p)} [\tau(i,l)]^\alpha \cdot [\eta(i,l)]^\beta} & \text{if } c_{ij} \in N(s^p) \\ 0 & \text{otherwise} \end{cases} \qquad (1.1)$$

where, $\tau(i,j)$ denotes the pheromone intensity at path (i,j), which is a connection between the edge i and j. Similarly, $\eta(i,j)$ is a heuristic value also known as *visibility* of path (i,j). Its value is generally set as the inverse of connection cost or distance. l is an edge not yet visited by ant k. Suppose $d(i,j)$ is representing the length of path (i,j) then the value of $\eta(i,j)$ can be calculated as $1/d(i,j)$. Furthermore, $N(s^p)$ is a set of feasible components or expedient neighbors of the kth ant on edge i. α and β are the controlling parameters to determine the relative importance of pheromone versus heuristic value where, $\alpha \in (0,1]$ and $\beta \in (0,1]$.

When all the ants of the colony have achieved their solution, the pheromone trail is updated for all the edges by the pheromone global updating rule, as given below.

$$\tau(i,j) \leftarrow \zeta.\tau(i,j) + \sum_{k=1}^{m} \Delta\tau_k(i,j) \qquad (1.2)$$

Here, the coefficient $\zeta \in (0,1]$ is presented in such as way that $(1-\zeta)$ would represent the evaporation of the pheromone trail between two steps/levels (the time taken to complete a cycle). The value of ζ is set to be less than unity to prohibit unlimited pheromone accumulation in the path. Similarly, the $\Delta\tau_k(i,j)$ is known as the quantity of pheromone laid by the kth ant, measured in per unit length of the path (i,j), which is expressed below

$$\Delta\tau_k(i,j) = \begin{cases} \frac{Q}{L_k} & \text{if ant } k \text{ travels through the path } (i,j) \\ 0 & \text{otherwise} \end{cases} \qquad (1.3)$$

where, Q and L_k are representing constant (usually, $Q = 1$) and the total travel length of the kth ant respectively. In this technique, the heuristic value presented in (1.1), $\eta(i, j)$ is usually adopted to favor the cost effective edges of the frame with a large amount of pheromone level, whereas in (1.2) the first term models the evaporated substance. This indicates the lower trail level due to pheromone decaying factors. It has to be noted that the trail for unflattering edges will be evaporated by time. The second term of (1.2) models the reinforcement of the pheromone trail. This will help newly joined ants in the ant colony system to effectively and easily sense the path of higher pheromone concentration. The amount of deposited substance on the path highly depends on the quality of solution achieved so far. The pheromone trail updating rule helps in simulating the change in intensity due to deposition of additional pheromone by new ants and the evaporated amount from the path.

For better understanding, a simple mathematical model is given in Fig. 1.2. It presents the traveling paths of two ants between the four points represented by bold and dashed lines. The starting and ending points of the ants

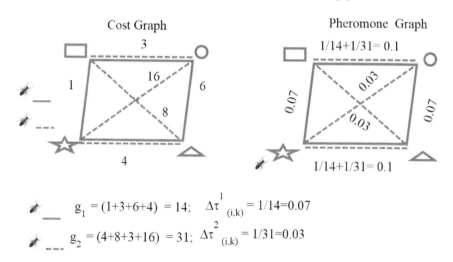

$$g_1 = (1+3+6+4) = 14; \quad \Delta\tau^1_{(i,k)} = 1/14 = 0.07$$

$$g_2 = (4+8+3+16) = 31; \quad \Delta\tau^2_{(i,k)} = 1/31 = 0.03$$

FIGURE 1.2
Demonstration of artificial ant following the shortest path with high pheromone intensity.

are considered at edges represented by *stars*. The cost graph represents the cost involved in traveling between two edges whereas, the pheromone graph presents the calculated pheromone value between two edges. Now, the probabilities of paths are calculated for selecting the shortest path by a new ant which starts its journey from the edge *star*. As per the calculation obtain from the figure, it can be analyzed that a new ant which has to start its journey from the edge *star* will move towards the edge represented by the *triangle* because this path has the highest concentration of the pheromone, i.e., 0.1, as compared to two other paths towards the nodes represented by the *square*, and *circle* with a pheromone concentration of 0.07 and 0.03 respectively. After reaching the *triangle* node, the ant will choose the path towards the *circle* as it has the highest substance concentration, i.e., 0.07. Similarly, it will move towards the *square* edge and then back to *star*. The flowchart of ACO optimization is presented in Fig. 1.3.

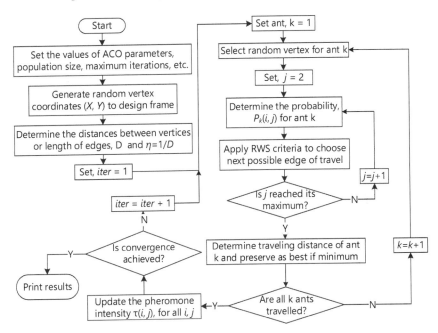

FIGURE 1.3
Flow chart of Ant Colony Optimization.

1.4 Source-code of ACO algorithm in Matlab

The Matlab source codes for fitness function, roulette wheel selection approach, and ACO algorithm are presented in Listings 1.1, 1.2 and 1.3 respectively. Here, the $OF(.)$ is representing the address of objective function.

```
1  % Distance calculation for TSP
2  function [Dist] = FUN_TSP(ant_path,nE,Distance)
3  % Calculation of distance traveled by ant 'k'
4  Dist=0;
5  for t=1:nE
6  Dist = Dist + Distance(ant_path(t),ant_path(t+1));
7  end
```

Listing 1.1
Fitness function for traveling salesman problem.

```
1  % Roulette wheel selection criteria
2  function loc = roulette(fitness)
3  partsum=0; j=1;
4  sumfit = sum(fitness); rn = rand*sumfit;
5  while j<=length(fitness)
6  partsum=partsum + fitness(j);
7  if partsum>=rn                 break;        end
8  j=j+1;
9  end
10 if j>length(fitness)     j=length(fitness); end
11 loc=j;
```

Listing 1.2
Roulette wheel selection (RWS) function.

```
1  % ACO algorithm
2  clc; clear;
3  Max_Iteration = 450;        % Max. no. of iterations
4  ant_pop = 50;               % Ant population size
5  Q = 1;                      % Constant
6  %
7  % Frame design...
8  % Coordinations of frame edges are generated randomly....
9  X = [1,4,10,30,11,23,28,30,15,11];  % X coordinates
10 Y = [42,36,10,46,30,44,8,38,19,4];  % Y coordinates
11 %
12 % Calculation of coordination matrix and visibility heuristic
13 %
14 nE = length(X);             % Number of cities or points...
15 Distance = zeros(nE,nE);
16 for i = 1:nE-1
17 for j = i+1:nE
18 % Euclidean distance between two points/coordinates
19 Distance(i,j) = sqrt((X(i)-X(j))^2 + (Y(i)-Y(j))^2);
20 Distance(j,i) = Distance(i,j);
21 end
22 end
23 ETA = (1./Distance);        % heuristic matrix
24 ETA(1:1+size(ETA,1):end) = 0;% remove inf and set values to 0
25 %
26 % Initialization
27 %
28 % ACO algorithm controlling parameters
29 alpha = 1; beta = 0.01; rho = 0.02;
30 TAU = 0.01.*rand(nE,nE);              % Initial values of pheromone
31 ant = zeros(ant_pop,length(X));      % Initial population of ants
32 fitness = inf.*ones(ant_pop,1);      % Initial fitness of ants.
33 best_fit = max(fitness);
34 Iteration = 0;                       % Initialize the iteration
35 while Iteration < Max_Iteration      % Iterations loop
36 Iteration = Iteration + 1;           % Iterations count
37 %
38 % Start ant journey from one point to other
39 %
40 for k = 1:ant_pop           % Ant k
41 ant0(1) = randi([1 nE],1,1);% Random start of ant k...
42 for l = 2:nE
43 i = ant0(end);     % Start the journey from chosen edge i
44 % numerator of transition probability
45 NUM = TAU(:,i).^alpha.*ETA(:,i).^beta;
46 DEN = sum(NUM);    % Denominator of transition probability
47 Prob = NUM/DEN;    % Transition probability
48 Prob(ant0)=0; %Set 0 Prob for path already visited by ant k
49 ant0(1) = roulette(Prob);  % Roulette wheel selection for roote
50 end
51 % Fitness calculation of ant k...
```

```matlab
52 fitness(k) = FUN_TSP([ant0 ant0(1)],nE,Distance);
53 if fitness(k) < best_fit      % Checking for best solution
54 best_ant = ant0;
55 best_fit = fitness(k);
56 end
57 ant(k,:) = ant0;              % Store ant0 to ant k
58 ant0 = [];                    % Restore ant0
59 end
60 %
61 % Update Phromone intensity at all the corners/paths
62 %
63 for k=1:ant_pop    % ant k
64 ANT0 = [ant(k,:) ant(k,1)];
65 for i=1:nE
66 % Amount of pheromone added at path(i,l) by ant k
67 TAU(ANT0(i),ANT0(i+1)) = TAU(ANT0(i),ANT0(i+1)) + Q/fitness(k);
68 end
69 end
70 % Pheromone evaporation with the time or iterations...
71 TAU=(1-rho)*TAU;
72 end
73 disp(best_ant);
74 disp(best_fit);
```

Listing 1.3
Source-code of ACO in Matlab.

1.5 Source-code of ACO algorithm in C++

Listing 1.4 presents the the source-code of ACO algorithm in C++.

```cpp
1 #include <iostream>
2 #include <math.h>
3 #include <ctime>
4 #include <limits>
5 #include <algorithm>
6 using namespace std;
7 double r() {return (double)(rand()%RAND_MAX)/RAND_MAX;}
8 double FUN_TSP(int ant_path[], int ant0, int nE, double *Distance)
9 {    int ant[nE+1];
10 for(int i=0; i<nE; i++)
11 {ant[i]=ant_path[i];}
12 ant[nE]=ant0;
13 double Dist=0;
14 for(int t=0; t<nE; t++)
15 {Dist=Dist+Distance[ant[t+1]*nE+ant[t]];}
16 return Dist;
17 }
18 int roulette(double fitness[], int nE)
19 {    double partsum=0, sumfit=0; int j=0;
20 for(int i=0; i<nE; i++)
21 {sumfit=sumfit+fitness[i];}
22 double rn=r()*sumfit;
23 while (j<nE)
24 {partsum=partsum+fitness[j];
25 if (partsum>=rn)
26 {return j;}
27 j++;}
28 if (j>nE-1) {j=nE-1;}
29 return j;
30 }
```

```cpp
31  int main()
32  {    srand(time(NULL));
33  int Max_Iteration = 500;
34  int ant_pop=50;
35  double Q=1;
36  int nE=10;
37  int X[nE]={1,4,10,30,11,23,28,30,15,11};
38  int Y[nE]={42,36,10,46,30,44,8,38,19,4};
39  double Distance[nE][nE]; double ETA[nE][nE]; double TAU[nE][nE];
40  int ant[ant_pop][nE]; double fitness[ant_pop]; double best_fit;
41  double alpha=1, beta=0.01, rho=0.02;
42  double NUM[nE], DEN, Prob[nE];
43  int ant0[nE], best_ant[nE], ANT0[nE+1], END, ix;
44  int Iteration=0;
45  for(int i=0; i<nE-1; i++)
46  {
47  for(int j=0; j<nE; j++)
48  {
49  Distance[i][j]=sqrt(pow(X[i]-X[j],2)+pow(Y[i]-Y[j],2));
50  ETA[i][j]=1/Distance[i][j];
51  if (i==j) {ETA[i][j]=0;}
52  TAU[i][j]=0.01*r();
53  }
54  }
55  for(int i=0; i<ant_pop; i++)
56  {fitness[i]=std::numeric_limits<double>::max();}
57  best_fit=*min_element(fitness,fitness+nE);
58  //Initialization
59  while (Iteration<Max_Iteration)
60  {
61  Iteration=Iteration+1;
62  for(int k=0; k<ant_pop; k++)
63  {    for(int s=0; s<nE; s++){ant0[s]=-1;}
64  END=0;
65  ant0[END]=rand()%nE;
66  for(int l=1; l<nE; l++)
67  {
68  for(int s=0; s<nE; s++){if (ant0[s]>-1){END=s;}}
69  ix=ant0[END];
70  double DEN=0;
71  for(int s=0; s<nE; s++)
72  {
73  NUM[s]=TAU[s][ix]*alpha*ETA[s][ix]*beta;
74  DEN=DEN+NUM[s];
75  }
76  for(int s=0; s<nE; s++)
77  {Prob[s]=NUM[s]/DEN;}
78  for(int s=0; s<=END; s++)
79  {
80  if (ant0[s]>-1)
81  {Prob[ant0[s]]=0;}
82  }
83  ant0[l]=roulette(Prob,nE);
84  }
85  fitness[k]=FUN_TSP(ant0, ant0[0], nE, *Distance);
86  if (fitness[k]<best_fit)
87  {
88  best_fit=fitness[k];
89  for(int s=0; s<nE; s++)
90  {best_ant[s]=ant0[s];}
91  }
92  for(int s=0; s<nE; s++)
93  {
94  ant[k][s]=ant0[s];
95  ant0[s]=-1;
96  }
97  }
```

```
98  //Update Pheromone
99  for(int k=0; k<ant_pop; k++)
100 {
101 for(int s=0; s<nE; s++)
102 {ANT0[s]=ant[k][s];}
103 ANT0[nE]=ant[k][0];
104 for(int i=0; i<nE; i++)
105 {TAU[ANT0[i]][ANT0[i+1]]=TAU[ANT0[i]][ANT0[i+1]]+(Q/fitness[k]);}
106 }
107 for(int i=0; i<nE; i++)
108 {
109 for(int j=0; j<nE; j++)
110 {TAU[i][j]=(1-rho)*TAU[i][j];}
111 }
112 cout<<best_fit<<endl;
113 }
114 //Printing best solution
115 cout<<"Best ant: ["<<best_ant[0];
116 for(int i=1; i<nE; i++)
117 {cout<<", "<<best_ant[i];}
118 cout<<"]"<<endl;
119 cout<<"Best fitness: "<<best_fit<<endl;
120 getchar();
121 return 0;
122 }
```

Listing 1.4
Source-code of ACO algorithm in C++.

1.6   Step-by-step numerical example of ACO algorithm

Example 1 *Solve the traveling salesman problem (TSP) for five cities.*

Solution: In order to solve the TSP problem for five cities, the coordinates or distances between the cities are needed. We assume the following coordinates for these cities:

$X = [79, 87, 13, 67, 63]$
$Y = [31, 90, 23, 82, 17]$

So, the dimension or number of variables $N = 5$ for this problem.

In the first step, we define the ACO algorithm parameters. For example, population size $= 4$, maximum iteration $= 1$, $\alpha = 1.0$, $\beta = 1.0$, $\tau_0 = 0.01$, and $\rho = 0.02$.

In the second step, the euclidean distances between these cities are determined. The distance between city '1' (79, 31) to city '2' (87, 90) is calculated as:

$$D(1,2) = D(2,1) = \sqrt{\left(X(1) - X(2)\right)^2 + \left(Y(1) - Y(2)\right)^2}$$

$$= \sqrt{\left(79 - 87\right)^2 + \left(31 - 90\right)^2} = 59.54$$

Similarly, others are also calculated and presented as:

$$D = \begin{bmatrix} 0.00 & 59.54 & 66.48 & 52.39 & 21.26 \\ 59.54 & 0.00 & 99.82 & 21.54 & 76.84 \\ 66.48 & 99.82 & 0.00 & 79.98 & 50.36 \\ 52.39 & 21.54 & 79.98 & 0.00 & 65.12 \\ 21.26 & 76.84 & 50.36 & 65.12 & 0.00 \end{bmatrix}$$

In the third step, heuristic η, also known as visibility of the path, is determined as:

$$\eta = 1/D = \begin{bmatrix} \infty & 0.0168 & 0.0150 & 0.0191 & 0.0470 \\ 0.0168 & \infty & 0.0100 & 0.0464 & 0.0130 \\ 0.0150 & 0.0100 & \infty & 0.0125 & 0.0199 \\ 0.0191 & 0.0464 & 0.0125 & \infty & 0.0154 \\ 0.0470 & 0.0130 & 0.0199 & 0.0154 & \infty \end{bmatrix} \quad \% \text{ element-wise}$$

inversion

In the fourth step, the initial pheromone intensity, τ is randomly allocated to edges, as suggested below.

$$\tau = \tau_0 \times rand(N, N) = \begin{bmatrix} 0.0059 & 0.0021 & 0.0077 & 0.0023 & 0.0013 \\ 0.0034 & 0.0086 & 0.0017 & 0.0062 & 0.0078 \\ 0.0034 & 0.0036 & 0.0054 & 0.0063 & 0.0008 \\ 0.0053 & 0.0085 & 0.0099 & 0.0079 & 0.0002 \\ 0.0038 & 0.0098 & 0.0038 & 0.0010 & 0.0079 \end{bmatrix}$$

In the fifth step, we start the iteration loop of the ACO algorithm. It is checked whether the maximum number of iterations is attained. If yes, we jump to the eighteenth step; otherwise move to the sixth step.

In the sixth step, a random vertex/node/city is chosen to start the journey of ant-1. Suppose it is city '3' then ant-1 will start its journey from node/city 3.

In the seventh step, we start the ants updating the loop. If the paths of all ants are covered then we jump to the fifteenth step, otherwise move to the eighth step.

In the eighth step, we determine the probabilities for ant-1 to take paths originating from city '3' to other cities, by using (1.1). The probability of ant-1 to visit city '1' is calculated as:

$$NUM = \tau(:, 3)^\alpha \star \eta(:, 3)^\beta \qquad \% \text{ numerator of probability function,}$$
component-wise multiplication

$$NUM = \begin{bmatrix} 0.0077 \\ 0.0017 \\ 0.0054 \\ 0.0099 \\ 0.0038 \end{bmatrix}^{\alpha=1} \star \begin{bmatrix} 0.0150 \\ 0.0100 \\ 0 \\ 0.0125 \\ 0.0199 \end{bmatrix}^{\beta=1} = \begin{bmatrix} 0.000116 \\ 0.000017 \\ 0.000000 \\ 0.000124 \\ 0.000075 \end{bmatrix}$$

$DEM = \sum \tau(:, 3)^\alpha \times \eta(:, 3)^\beta == 0.000116 + 0.000017 + 0.000000 + 0.000124 + 0.000075 = 0.000332$ \quad \% denominator of probability function

Now, we determine the probability for ant-1 to visit other cities as follows:

$P_3^r = NUM/DEM = \begin{bmatrix} 0.3488 & 0.0513 & 0 & 0.3727 & 0.2272 \end{bmatrix}^T$ % component wise division.

In the ninth step, we apply the roulette wheel selection (RWS) criteria on P_3^r to select the most probable city to be visited by ant-1. Suppose city '4' is selected because it has the highest probability then the root of ant-1 is updated as:

$ant(1,:) = [3, 4]$

In the tenth step, the eighth step is repeated to determine the probability of ant-1 to visit other cities from current city '4', as calculated below:

$P_4^r = \begin{bmatrix} 0.1031 & 0.6759 & 0.1850 & 0 & 0.0361 \end{bmatrix}^T$

In the eleventh step, the probability of already visited cities is set to zero, as suggested below. This will avoid the revisits of this ant to already visited cities.

$P_4^r = \begin{bmatrix} 0.1031 & 0.6759 & 0 & 0 & 0.0361 \end{bmatrix}^T$

Suppose ant-1 is visiting city '2' (assumed from RWS); therefore the completed path of this ant can be updated as:

$ant(1,:) = [3, 4, 2]$

Similarly other paths are also determined and ant-1 returns to city '3' after visiting all cities, as given below:

$ant(1,:) = [3, 4, 2, 5, 1, 3]$

In the twelfth step, we calculate the fitness or distance traveled by this ant, as shown below.

$D(3,4) = 79.98;\ D(4,2) = 21.54;\ D(2,5) = 76.84;\ D(5,1) = 21.26;\ D(1,3) = 66.48;$

Therefore the fitness of ant-1 or distance traveled is calculated as:
$Fit(1) = 79.98 + 21.54 + 76.84 + 21.26 + 66.48 = 266.10$

By repeating the steps eighth to twelfth, the path traveled by other ants and respective distances are also calculated, as shown below.

$ant(1,:) = [3, 4, 2, 5, 1, 3]$ $Fit(1) = 266.10$
$ant(2,:) = [2, 4, 1, 5, 3, 2]$ $Fit(2) = 245.38$
$ant(3,:) = [5, 3, 4, 2, 1, 5]$ $Fit(3) = 232.68$
$ant(4,:) = [2, 5, 1, 4, 3, 2]$ $Fit(4) = 330.30$

In the thirteenth step, we compare the fitness of each ant with the current best ant and preserve the best ant that traveled through the shortest path, e.g., $best\_ant = ant(3,:) = [5, 3, 4, 2, 1, 5]$ and $best\_fit = 232.68$.

In the fourteenth step, we return to the seventh step.

In the fifteenth step, we update the pheromone intensity τ on the paths followed by these ants. The substance laid by ant k on a path connecting cities i and j is updated as:

$$\tau(i,j) = \tau(i,j) + 1/Fit(k)$$

The pheromone laid by ant-1 is updated as:

$$\tau(3,4) = \tau(3,4) + 1/Fit(1) = 0.0063 + 1/(266.10) = 0.0063 + 0.0038 = 0.0101$$
$$\tau(4,2) = \tau(4,2) + 1/Fit(1) = 0.0085 + 1/(266.10) = 0.0085 + 0.0038 = 0.0123$$
$$\tau(2,5) = \tau(2,5) + 1/Fit(1) = 0.0078 + 1/(266.10) = 0.0078 + 0.0038 = 0.0116$$
$$\tau(5,1) = \tau(5,1) + 1/Fit(1) = 0.0038 + 1/(266.10) = 0.0038 + 0.0038 = 0.0076$$
$$\tau(1,3) = \tau(1,3) + 1/Fit(1) = 0.0077 + 1/(266.10) = 0.0077 + 0.0038 = 0.0115$$

Similarly, the pheromone laid by other ants is also updated on these paths, as presented below:

$$\tau = \begin{bmatrix} 0.0059 & 0.0021 & 0.0115 & 0.0053 & 0.0097 \\ 0.0077 & 0.0086 & 0.0017 & 0.0103 & 0.0146 \\ 0.0034 & 0.0107 & 0.0054 & 0.0144 & 0.0008 \\ 0.0094 & 0.0166 & 0.0129 & 0.0079 & 0.0002 \\ 0.0106 & 0.0098 & 0.0122 & 0.0010 & 0.0079 \end{bmatrix}$$

In the sixteenth step, we evaporate the ρ amount of pheromone. The remaining pheromone intensity is expressed as:

$$\tau = (1 - \rho) * \tau = \begin{bmatrix} 0.0058 & 0.0021 & 0.0112 & 0.0052 & 0.0095 \\ 0.0075 & 0.0084 & 0.0017 & 0.0101 & 0.0143 \\ 0.0033 & 0.0105 & 0.0053 & 0.0141 & 0.0008 \\ 0.0092 & 0.0162 & 0.0127 & 0.0077 & 0.0002 \\ 0.0104 & 0.0096 & 0.0119 & 0.0010 & 0.0077 \end{bmatrix}$$

In the seventeenth step, we return to the fifth step.

In the eighteenth step, we print the best ant, i.e. *best_ant*, and its fitness, *best_fit*.

1.7   Conclusion

This chapter presents the standard version of the ACO algorithm. The optimal characteristics of ant colony are mathematically modeled into some set of equations. The artificial ant follows the shortest route between nest and food source. Matlab and C++ source codes are also presented in this chapter for hands-on practice. Furthermore, the step-by-step example is presented for beginners to understand the concept of basic ACO.

Acknowledgment

This work was supported by Marie Sklodowska-Curie Fellowship (COFUND-Multiply) received from the European Union's Horizon 2020 research and innovation programme under the Marie Sklodowska-Curie Grant Agreement no. 713694.

References

1. Colorni A, Dorigo M, Maniezzo V. "Distributed optimization by ant colonies" in Proceedings of the First European Conference on Artificial Life 1992 Dec (vol. 142, pp. 134-142).

2. Dorigo M, Birattari M, Sultzle T. "Ant colony optimization: Artificial ants as a computational intelligence technique", *IEEE Computational Intelligence Magazine*, vol 1, no. 4, 2006, pp. 28-39

3. Dorigo M, Di Caro G. "Ant colony optimization: a new meta-heuristic" in *Proceedings of IEEE Congress on Evolutionary Computation, (CEC-99)*, the IEEE; vol. 2. 6-9 July 1999. pp. 1470-1477

4. Blum C, Dorigo M. "The hyper-cube framework for ant colony optimization". *IEEE Trans Systems Man Cybernetics* B 2004; 34(2), pp. 1161-1172

5. Dorigo M, Maniezzo V, Colorni A. "Ant system: optimization by a colony of cooperating agents". *IEEE Trans Syst Man Cybern B Cybern*, 1996, doi:10.1109/3477.484436

6. Dorigo M. "Optimization, learning and natural algorithms". Ph.D. dissertation, Dipartimento di Elettronica, Politecnico di Milano, Italy, 1992.

7. Dorigo M, Gambardella LM. "Ant colony system: A cooperative learning approach to the traveling salesman problem". *IEEE Trans Evol Computat*, 1997; 1(1), pp. 53-66.

8. Lee K.Y., El-Sharkawi M.A. *Modern Heuristic Optimization Techniques: Theory and Applications to Power Systems*, John Wiley & Sons, vol. 39, 2008.

2

Artificial Bee Colony Algorithm

Bahriye Akay

Department of Computer Engineering
Erciyes University, Melikgazi, Kayseri, Turkey

Dervis Karaboga

Department of Computer Engineering
Erciyes University, Melikgazi, Kayseri, Turkey

CONTENTS

2.1 Introduction

Swarm intelligence models the collective behaviour of social creatures. The collective intelligence arises from task division and self-organization which is adaptation to new conditions without a global supervision. Self-organization is achieved by repeating the rewarding actions (positive feedback), abandoning repetitive behaviour patterns (negative feedback), communicating to neighbouring agents (multiple interactions) and exploring undiscovered patterns to avoid stagnation states (fluctuation).

Honey bees are social creatures that exhibit swarm intelligence in some of their activities such as nest site selection, mating, foraging etc. As well as the other activities, there is a task division among the bees in the foraging performed to find profitable sources to maximize the nectar amount transferred to the hive. To achieve this crucial task efficiently, the bees are divided into three categories: employed bees, onlooker bees and scouts. The employed bees are responsible for bringing the nectar from discovered flowers to the hive and they dance to give profit and location information about the source to waiting bees in the hive which are called onlookers. The onlooker bees watch the dances and fly to potentially good flowers chosen according to the information gathered from dances. A potentially good solution has high probability to be chosen by an onlooker bee. This positive feedback in foraging and dancing creates interaction between the bees. When a flower is selected by a bee, its nectar decreases by each exploitation and finally exhausts. The exhausted source is abandoned (negative feedback) and a search to find an undiscovered source (fluctuation) is performed by a scout bee. The positive and negative feedback, interaction and fluctuation properties show that a bee swarm is self-organizing and adaptable to internal and environmental conditions without a higher level guidance.

In 2005, Karaboga was inspired by the foraging behaviour of honey bees and proposed the Artificial Bee Colony (ABC) algorithm which simulates the task division and self-organization in a bee colony [1]. The algorithm has three phases: employed bees phase, onlooker bees phase and scout bee phase. The food sources (solutions) in an environment (search space) are searched by the bees to maximize the nectar amount (fitness of the solutions). As in real bees, the employed bees phase searches the vicinity of the sources discovered so far while the onlooker bees phase recruits the waiting bees to quality sources based on the information gathered from the employed bees. The scout bee phase tries to find new undiscovered flowers. Performance of the algorithm has been investigated in single-objective unconstrained [2, 3], constrained optimization [4], multi-objective optimization [5] and the algorithm have been used in many research areas successfully [6, 7]. According to the studies, the ABC algorithm is a simple and powerful meta-heuristic that can be used efficiently especially on high-dimensional and multi-modal problems. On hybrid and composite problems, the algorithm was modified to improve its local search capability and convergence rate. On constrained problems, its selection strategy was modified based on Deb's rules and on multi-objective problems, its greedy selection strategy was replaced with non-dominated sorting to rank solutions and build a population of dominated and non-dominated solutions. It is a good alternative in swarm intelligence algorithms with the advantage of having a smaller number of control parameters and balanced exploration/exploitation capability.

This chapter is organized as follows: in Section 2.2 the pseudo-code for the ABC algorithm is provided and each phase of the algorithm is explained in detail. Control parameters of the algorithm are introduced and their effects are discussed. In Sections 2.3 and 2.4, the ABC algorithm implementations using Matlab and C++ programming languages are presented, respectively. In Section 2.5 a step-by-step procedure is given to show the behaviour of the algorithm and finally it is concluded in Section 2.6.

2.2   The original ABC algorithm

The Artificial Bee Colony [1] algorithm is an optimization tool that exhibits the swarm intelligence of honey bees in the foraging task. In the algorithm, each solution represents a food source position and the algorithm tries to find the source with the maximum nectar amount. The pseudo-code of the ABC algorithm is given in Alg. 1.

Algorithm 1 Main steps of ABC algorithm.

1: Set values for the control parameters
2: SN: Number of Food Sources,
3: MCN: Termination Criteria, Maximum Number of Cycles,
4: $limit$: Maximum number of exploitations for a solution
　　　　　　　　　　　　　　　　　　　　　　　　▷ //Initialization
5: **for** $i = 1$ to SN **do**
6: 　　$x(i) \longleftarrow$ a random food source location by Eq. 2.1
7: 　　$trial(i) = 0$
8: **end for**
9: $cyc = 1$
10: **while** $cyc < MCN$ **do**　　　　　　　　　　▷ //Employed Bees' Phase
11: 　　**for** $i = 1$ to SN **do**
12: 　　　　$\hat{x} \longleftarrow$ a neighbour food source location generated by Eq. 2.2
13: 　　　　**if** $f(\hat{x}) < f_i$ **then**
14: 　　　　　　$x(i) = \hat{x}$
15: 　　　　　　$trial(i) = 0$
16: 　　　　**else**
17: 　　　　　　$trial(i) = trial(i) + 1$
18: 　　　　**end if**
19: 　　**end for**
20: 　　$p \longleftarrow$ assign probability by Eq. 2.3
　　　　　　　　　　　　　　　　　　　　　▷ //Onlooker Bees' Phase
21: 　　$i = 0$
22: 　　$t = 0$
23: 　　**while** $t < SN$ **do**
24: 　　　　**if** $rand(0, 1) < p(i)$ **then**
25: 　　　　　　$t = t + 1$
26: 　　　　　　$\hat{x} \longleftarrow$ a neighbour food source location generated by Eq. 2.2
27: 　　　　　　**if** $f(\hat{x}) < f_i$ **then**
28: 　　　　　　　　$x(i) = \hat{x}$
29: 　　　　　　　　$trial(i) = 0$
30: 　　　　　　**else**
31: 　　　　　　　　$trial(i) = trial(i) + 1$
32: 　　　　　　**end if**
33: 　　　　**end if**
34: 　　　　$i = (i + 1) \mod (SN - 1)$
35: 　　**end while**
36: 　　Memorize the best solution　　　　　　　　　▷ //Scout bee phase
37: 　　$si = \{i : trial(i) = max(trial)\}$
38: 　　**if** $trial(si) > limit$ **then**
39: 　　　　$x(si) \longleftarrow$ a random food source location by Eq. 2.1
40: 　　　　$trial(si) = 0$
41: 　　**end if**
42: 　　$cyc + +$
43: **end while**

In the initialization phase of ABC, SN number of solutions (food sources) are generated by Eq. 2.1.

$$x_{ij} = x_j^{lb} + rand(0, 1)(x_j^{ub} - x_j^{lb}) \tag{2.1}$$

where $i \in \{1, \ldots, SN\}$, $j \in \{1, \ldots, D\}$, D is the problem dimension, x_j^{lb} and x_j^{ub} are the lower and upper bounds in jth dimension of design parameters, respectively.

The food source population is evolved by the employed bees, onlooker bees and scout bee phases based on swarm intelligence characteristics. In the employed bees phase, the nectar exploitation behaviour is simulated by a local search around each source i by Eq. 2.2.

$$\hat{x}_{ij} = x_{ij} + \phi_{ij}(x_{ij} - x_{kj}) \tag{2.2}$$

where k is a randomly chosen neighbour of ith solution and $k \in \{1, 2, \ldots CS\}$: $k \neq i$ and ϕ_{ij} is a real random number drawn from uniform distribution within range [-1,1]. j is a random dimension which is going to be perturbed.

Choosing profitable sources based on the information gained from the employed bees is simulated by a selection strategy in which each solution is assigned a probability proportional to its quality. The quality is measured by the solution fitness (Eq. 2.4) and the probability can be defined by (Eq. 2.3):

$$p_i = 0.1 + 0.9 * \frac{fitness_i}{max(\vec{fitness})} \tag{2.3}$$

where $fitness_i$ (Eq. 2.4) is inversely proportional to the cost function for minimization purposes. When an onlooker bee selects a food source, the bee conducts a local search around the source by Eq. 2.2.

$$fitness_i = \begin{cases} \frac{1}{1+f_i}, if\ f_i \geq 0 \\ 1 + |f_i|, if\ f_i < 0 \end{cases} \tag{2.4}$$

In both the employed bees and onlooker bees phases, if the current source cannot be improved by a new source produced by the local search, its counter corresponding the number of exploitations is incremented by 1. If the counter exceeds a predefined number (control parameter, limit), the solution is discarded from the population by simulating food source exhaustion. As in real scout bees, the bee abandons the exhausted source and searches for a new source by Eq. 2.1.

2.3   Source-code of ABC algorithm in Matlab

```
1 %/* Control Parameters of ABC algorithm*/
2 FoodNumber=20; %/*The number of food sources*/
3 limit=100; %/*used to decide whether a food source will be abandoned*/
4 maxCycle=1000; %/*The maximum number of cycles */
5 %/* Problem specific variables*/
```

```matlab
 6 objfun='rastrigin'; %cost function to be optimized
 7 D=20; %/*The number of parameters of the problem to be optimized*/
 8 ub=ones(1,D)*5.12; %/*lower bounds of the parameters. */
 9 lb=ones(1,D)*(-5.12);%/*upper bound of the parameters.*/
10 %Foods  [FoodNumber][D]; /*population of food sources.*/
11 %ObjVal[FoodNumber]; /*objective function values associated with food
        sources */
12 %Fitness[FoodNumber]; /*fitness vector*/
13 %trial[FoodNumber]; /*the number of exploitations of each source*/
14 %prob[FoodNumber]; /* probability of each source */
15 %solution [D]; /*New solution (neighbour)*/
16 %ObjValSol; /*Objective function value of new solution*/
17 %FitnessSol; /*Fitness value of new solution*/
18 %GlobalParams[D]; /*the optimum solution vector*/
19 %GlobalMin; /*the optimum function value*/
20 %%Initialization
21 Range = repmat((ub-lb),[FoodNumber 1]);
22 Lower = repmat(lb, [FoodNumber 1]);
23 Foods = rand(FoodNumber,D) .* Range + Lower;
24 ObjVal=feval(objfun,Foods);
25 Fitness=calculateFitness(ObjVal);
26 trial=zeros(1,FoodNumber); %reset trial counters
27 [GlobalMin,GlobalParams]=MemorizeBestSolution(Foods,ObjVal,ObjVal(1),
        Foods(1,:));
28 cycle=1;
29 while ((cycle <= maxCycle))
30 %%%% EMPLOYED BEE PHASE %
31     for i=1:(FoodNumber)
32         [sol,ObjValSol,FitnessSol]=GenerateNewSolution(Foods,i,objfun,
        lb,ub);
33         % /*greedy selection*/
34         if (FitnessSol>Fitness(i))
35             Foods(i,:)=sol;
36             Fitness(i)=FitnessSol;
37             ObjVal(i)=ObjValSol;
38             trial(i)=0;
39         else
40             trial(i)=trial(i)+1;
41         end
42     end
43 %%%%CalculateProbabilities %
44 prob=(0.9.*Fitness./max(Fitness))+0.1;
45 %%%% ONLOOKER BEE PHASE %
46 i=1;
47 t=0;
48 while(t<FoodNumber)
49     if(rand<prob(i))
50         t=t+1;
51         [sol,ObjValSol,FitnessSol]=GenerateNewSolution(Foods,i,objfun,
        lb,ub);
52         % /*greedy selection*/
53         if (FitnessSol>Fitness(i))
54             Foods(i,:)=sol;
55             Fitness(i)=FitnessSol;
56             ObjVal(i)=ObjValSol;
57             trial(i)=0;
58         else
59             trial(i)=trial(i)+1;
60         end
61     end
62     i=i+1;
63     if (i==(FoodNumber)+1)
64         i=1;
65     end
66 end
67 [GlobalMin,GlobalParams]=MemorizeBestSolution(Foods,ObjVal,GlobalMin,
        GlobalParams);
```

```matlab
68  %%% SCOUT BEE PHASE %
69  ind=find(trial==max(trial));
70  ind=ind(end);
71  if (trial(ind)>limit)
72      trial(ind)=0;
73      Foods(ind,:)=(ub-lb).*rand(1,D)+lb;
74      ObjVal(ind)=feval(objfun,sol);
75      Fitness(ind)=calculateFitness(ObjValSol);
76  end
77  fprintf('Cycle=%d ObjVal=%g\n',cycle,GlobalMin);
78  cycle=cycle+1;
79  end % End of ABC
80  function [sol,ObjValSol,FitnessSol]=GenerateNewSolution(Foods,i,objfun
        ,lb,ub)
81  [FoodNumber,D]=size(Foods);
82  %/*The parameter to be changed is determined randomly*/
83      Param2Change=fix(rand*D)+1;
84      %/*A randomly chosen solution is used in producing a mutant
        solution of the solution i*/
85      neighbour=fix(rand*(FoodNumber))+1;
86      %/*Randomly selected solution must be different from the
        solution i*/
87          while(neighbour==i)
88              neighbour=fix(rand*(FoodNumber))+1;
89          end
90      sol=Foods(i,:);
91      %  /*v_{ij}=x_{ij}+\phi_{ij}*(x_{kj}-x_{ij}) */
92      sol(Param2Change)=Foods(i,Param2Change)+(Foods(i,Param2Change)-
        Foods(neighbour,Param2Change))*(rand-0.5)*2;
93      %  /*if generated parameter value is out of boundaries, it is
        shifted onto the boundaries*/
94          ind=find(sol<lb);
95          sol(ind)=lb(ind);
96          ind=find(sol>ub);
97          sol(ind)=ub(ind);
98          %evaluate new solution
99          ObjValSol=feval(objfun,sol);
100         FitnessSol=calculateFitness(ObjValSol);
101 end
102 function [GlobalMin,GlobalParams]=MemorizeBestSolution(Foods,ObjVal,
        GlobalMin,GlobalParams)
103 %/*The best food source is memorized*/
104 BestInd=find(ObjVal==min(ObjVal));
105 BestInd=BestInd(end);
106 if(GlobalMin>ObjVal(BestInd))
107 GlobalMin=ObjVal(BestInd);
108 GlobalParams=Foods(BestInd,:);
109 end
110 end
111 function fFitness=calculateFitness(fObjV)
112 fFitness=zeros(size(fObjV));
113 ind=find(fObjV>=0);
114 fFitness(ind)=1./(fObjV(ind)+1);
115 ind=find(fObjV<0);
116 fFitness(ind)=1+abs(fObjV(ind));
117 end
118 function ObjVal=Sphere(Colony,xd)
119 S=Colony.*Colony;
120 ObjVal=sum(S');
121 end
```

Listing 2.1

ABC Source-code implemented using Matlab.

2.4 Source-code of ABC algorithm in C++

```cpp
#include <iostream>
using namespace std;
#include <stdio.h>
#include <stdlib.h>
#include <math.h>
#include <conio.h>
#include <time.h>
/* Control Parameters of ABC algorithm*/
#define FoodNumber 20
#define limit 100
#define maxCycle 3000
/* Problem specific variables*/
#define D 50
#define lb -5.12
#define ub 5.12
double Foods[FoodNumber][D];
double f[FoodNumber];
double fitness[FoodNumber];
double trial[FoodNumber];
double prob[FoodNumber];
double solution[D];
double ObjValSol;
double FitnessSol;
int neighbour, param2change;
double GlobalMin;
double GlobalParams[D];
double r; /*a random number in the range [0,1)*/
/*benchmark functions */
double sphere(double sol[D]);
typedef double (*FunctionCallback)(double sol[D]);
FunctionCallback function = &sphere;
double CalculateFitness(double fun){
  double result = 0;
  if (fun >= 0)
    result = 1 / (fun + 1);
  else
    result = 1 + fabs(fun);
  return result;
}
void MemorizeBestSource(){
  int i, j;

  for (i = 0; i<FoodNumber; i++)
    if (f[i]<GlobalMin)
    {
      GlobalMin = f[i];
      for (j = 0; j < D; j++)
        GlobalParams[j] = Foods[i][j];
    }
}
void init(int index){
  int j;
  for (j = 0; j<D; j++){
    r = ((double)rand() / ((double)(RAND_MAX)+(double)(1)));
    Foods[index][j] = r*(ub - lb) + lb;
    solution[j] = Foods[index][j];
  }
  f[index] = function(solution);
  fitness[index] = CalculateFitness(f[index]);
  trial[index] = 0;
}
void initial(){
```

```
63    int i;
64    for (i = 0; i<FoodNumber; i++)
65      init(i);
66
67    GlobalMin = f[0];
68    for (i = 0; i < D; i++)
69      GlobalParams[i] = Foods[0][i];
70  }
71  void SendEmployedBees(){
72    int i, j;
73    for (i = 0; i<FoodNumber; i++){
74      r = ((double)rand() / ((double)(RAND_MAX)+(double)(1)));
75      param2change = (int)(r*D);
76      r = ((double)rand() / ((double)(RAND_MAX)+(double)(1)));
77      neighbour = (int)(r*FoodNumber);
78      while (neighbour == i){
79        r = ((double)rand() / ((double)(RAND_MAX)+(double)(1)));
80        neighbour = (int)(r*FoodNumber);
81      }
82      for (j = 0; j < D; j++)
83        solution[j] = Foods[i][j];
84      /*v_{ij}=x_{ij}+\phi_{ij}*(x_{kj}-x_{ij}) */
85      r = ((double)rand() / ((double)(RAND_MAX)+(double)(1)));
86      solution[param2change] = Foods[i][param2change] + (Foods[i][
          param2change] - Foods[neighbour][param2change])*(r - 0.5) * 2;
87      if (solution[param2change]<lb)
88        solution[param2change] = lb;
89      if (solution[param2change]>ub)
90        solution[param2change] = ub;
91      ObjValSol = function(solution);
92      FitnessSol = CalculateFitness(ObjValSol);
93      /*greedy selection*/
94      if (FitnessSol>fitness[i]){
95        trial[i] = 0;
96        for (j = 0; j < D; j++)
97          Foods[i][j] = solution[j];
98        f[i] = ObjValSol;
99        fitness[i] = FitnessSol;
100     }
101     else
102       trial[i] = trial[i] + 1;
103   }
104 }
105 void CalculateProbabilities(){
106   int i;
107   double maxfit;
108   maxfit = fitness[0];
109   for (i = 1; i<FoodNumber; i++)
110     if (fitness[i]>maxfit)
111       maxfit = fitness[i];
112   for (i = 0; i<FoodNumber; i++)
113     prob[i] = (0.9*(fitness[i] / maxfit)) + 0.1;
114 }
115 void SendOnlookerBees(){
116   int i, j, t;
117   i = 0;
118   t = 0;
119   while (t<FoodNumber){
120     r = ((double)rand() / ((double)(RAND_MAX)+(double)(1)));
121     if (r<prob[i]) /*choose a food source depending on its probability
          to be chosen*/
122     {
123       t++;
124       r = ((double)rand() / ((double)(RAND_MAX)+(double)(1)));
125       param2change = (int)(r*D);
126
127       r = ((double)rand() / ((double)(RAND_MAX)+(double)(1)));
```

```cpp
128        neighbour = (int)(r*FoodNumber);
129
130        while (neighbour == i){
131          r = ((double)rand() / ((double)(RAND_MAX)+(double)(1)));
132          neighbour = (int)(r*FoodNumber);
133        }
134        for (j = 0; j < D; j++)
135          solution[j] = Foods[i][j];
136
137        /*v_{ij}=x_{ij}+\phi_{ij}*(x_{kj}-x_{ij}) */
138        r = ((double)rand() / ((double)(RAND_MAX)+(double)(1)));
139        solution[param2change] = Foods[i][param2change] + (Foods[i][
       param2change] - Foods[neighbour][param2change])*(r - 0.5) * 2;
140        if (solution[param2change]<lb)
141          solution[param2change] = lb;
142        if (solution[param2change]>ub)
143          solution[param2change] = ub;
144        ObjValSol = function(solution);
145        FitnessSol = CalculateFitness(ObjValSol);
146        /*greedy selection*/
147        if (FitnessSol>fitness[i]){
148          trial[i] = 0;
149          for (j = 0; j < D; j++)
150            Foods[i][j] = solution[j];
151          f[i] = ObjValSol;
152          fitness[i] = FitnessSol;
153        }
154        else
155          trial[i] = trial[i] + 1;
156      }
157      i++;
158      if (i == FoodNumber)
159        i = 0;
160    }/*while*/
161 }
162 void SendScoutBees(){
163    int maxtrialindex, i;
164    maxtrialindex = 0;
165    for (i = 1; i<FoodNumber; i++)
166      if (trial[i]>trial[maxtrialindex])
167        maxtrialindex = i;
168    if (trial[maxtrialindex] >= limit)
169      init(maxtrialindex);
170 }
171 int main(){
172    int cycle, j;
173    double mean;
174    mean = 0;
175    srand(time(NULL));
176    initial();
177    MemorizeBestSource();
178    for (cycle = 0; cycle<maxCycle; cycle++){
179      SendEmployedBees();
180      CalculateProbabilities();
181      SendOnlookerBees();
182      MemorizeBestSource();
183      SendScoutBees();
184    }
185    for (j = 0; j < D; j++)
186      cout << "GlobalParam[" << j + 1 << "]: " << GlobalParams[j] <<
       endl;
187    cout << "GlobalMin =" << GlobalMin << endl;
188 return 0;
189 }
190 double sphere(double sol[D]){
191    int j;
192    double top = 0;
```

```
193    for (j = 0; j<D; j++)
194        top = top + sol[j] * sol[j];
195    return top;
196 }
```

Listing 2.2
ABC Source-code implemented using C++

2.5 Step-by-step numerical example of the ABC algorithm

In this chapter, a numerical example of how the ABC algorithm works is given step by step to solve the Sphere problem. Assume that the problem dimension (D) is 5 and the search space range is $[-5.12, 5.12]$ in all dimensions. The value of the *limit* parameter is 2 and the number of food sources (SN) is 6. $\vec{X}_i$ $(i = 1, 2, \ldots, SN)$ solutions are initialized by Eq. 2.1 and when each food source location is evaluated in the objective function (Sphere), the following $ObjVal_i$ values are obtained.

$\vec{X}_1 = \{4.1460, 0.97170, -2.0820, 3.0824, -2.6902\}, ObjVal_1 = 39.2066$
$\vec{X}_2 = \{4.9126, -2.4350, -1.8557, -4.8208, -0.4214\}, ObjVal_2 = 56.9242$
$\vec{X}_3 = \{-0.6260, 1.0531, -0.7765, 4.3914, 4.7420\}, ObjVal_3 = 43.8748$
$\vec{X}_4 = \{-3.9821, 2.1628, 0.0805, 2.3585, 0.4792\}, ObjVal_4 = 26.3335$
$\vec{X}_5 = \{-2.4774, -2.8493, -4.2443, -0.1166, 0.2164\}, ObjVal_5 = 32.3305$
$\vec{X}_6 = \{-0.9347, -3.9176, -2.4321, 0.8040, -2.7484\}, ObjVal_6 = 30.3365$

In the initialization phase, the number of trial values is set to 0 for all solutions. $\vec{trial} = \{0, 0, 0, 0, 0, 0\}$. $GlobalMin = 26.3335$ and the corresponding $GlobalParams = \{-3.9821, 2.1628, 0.0805, 2.3585, 0.4792\}$.

After initialization, ABC starts to iterate the phases in each cycle. In the employed bee phase, for each solution, a new solution is generated and a greedy selection is applied. At the first cycle, for $\vec{X}_1$, assume that $k = 4, j = 3$ and $\phi = 0.3582$ values are drawn randomly. It means that a new solution is generated by changing the 3rd parameter of $\vec{X}_1$ by substituting k and ϕ values in Eq. 2.2. $New_{1,3} = x_{1,3} + 0.3582 * (x_{1,3} - x_{4,3}) = -2.8566$. The other parameters of $\vec{New}_1$ but for the 3rd parameter are copied from $\vec{X}_1$. Corresponding objective function and fitness values are calculated as 43.0321 and 0.0227, respectively. The same procedure is applied for all solutions as shown in Table 2.1. Each new solution is compared to its basis and the better one is kept in the population. If the basis solution is retained, its counter is incremented by 1; otherwise it is set to 0. Because $\vec{X}_1$ is better than New_1, $\vec{X}_1$ is selected and its counter is incremented by 1. The food source population and trial counters after selection are shown in Table 2.2.

TABLE 2.1
New Food Source Locations Generated by Employed Bee Phase at cycle=1.

	K	j	ϕ	x1	x2	x3	x4	x5	ObjValSol	FitnessSol
New_1	4	3	0.3582	4.1460	0.9717	-2.8566	3.0824	-2.6902	43.0321	0.0227
New_2	3	2	0.9759	4.9126	-5.1200	-1.8557	-4.8208	-0.4214	77.2094	0.0128
New_3	6	1	0.8265	-0.3709	1.0531	-0.7765	4.3914	4.7420	43.6205	0.0224
New_4	1	4	-0.4762	-3.9821	2.1628	0.0805	2.7032	0.4792	28.0783	0.0344
New_5	1	2	0.4424	-2.4774	-4.5397	-4.2443	-0.1166	0.2164	44.8210	0.0218
New_6	4	1	-0.0116	-0.9700	-3.9176	-2.4321	0.8040	-2.7484	30.4038	0.0318

TABLE 2.2
The Food Source Population and Trial Counters after Selection in the Employed Bee Phase at Cycle=1.

	Selected	x1	x2	x3	x4	x5	ObjVal	Fitness	Trial
$\vec{X}_1$	$\vec{X}_1$	4.1460	0.97170	-2.0820	3.0824	-2.6902	39.2066	0.0249	1
$\vec{X}_2$	$\vec{X}_2$	4.9126	-2.4350	-1.8557	-4.8208	-0.4214	56.9242	0.0173	1
$\vec{X}_3$	New_3	-0.3709	1.0531	-0.7765	4.3914	4.7420	43.6205	0.0224	0
$\vec{X}_4$	$\vec{X}_4$	-3.9821	2.1628	0.0805	2.3585	0.4792	26.3335	0.0366	1
$\vec{X}_5$	$\vec{X}_5$	-2.4774	-2.8493	-4.2443	-0.1166	0.2164	32.3305	0.0300	1
$\vec{X}_6$	$\vec{X}_6$	-0.9347	-3.9176	-2.4321	0.8040	-2.7484	30.3365	0.0319	1

Based on the fitness values calculated by Eq. 2.4, the probability values are assigned by Eq. 2.3 as $\vec{P} = \{0.7118, 0.5247, 0.6513, 1.0000, 0.8381, 0.8850\}$. In the onlooker bees phase, for each source a random real number is drawn and if this number is lower than the probability value, an onlooker bee flies to search the vicinity of that solution by Eq. 2.2. For example, assume that the random number is 0.2548 for $\vec{X}_1$. Since this number is lower than the probability of $\vec{X}_1$ (0.7118), a new solution is generated in the vicinity of $\vec{X}_1$ and a greedy selection is applied between $\vec{X}_1$ and its mutant. For $\vec{X}_2$, assume that 0.5687 is generated randomly. Because this number is higher than the probability of $\vec{X}_2$, an onlooker bee is not recruited to this source. This process is iterated until the number of onlooker bees recruited to the sources is equal to the number of food sources ($SN = 6$ in this case). All the iterations at the first cycle of onlooker bee phase are given step-by-step in Table 2.3. The food

TABLE 2.3
The Search and Selection in the Onlooker Bee Phase at Cycle=1.

i	P_i	random	Fly	K	j	ϕ	x1	x2	x3	x4	x5	OF	Fit	Selected	trial
1	0.7118	0.2548	Yes	3	5	0.4455	4.1460	0.9717	-2.0820	3.0824	-5.1200	58.1838	0.0169	$\vec{X}_1$	2
2	0.5247	0.5687	No												1
3	0.6513	0.9037	No												0
4	1.0000	0.8909	Yes	5	2	-0.6044	-3.9821	-0.8665	0.0805	2.3585	0.4792	22.4066	0.0427	$\vec{X}_4 = New4$	0
5	0.8381	0.3054	Yes	4	4	-0.0402	-2.4774	-2.8493	-4.2443	-0.0171	0.2164	32.3172	0.0300	$\vec{X}_5 = New5$	0
6	0.8850	0.9047	No												1
1	0.7118	0.6099	Yes	6	4	0.6110	4.1460	0.9717	-2.0820	4.4745	-2.6902	49.7266	0.0197	$\vec{X}_1$	3
2	0.5247	0.5767	No												1
3	0.6513	0.1829	Yes	2	2	-0.9427	-0.3709	-2.2351	-0.7765	4.3914	4.7420	47.5073	0.0206	$\vec{X}_3$	1
4	1.0000	0.4899	Yes	6	1	0.4254	-5.1200	-0.8665	0.0805	2.3585	0.4792	32.7639	0.0296	$\vec{X}_4$	1

TABLE 2.4

The Food Source Population and Trial Counters after the Onlooker Bee Phase at Cycle=1.

	x1	x2	x3	x4	x5	ObjVal	Fitness	Trial
$\vec{X}_1$	4.1460	0.97170	-2.0820	3.0824	-2.6902	39.2066	0.0249	3
$\vec{X}_2$	4.9126	-2.4350	-1.8557	-4.8208	-0.4214	56.9242	0.0173	1
$\vec{X}_3$	-0.3709	1.0531	-0.7765	4.3914	4.7420	43.6205	0.0224	1
$\vec{X}_4$	-3.9821	-0.8665	0.0805	2.3585	0.4792	22.4066	0.0427	1
$\vec{X}_5$	-2.4774	-2.8493	-4.2443	-0.0171	0.2164	32.3172	0.0300	0
$\vec{X}_6$	-0.9347	-3.9176	-2.4321	0.8040	-2.7484	30.3365	0.0319	1

TABLE 2.5

The Food Source Population and Trial Counters after the Scout Phase at Cycle=1.

	x1	x2	x3	x4	x5	ObjVal	Fitness	Trial
$\vec{X}_1$	3.2228	4.1553	-3.8197	4.2330	1.3554	32.7639	0.0296	0
$\vec{X}_2$	4.9126	-2.4350	-1.8557	-4.8208	-0.4214	56.9242	0.0173	1
$\vec{X}_3$	-0.3709	1.0531	-0.7765	4.3914	4.7420	43.6205	0.0224	1
$\vec{X}_4$	-3.9821	-0.8665	0.0805	2.3585	0.4792	22.4066	0.0427	1
$\vec{X}_5$	-2.4774	-2.8493	-4.2443	-0.0171	0.2164	32.3172	0.0300	0
$\vec{X}_6$	-0.9347	-3.9176	-2.4321	0.8040	-2.7484	30.3365	0.0319	1

source population and the number of trials after the onlooker bee phase is completed at the first cycle are given in Table 2.4.

Once the onlooker bee phase is completed, it is checked whether $GlobalMin$ and $GlobalParams$ need to be updated. Since the objective function value of $\vec{X}_2$ is lower than current $GlobalMin$, $GlobalMin$ is assigned 22.4066 (objective value of $\vec{X}_2$) and $GlobalParams = \{-3.9821, -0.8665, 0.0805, 2.3585, 0.4792\}$.

The scout bee phase checks whether any trial counter exceeds the *limit* control parameter. Since the value of *limit* is predetermined as 2 and the counter associated with $\vec{X}_1$ is 3, the food source $\vec{X}_1$ is assumed to be exhausted. $\vec{X}_1$ is replaced with a new random food source location generated by Eq. 2.1 and its counter is set to 0. The food source population and the number of trials after the scout bee phase is completed at the first cycle are given in Table 2.5.

When the algorithm was iterated through 50 cycles using the control parameter values mentioned before, the following values were obtained: $GlobalMin = 0.0130748$, $GlobalParams = \{-0.0524, 0.0679, -0.0620, -0.0037, -0.0432\}$. The convergence of $GlobalMin$ by the ABC algorithm is shown in Fig. 2.1:

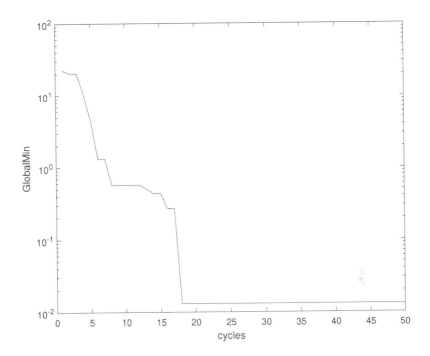

FIGURE 2.1
Convergence of *GlobalMin* by ABC algorithm, $SN = 6, limit = 2$.

2.6 Conclusions

In this chapter, basic principles of the Artificial Bee Colony Algorithm are provided and the source codes of the ABC algorithm in Matlab and C++ programming languages are also given for the researchers who want to implement ABC and solve their optimization problems. For the readers who need a better understanding on how the ABC algorithm iterates the phases and cycles, a step-by-step procedure on a numerical example is presented. We hope that this chapter may be helpful to the researchers interested in the Artificial Bee Colony algorithm.

References

1. D. Karaboga. An idea based on honey bee swarm for numerical optimization. Technical Report TR06, Erciyes University, Engineering Faculty, Computer Engineering Department, 2005.

2. D. Karaboga, B. Basturk. A powerful and efficient algorithm for numerical function optimization: Artificial bee colony (abc) algorithm. *Journal of Global Optimization*, 39(3):459-471, 2007.

3. D. Karaboga, B. Akay. A comparative study of artificial bee colony algorithm. *Applied Mathematics and Computation*, 214:108-132, 2008.

4. D. Karaboga, B. Akay. A modified artificial bee colony (abc) algorithm for constrained optimization problems. *Applied Soft Computing*, 11(3):3021-3031, 2011.

5. B. Akay. Synchronous and asynchronous pareto-based multi-objective artificial bee colony algorithms. *Journal of Global Optimization*, 57(2):415-445, October 2013.

6. D. Karaboga, B. Akay. A survey: Algorithms simulating bee swarm intelligence. *Artificial Intelligence Review*, 31(1):68-55, 2009.

7. D. Karaboga, B. Gorkemli, C. Ozturk, N. Karaboga. A comprehensive survey: artificial bee colony (abc) algorithm and applications. *Artificial Intelligence Review*, 42(1):21-57, 2014.

3

Bacterial Foraging Optimization

Sonam Parashar

Department of Electrical Engineering
Malaviya National Institute of Technology, Jaipur, India

Nand K. Meena

School of Engineering and Applied Science
Aston University, Birmingham, B4 7ET, United Kingdom

Jin Yang

School of Engineering and Applied Science
Aston University, Birmingham, United Kingdom

Neeraj Kanwar

Department of Electrical Engineering
Manipal University Jaipur, Jaipur, India

CONTENTS

3.1 Introduction

The bacterial foraging optimization (BFO) is a swarm intelligence based optimization method introduced by Kevin Passino in 2002. The method is inspired by the foraging theory of animals and mathematically models the foraging behavior of the bacteria present in the human intestine, named as *E. coli*, for the optimization purpose [1]. The foraging theory represents the idea that every animal or bacteria searches for food or nutrients in such a way that maximizes their energy intake per unit time spent on foraging subjected to some geographical and physiological constraints. The algorithm is further modified and some modifications in algorithm structure are suggested for the simplification of the algorithm such as elimination of the swim step parameter, a clear adaptation rule for the step size, the use of a uniform distribution, the position initialization, the inclusion of best individual information in the movement equation and the removal of cell-to-cell communication [2]. An adaptive BFO dynamically adjusts the run-length unit parameter to balance the trade-off between exploration and exploitation of the search space [3]. BFO can optimize the single objective functions efficiently.

However, for the multi-objective functions, a multi-objective variant of BFO is also introduced. The multi-objective bacteria foraging algorithm (MBFO) gives the pareto optimal solutions by taking decisions based on the idea of integration between the health sorting approach and pareto dominance mechanism [4]. In addition, a hybrid approach known as bacterial foraging algorithm-genetic algorithm (BFO-GA) is also proposed to improve the performance of the algorithm for solving the multi-objective optimization problems [5]. This chapter presents the insights of the BFO algorithm and is organized as follows: Section 3.1 presents the introduction of BFO followed by its algorithm in Section 3.2. In Section 3.3, the pseudo-codes of the BFO algorithm are presented. The Matlab source-code is given in Section 3.4 followed by some solved examples and conclusions in Sections 3.5 and 3.6 respectively.

3.2   Bacterial foraging optimization algorithm

The *E. coli* bacterium has a structure composed of the plasma membrane, cell wall, and capsule that contains the cytoplasm and nucleoid. The pili (singular, pilus) are used for a type of gene transfer to other *E. coli* bacteria, and flagella (singular, flagellum) are used for locomotion. The bacterium follows the foraging theory, always searching for high nutrient gradient to maximize the energy intake per unit time spent on foraging by using some intelligence. This process is achieved through two types of motions i.e. tumble and run or swim. The bacterium swims or runs in a nutrient gradient area and keep searching for the high nutrient gradient with the help of locomotion in a counterclockwise direction which groups the bacterium. If the bacterium finds no nutrient or noxious substance, the flagellum moves in the clockwise direction which means it pulls on the cell and results in a little displacement of the bacterium with undefined direction known as tumble motion of the bacterium. For this switching between the two motions tumble and run, firstly, the cell slows down or stops and then starts to move slowly.

The BFO algorithm mainly comprises the four main steps, Chemotaxis, Swarming, Reproduction and Elimination or Dispersal. These steps are explained in the following sections:

3.2.1 Chemotaxis

Chemotaxis is a phenomenon which describes the motion patterns of the bacterium over the nutrient gradient. Generally, the bacterium follows the Brownian motion and chooses its direction of motion for searching the food or nutrient gradient with some self-intelligence of locomotion. The motion of a bacterium at ith iteration can be defined as

$$\theta_i(j+1, k, l) = \theta_i(j, k, l) + C_i \phi_j \tag{3.1}$$

where, θ_i is the location of ith bacterium, j is the index for the chemotaxis step, k is the index for the reproduction steps, l is the index for the elimination and dispersal step, C_i is a unit run length of the bacterium and ϕ_j is the direction angle of the jth step, expressed as

$$\phi_j = \frac{\Delta_i}{\sqrt{\Delta_i^T \cdot \Delta_i}} \tag{3.2}$$

where, Δ_i represents i^{th} value of direction vector in j^{th} chemotactic step, $\Delta_i \epsilon R^P$, R^P is a random vector of P dimension having values between $[-1, 1]$ and is used for tumble.

3.2.2 Swarming

This step is about the communication of each bacterium. The swarm of bacteria communicates by releasing some cell attractants and repellents. The attractant responses are obtained from the serine or aspartate whereas repellent responses are obtained from metal ions such as Ni and Co. The cell-to-cell communication in the presence of attractant and repellent is obtained from the following expression

$$J_{cc}(\theta, P(j, k, l)) = \sum_{i=1}^{S} J_{cc}^j(\theta, \theta_i(j, k, l))$$

$$= \sum_{i=1}^{S} \left[-d_{attract} \cdot \exp\left(-w_{attract} \sum_{m=1}^{p} (\theta_m - \theta_m^i)^2 \right) \right] \tag{3.3}$$

$$+ \left[h_{repellent} \cdot \exp\left(-w_{repellent} \sum_{m=1}^{p} (\theta_m - \theta_m^i)^2 \right) \right]$$

where, $J_{cc}(\theta, P(j, k, l))$ is the objective function value to be added to the actual objective function, S is the total number of bacteria, P is the number of variables to be optimized which are present in each bacterium,

$\theta = [\theta_1, \theta_2, \theta_3, \ldots, \theta_p]^T$ denotes a point in the P-dimensional search space, θ^i_m is the mth component of the ith bacterium position, $d_{attract}$ is the depth of the attractant released by the cell, $w_{attract}$ is a measure of the width of attractant signal, $h_{repellant}$ is the height of the repellant effect (i.e., $h_{repellant} = d_{attract}$), and $w_{repellant}$ is the measure of the width of the repellant.

3.2.3   Reproduction

In this process, the members of the bacteria's swarm having high nutrients will only survive and reproduce new ones by splitting themselves into two. For this purpose, the population first sorted into ascending order according to the fitness of the objective function, the members responsible for poor fitness of the objective function will be considered as the members with nutrient deficiency or the weak members which will further be replaced by the newly reproduce members from the healthy bacteria responsible for the best fitness of the objective function. This replacement balances the number of bacteria members in the swarm.

3.2.4   Elimination and dispersal

After some reproduction steps, elimination and dispersal take place to avoid the local minima. At this stage, each bacterium will be assigned a probability P_{ed}. The bacterium is then selected according to their probability to change their location in the search space. The bacterium having highest probability will change its location first.

3.3   Pseudo-code of bacterial foraging optimization

In this section, the pseudo-codes for the basic version of BFO is presented in Algorithm 2.

Algorithm 2 Pseudo-code of basic Bacterial Foraging Optimization.

1: Determine the objective function $OF(.)$
2: set the algorithm parameters such as p=dimensions; N_c=chemotactic steps; N_s=swimming length; S=bacterium; N_{ed}=elimination; N_r=reproduction steps; p_{ed}=elimination probability etc.
3: set the lower and upper bounds $[LB, UB]$ for each variable/dimension p
4: randomly generate initial population of bacteria, as follows
5: **for** each k-th reproduction step **do**
6:     **for** each j-th chemotactic step **do**
7:         **for** each i-th bacterium **do**
8:             **for** each d-th dimension **do**

9: $$pp(d, k, j, i) = LB(d) + rand * (UB(d) - LB(d))$$

10: **end for**

11: apply correction algorithm to i-th bacterium, if infeasible

12: calculate the fitness value of i-th bacterium

13: **end for**

14: **end for**

15: **end for**

16: Start the elimination or dispersal step...

17: **for** each l-th elimination-dispersal step **do**

18: **for** each k-th reproduction step **do**

19: **for** each j-th chemotactic step **do**

20: **for** each i-th bacterium **do**

21: Update the i-th bacterium by using (3.2) and (3.3), as

22: $$pp_i(j + 1, k, l) = pp_i(j, k, l) + C_i \cdot \frac{\Delta_i}{\sqrt{\Delta_i^T \cdot \Delta_i}}$$

23: apply correction algorithm to i-th bacterium, if infeasible

24: calculate the fitness value of newly generated bacterium i

25: **end for**

26: apply the swimming operation for Ns bacteria

27: apply reproduction-elimination by sorting bacteria according to their fitness values and then split the top 50% and eliminate the remaining

28: **end for**

29: **end for**

30: **end for**

31: return the best result.

3.4 Matlab source-code of bacterial foraging optimization

The Matlab source-code of basic BFO is presented in Listing 3.1.

```matlab
% Bacterial Foraging Optimization Algorithm...
clc ;
clear ;
%Algorithm parameters....
N_bact=60;              %Total number of bacteria,
N_die=round(N_bact/2); %Number of weak bacteria died during
    reproduction,
N_chemo=30;            %The number of chemotactic steps,
Ns=7;                  %The swimming length,
N_repro=10;            %The number of reproduction steps,
N_disp=5;              %The number of elimination-dispersal events,
P_disp=0.5;            %Elimination-dispersal probability,
LB=[-3*pi -3*pi];      %lower bounds of variables,
UB=[3*pi 3*pi];        %upper bounds of variables,
p=length(UB);          %number of variables or dimensions,
C=1;                   %unit length step size for a run/tumble
%
%initialization, random but feasible population
for r=1:N_repro        %reproduction starts...
for c=1:N_chemo        %chemotaxis starts...
```

```
20 for b=1:N_bact        %individual bacteria...
21 for v=1:p             %variable....
22 pp(v,b,c,r)=LB(v)+rand*(UB(v)-LB(v));
23 end
24 Fit(b,c,r)=OF(pp(:,b,c,r));
25 end
26 end
27 end
28 best_fit=min(Fit(:));
29 loc=find(best_fit==Fit);
30 best_sol=pp(:,loc);
31 % % %
32 % %dispersal starts here....
33 for d=1:N_disp
34 for r=1:N_repro        %reproduction starts...
35 for c=1:N_chemo        %chemotaxis starts...
36 for b=1:N_bact         %individual bacteria...
37 FitLast=Fit(b,c,r);
38 dd(:,b)=unifrnd(-1,1,[p,1]);
39 %add on the cell-to-cell attractant
40 pp(:,b,c,r)=pp(:,b,c,r)+C*dd(:,b)/sqrt(dd(:,b)'*dd(:,b));
41 %New fitness calculations....
42 Fit(b,c,r)=OF(pp(:,b,c,r));
43 %swimming behaviour of bacteria
44 sw = 1;
45 while sw < Ns
46 if Fit(b,c,r)<FitLast
47 %addonthecell-to-cell attractant
48 pp(:,b,c,r)=pp(:,b,c,r)+C*dd(:,b)/sqrt(dd(:,b)'*dd(:,b));
49 Fit(b,c,r)=OF(pp(:,b,c,r)); %New fitness calculations....
50 FitLast=Fit(b,c,r); %Restore the fittest bacterium..
51 else
52 sw=Ns;
53 end
54 sw=sw+1;
55 end
56 end
57 end
58 %Reprodution...
59 Fit_health=sum(Fit(:,:,r),2); %Health of bacteria, 1 to N_bact
60 [Health,loc]=sort(Fit_health);
61 %spliting the top (N_bact/2) bacteria with better fitness
62 g=0;                   %counter for spliting bateria...
63 for sp=N_die+1:N_bact
64 g=g+1;
65 pp(:,loc(sp),1,r)=pp(:,loc(g),1,r);
66 Fit(loc(sp),1,r)=OF(pp(:,loc(sp),1,r));
67 end
68 end
69 %elimination/dispersal operations...
70 for u=1:N_bact
71 rr=rand;
72 if rr<P_disp
73 for v=1:p  %variable....
74 pp(v,u,1,1)=LB(v)+rand*(UB(v)-LB(v));
75 end
76 Fit(u,1,1)=OF(pp(:,u,1,1));
77 end
78 end
79 % Preserve the best...
80 best_fit2=min(Fit(:));
81 if best_fit2<best_fit
82 loc2=find(best_fit2==Fit);
83 best_sol=pp(:,loc2);
84 best_fit=best_fit2;
85 end
86 end
```

```
87 %print the solution
88 disp(best_fit);
89 disp(best_sol);
```

Listing 3.1
Source-code of BFO in Matlab.

3.5 Numerical examples

In this section, the application of the BFO algorithm is demonstrated by a step-by-step example.

Example 2 *Determine the global minima of the Griewank function expressed as*

$$f(x) = 1 + \frac{1}{4000} \sum_{i=1}^{n} x_i^2 - \prod_{i=1}^{n} cos\left(\frac{x_i}{\sqrt{i}}\right) \quad (3.4)$$

where, $x \epsilon [-3\pi, 3\pi]$

Solution: A contour plot of the Griewank function is presented in Fig. 3.1. It shows that this function is non-convex with multiple minima.

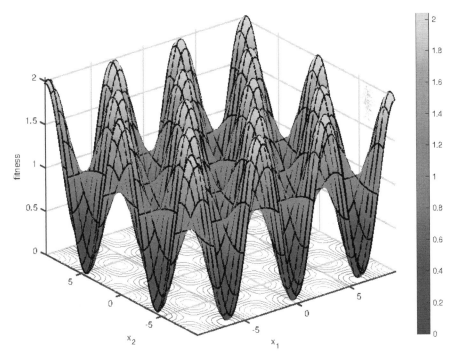

FIGURE 3.1
Griewank function for $n=2$.

In Listing 3.2, the Matlab source code is presented for the Griewank function, $f(x)$.

```
1  %Griewank function
2  function [Fit] = OF(x)
3  n=length(x);
4  sum=0;
5  product=1;
6  for i=1:n
7  y=x(i);
8  sum=sum + y^2/4000;
9  product=product*cos(y/sqrt(i));
10 end
11 Fit=1+sum-product;
```

Listing 3.2
Definition of Griewank function $OF(.)$ in Matlab.

Now, we demonstrate how BFO can help to determine the optimal values of x such that $f(x)$ is minimum. In the first step, assume that $n = 2$, upper and lower limits [LB UB], i.e., [-3π 3π] along with all algorithm parameters are provided. For example, $N_{bact} = 2$, $N_{die} = round(N_{bact}/2)$, $N_{chemo} = 3$, $N_s = 3$, $N_{repro} = 1$, $N_{disp} = 1$, $P_{disp} = 0.5$, $p = length(UB)$, $C = 1$ etc.

In step two, a random population of $N_{bact} = 2$ bacteria, with $n = 2$ dimensions, are generated for each chemotactic step, as shown below.

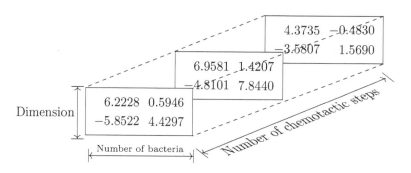

In the third step, determine the fitness of each bacterium by using $OF(.)$. For example, the first bacterium of the first chemotactic step is $x(:, 1, 1) = [6.2228 \ -5.8522]^T$; thus its fitness value will be $OF(x(:, 1, 1)) = 1.5605$. The second bacterium of the first chemotactic step is $x(:, 2, 1) = [0.5946 \ 4.4297]^T$ and its fitness $OF(x(:, 2, 1)) = 1.8333$. Similarly, fitness values of all bacteria in chemotactic steps two and three are determined and presented below.

In the fourth step, we retain the best bacterium based on fitness value. In the proposed work, we are minimizing objective function $OF(.)$; therefore a bacterium with minimum fitness value will be the best. For example, minimum fitness value is 0.6064 which corresponds to the bacterium $[-0.4830\ 1.5690]^T$ in the third chemotactic step so $x_{best} = [-0.4830\ 1.5690]^T$ with $Fit_{best} = 0.6064$.

In the fifth step, if the maximum number of reproduction steps is reached then jump to the sixteenth step; otherwise move to step five.

In the sixth step, we store the fitness of bacterium $x(:,1,1)$ or '1' into $FitLast = OF(x(:,1,1)) = 1.5605$ and a random vector Δ_1 is generated for bacterium '1', e.g., $\Delta_1 = [-0.4462\ -0.9077]^T$.

In the seventh step, we update the position of the first bacterium by using (3.1) as $x(:,1,1) = x(:,1,1) + C \cdot \dfrac{\Delta_1}{\sqrt{\Delta_1^T \cdot \Delta_1}} = [5.7816\ -6.7496]^T$.

In the eighth step, we determine the fitness value of the updated bacterium by using $OF(.)$, i.e., $OF(x(:,1,1)) = 0.9669$.

In the ninth step, the swimming behaviour of the bacterium is executed. To do this, a swimming counter is set, i.e., $sw = 1$. While $sw < N_s$ then $FitLast$ is compared with $OF(x(:,1,1))$. If $OF(x(:,1,1)) < FitLast$ then $x(:,1,1)$ is further updated as $x(:,1,1) = x(:,1,1) + C \cdot \dfrac{\Delta_1}{\sqrt{\Delta_1^T \cdot \Delta_1}} = [5.3405\ -7.6471]^T$, where $OF(x(:,1,1)) = 0.6455$. If $OF(x(:,1,1)) > FitLast$ then set $sw = N_s$. Similarly, other bacteria of all chemotactic steps are updated and their fitness values are also evaluated, as shown below.

Updated bacteria in all chemotactic steps

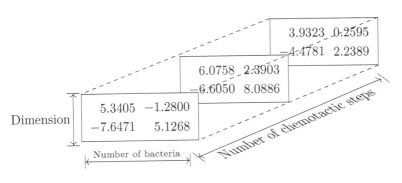

Fitness values of all new bacteria

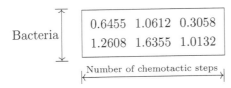

In the tenth step, reproduction is performed. First, we determine the sum of bacteria fitness in all chemotactic steps and then the population is sorted according to the evaluated sum. Assume that sum is called $FitHealth$ and sorted fitness values are referred as $Health$, as demonstrated below.

$FitHealth = [(0.6455 + 1.0612 + 0.3058) \quad (1.2608 + 1.6355 + 1.0132)]^T = [2.0125 \quad 3.9095]^T$ and $Health = [2.0125 \quad 3.9095]^T$.

In the eleventh step, we eliminate the bottom $N_{bact}/2$ bacteria and split the top $N_{bact}/2$ bacteria population, with better fitness values, to keep the population constant. From $FitHealth$, it is observed that bacteria two has poor fitness; therefore we eliminate this poor bacteria by replacing with fittest bacteria, i.e. bacteria one, in the first chemotactic step. The updated bacteria of the first chemotactic step are given below; in the second and third chemotactic steps, bacteria will remain unaltered.

$$\text{Dimension} \begin{bmatrix} 5.3405 & 5.3405 \\ -7.6471 & -7.6471 \end{bmatrix}$$

$$\underset{\longleftarrow \text{Bacteria} \longrightarrow}{}$$

In the twelfth step, the elimination-dispersal operation is performed. To do this, a random number, $rr \epsilon [0\ 1]$ is generated for each bacterium. If $rr < P_{disp}$ then that bacterium is randomly updated. For example, $rr = 0.9963$ which is greater than P_{disp} therefore, bacterium one, i.e. $x(:, 1, 1) = [5.3405\ -7.6471]^T$ will remain unaltered. Similarly, rr is generated for bacterium two, e.g., $rr = 0.3469$ which is less than P_{disp} therefore updated randomly as $x(:, 2, 1) = LB(2) + (UB(2) - LB(2)). * rand(2, 1) = [0.3203\ 1.0687]^T$. The updated bacteria of the first chemotactic step are shown below; the others remain the same.

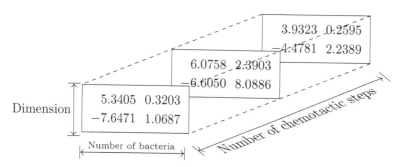

In the thirteenth step, we determine the fitness values of updated bacteria and then store in the fitness matrix as

$$\text{Bacteria} \begin{bmatrix} 0.6455 & 1.7725 & 0.7354 \\ 0.3095 & 0.9051 & 0.6064 \end{bmatrix}$$

$$\underset{\longleftarrow \text{Number of chemotactic steps} \longrightarrow}{}$$

In the fourteenth step, we determine the best bacterium of the new population and then compare with the best of the old one. If the new best is better, we replace with the new; otherwise not. For example, the best fitness of the updated population is $Fit_{best}^{new} = 0.3095$, which corresponds to bacterium $x_{best}^{new} = [0.3203 \ 1.0687]^T$. Here we are minimising the function $OF(.)$ and $Fit_{best}^{new} < Fit_{best}$ thus $x_{best} = x_{best}^{new} = [0.3203 \ 1.0687]^T$.

In the fifteenth step, we check whether the maximum number of elimination-dispersal events is reached. If yes, move to the sixteenth step; otherwise return to the fifth step.

In the sixteenth step, we print the best solution x_{best} with its corresponding fitness Fit_{best}.

3.6 Conclusions

In this chapter, the standard version of the BFO algorithm, inspired by the foraging behavior of bacteria, is discussed. The pseudo-codes of the algorithm are presented and explained. In order to understand the coding of this optimization method, Matlab source-code is also provided. The applicability of this method to mathematical optimization problems is demonstrated by step-by-step example in which the benchmark Griewank function is minimized by using the BFO algorithm.

3.7 Acknowledgement

This work was supported by the Engineering and Physical Sciences Research Council (EPSRC) of the United Kingdom (Reference Nos.: EP/R001456/1 and EP/S001778/1).

References

1. K.M. Passino. "Biomimicry of bacterial foraging for distributed optimization and control". *IEEE Control Systems*, vol. 22(3), 2002, pp.52-67.

2. M. A. Munoz, S. K. Halgamuge, W. Alfonso and E. F. Caicedo. "Simplifying the Bacteria Foraging Optimization Algorithm". *IEEE Congress on Evolutionary Computation*, Barcelona, 2010, pp. 1-7. doi: 10.1109/CEC.2010.5586025

3. H. Chen, Y. Zhu, and K. Hu. "Adaptive bacterial foraging optimization". *Abstract and Applied Analysis, Hindawi*, vol. 2011. doi:10.1155/2011/108269

4. B. Niu, H. Wang, J. Wang, and L. Tan. "Multi-objective bacterial foraging optimization". *Neurocomputing*, vol. 116, 2013, pp.336-345.

5. P. Manikandan and D. Ramyachitra. "Bacterial Foraging Optimization-Genetic Algorithm for Multiple Sequence Alignment with Multi-Objectives". *Scientific Reports*, vol. 7(1), 2017, Article number: 8833.

4

Bat Algorithm

Xin-She Yang

School of Science and Technology
Middlesex University, London, United Kingdom

Adam Slowik

Department of Electronics and Computer Science
Koszalin University of Technology, Koszalin, Poland

CONTENTS

4.1 Introduction

The original bat algorithm (BA) was first developed by Xin-She Yang in 2010 [1], which attempts to mimic the main characteristics of echolocation of microbats. Due to its good convergence, it has been applied to solve various optimization problems [2], and it has also been extended to solve a multi-objective optimization problem [3] and chaotic variants [4]. Its applications are diverse [5, 6], and a detailed review about the earlier literature was first given by Yang and He in 2013 [2]. In addition, many variants have been developed to improve its efficiency in various ways, including a hybrid bat algorithm [7], directional bat algorithm [8, 9], discrete bat algorithm [10], and others.

The global convergence of the standard bat algorithm has recently been proved mathematically using both a Markovian framework and dynamical system theory [11].

The rest of this chapter provides all the fundamentals of the bat algorithm with the main pseudo-code, Matlab code and C++ code, followed by some simple demonstration of how it works.

4.2   Original bat algorithm

Bats, especially microbats, use echolocation for navigation, which consists of emitting a series of short, ultrasonic pulses that typically last a few milliseconds with frequencies in the range of 25 kHz to about 150 kHz. The loudness of such bursts can be up to 110 dB. When homing for prey, microbats typically increase their pulse emission rates and frequencies, which is also accompanied by variation of their loudness. The main purpose of such frequency tuning and echolocation is for navigation and hunting so as to increase the accuracy of detection and success rate of capturing the prey. Such characteristics are simulated in the bat algorithm.

4.2.1   Description of the bat algorithm

For a population of n bats, each bat (say, bat i) is associated with a location or position vector $\boldsymbol{x}_i$ and a corresponding velocity vector $\boldsymbol{v}_i$. During the iterations, each bat can vary its pulse emission rate r_i, loudness A_i and its frequency f_i.

In general, a position vector is considered as a solution vector to an optimization problem in a D-dimensional search space with D independent design variables

$$\boldsymbol{x} = [x_1, \ x_2, \ x_3, \ ..., \ x_D], \tag{4.1}$$

which will vary with the iteration or a pseudo-time counter t. So, we use the notation $\boldsymbol{x}_i^t$ to denote the position of bat i at iteration t. Its corresponding velocity is thus denoted by $\boldsymbol{v}_i^t$.

Frequency tuning can be carried out by

$$f_i = f_{\min} + \beta(f_{\max} - f_{\min}), \tag{4.2}$$

where $f_{\min}$ and $f_{\max}$ are the minimum and maximum ranges, respectively, of the frequency f_i for each bat i. Here, β is a uniformly distributed number drawn from $[0, 1]$. In most applications, we should use $f_i = O(1)$. For example, we can start with $f_{\min} = 0$ and $f_{\max} = 2$.

This frequency variation is then used for updating the velocity of each bat

$$v_i^{t+1} = v_i^t + (x_i^t - x_*)f_i, \tag{4.3}$$

where x_* is the best solution found at iteration t.

The position or solution vector x_i is updated by

$$x_i^{t+1} = x_i^t + (\Delta t)v_i^{t+1}, \tag{4.4}$$

where Δt is the iteration or time increment. As all iterative algorithms are updated in a discrete manner, we usually set $\Delta t = 1$. Thus, we can simply consider the vectors without any physical units, and then write the update equation as

$$x_i^{t+1} = x_i^t + v_i^{t+1}. \tag{4.5}$$

The pulse emission rate r_i can monotonically increase from a lower value $r_i^{(0)}$, while the loudness can reduce from a higher value $A^{(0)} = 1$. We have

$$A_i^{t+1} = \alpha A_i^t, \quad r_i^{t+1} = r_i^{(0)}[1 - \exp(-\gamma t)], \tag{4.6}$$

where $0 < \alpha < 1$ and $\gamma > 0$ are constants. When t is large enough ($t \to \infty$), we have $A_i^t \to 0$ and $r_i^t \to r_i^{(0)}$. To start with the simple demonstration later, we can use $\alpha = 0.97$ and $\gamma = 0.1$.

For the local modification around a member of the best solution, we can use

$$x_{\text{new}} = x_{\text{old}} + \sigma \epsilon_t A^{(t)}, \tag{4.7}$$

where ϵ_t is a random number drawn from a normal distribution N(0,1), and σ is the standard deviation acting as a scaling factor. For simplicity, we can use $\sigma = 0.1$ in our later implementation. Here, $A^{(t)}$ is the average loudness at iteration t.

4.2.2 Pseudo-code of BA

Based on the above descriptions, the main steps of the BA consists of a single loop with some probabilistic switching during the iteration. Thus, the procedure of the bat algorithm is summarized in the pseudo-code shown in Algorithm 3.

4.2.3 Parameters in the bat algorithm

Apart from the population size n, there are two parameters and four initialization values in the bat algorithm. They are $\alpha, \gamma, f_{\min}, f_{\max}, A_i^{(0)}$ and $r_i^{(0)}$. In our implementation, we have used $\alpha = 0.97$, $\gamma = 0.1$, $f_{\min} = 0$, $f_{\max} = 2$, and $A_i^{(0)} = r_i^{(0)} = 1$. Though n should vary for different applications, we have used $n = 10$ here.

Algorithm 3 Pseudo-code of the bat algorithm.

1: Define the objective function $f(\boldsymbol{x})$
2: Initialize the bat population $\boldsymbol{x}_i$ and $\boldsymbol{v}_i (i = 1, 2, ..., n)$
3: Initialize frequencies f_i, pulse rates r_i and loudness A_i
4: Set the iteration counter $t = 0$
5: **while** $(t < t_{\max})$ **do**
6:     Vary r_i and A_i
7:     Generate new solutions by adjusting frequencies
8:     Update velocities and locations/solutions via Eqs (4.3) and (4.5)
9:     **if** rand $> r_i$ **then**
10:         Select a solution among the best solutions
11:         Generate a local solution around the selected best solution
12:     **end if**
13:     Generate a new solution by flying randomly
14:     **if** rand $> A_i$ and $f(\boldsymbol{x}_i) < f(\boldsymbol{x}_j)$ **then**
15:         Accept the new solution
16:     **end if**
17:     Rank the bats and find the current best solution $\boldsymbol{x}_*$
18: **end while**

4.3   Source code of bat algorithm in Matlab

The implementation of this algorithm is relatively straightforward using Matlab. Here, we present a detailed description of a simple Matlab implementation to find the minimum of the function $f(\boldsymbol{x})$

$$\text{minimize} f(\boldsymbol{x}) = (x_1 - 1)^2 + (x_2 - 1)^2 + ... + (x_D - 1)^2, \quad x_i \in \mathbb{R}, \quad (4.8)$$

which has a global minimum of $\boldsymbol{x}_* = (1, 1, ..., 1)$. In theory, the algorithm can search the whole domain; we always impose some limits in practice. Here, for simplicity, we use the following lower bound (Lb) and upper bound (Ub):

$$Lb = [-5, -5, ..., -5], \quad Ub = [+5, +5, ..., +5]. \quad (4.9)$$

The Matlab code for the bat algorithm here consists of three parts: initialization, the main loop, and the objective function. The whole codes should consist of all the lines of codes in a sequential order. However, for ease of understanding, we split them into three parts.

The first part is mainly initialization of parameter values such as α and γ, as well as the generation of the initial population of $n = 10$ bats. The cost function is the objective function to be given later in a Matlab function.

It is worth pointing out that the simple bounds should be checked at each iteration when new solutions are generated. However, for simplicity, the implementation here does not carry out this check, and care should be taken when extending the simple computer code here.

```
1  function [best,fmin]=bat_algorithm
2  % Default parameters
3  n=10;                 % Population size, typically 10 to 25
4  A=1;                  % Loudness (constant or decreasing)
5  r=1;                  % Pulse rate (constant or decreasing)
6  alpha=0.97;           % Parameter alpha
7  gamma=0.1;            % Parameter gamma
8  % Frequency range
9  Freq_min=0;           % Frequency minimum
10 Freq_max=2;           % Frequency maximum
11 % Max number of iterations
12 t_max=1000;
13 t=0;                  % Initialize iteration counter
14 % Dimension of the search variables
15 d=5;
16 % Initialization of arrays
17 Freq=zeros(n,1);      % Frequency
18 v=zeros(n,d);         % Velocities
19 Lb=-5*ones(1,d);      % Lower bounds
20 Ub=5*ones(1,d);       % Upper bounds
21 % Initialize the population/solutions
22 for i=1:n,
23    Sol(i,:)=Lb+(Ub-Lb).*rand(1,d);
24    Fitness(i)=Fun(Sol(i,:));
25 end
26 % Find the best solution of the initial population
27 [fmin,I]=min(Fitness);
28 best=Sol(I,:);
```

Listing 4.1
BA demo initialization.

The second part is the main part, consisting a loop over the whole population at each iteration. The pulse emission rate and loudness are varied first, then new solutions are generated and evaluated. The new population is checked and the current best solution is found and updated. Though each bat can have its own pulse emission rate r_i and loudness A_i, we have used the same values of r and A for all bats. This means that the average of A_i is now simply A itself.

```
1  % Start the iterations -- Bat Algorithm
2  while (t<t_max)
3     % Varying parameters
4     r=r*(1-exp(-gamma*t));
5     A=alpha*A;
6     % Loop over all bats/solutions
7     for i=1:n,
8        Freq(i)=Freq_min+(Freq_max-Freq_min)*rand;
9        v(i,:)=v(i,:)+(Sol(i,:)-best)*Freq(i);
10       S(i,:)=Sol(i,:)+v(i,:);
11       % Check a switching condition
12       if rand>r,
13       S(i,:)=best+0.1*randn(1,d)*A;
14       end
15
16       % Evaluate new solutions
17       Fnew=Fun(S(i,:));
18       % If the solution improves or not to loud
19        if (Fnew<=Fitness(i)) & (rand>A) ,
20           Sol(i,:)=S(i,:);
```

```
21        Fitness(i)=Fnew;
22      end
23
24    % Update the current best solution
25      if Fnew<=fmin,
26        best=S(i,:);
27        fmin=Fnew;
28      end
29    end % end of for
30    t=t+1;  % Update iteration counter
31  end
32  % Output the best solution
33  disp([ 'Best =',num2str(best),' fmin=',num2str(fmin)]);
```

Listing 4.2
Main iterations for the bat algorithm.

The third part is the objective function and ways for implementing the lower and upper bounds. This will ensure the solution vector should be within the regular bounds; however, we have not implemented the bounds here for simplicity. A new solution is evaluated by calling the objective or cost function.

```
1  %% Cost or objective function
2  function z=cost(x)
3  z=sum((x-1).^2);  % Solutions should be (1,1,...,1)
```

Listing 4.3
The objective function.

If we run this code with a maximum number of 1000 iterations, we can get the best minimum value as $f_{best} = 3.49 \times 10^{-29}$. Obviously, due to the random numbers used, these results are not exactly repeatable, but the order of the magnitude is easily reachable in the simulation.

4.4   Source code in C++

As the bat algorithm is simple, it can be implemented easily in any other programming language. For the same algorithm to solve the same problem as given above in Matlab, we can implement it in C++ as follows:

```
1  #include <iostream>
2  #include <time.h>
3  #include <random>
4  using namespace std;
5  //Definition of the objective function OF(.)
6  float Fun(float x[], int size_array)
7  {   float t=0;
8          for(int i=0; i<size_array; i++)
9          {t=t+(x[i]-1)*(x[i]-1);}
10      return t;
11  }
12  //Generate pseudo random value from the range [0, 1)
13  float ra() {return (float)rand()/RAND_MAX;}
14  int main()
15  {
```

```cpp
16  //Normal distribution generator
17  std::default_random_engine generator;
18  std::normal_distribution<double> distribution(0,1.0);
19  //Initialization of pseudo random generator
20  srand(time(NULL));
21  int n=10;          //Population size, typically 10 to 25
22  float A=1.0;       //Loudness (constant or decreasing)
23  float r=1.0;       //Pulse rate (constant or increasing)
24  float alpha=0.97;  //Parameter alpha
25  float gamma=0.1;   //Parameter gamma
26  //Frequency range
27  float Freq_min=0.0;  //Frequency minimum
28  float Freq_max=2.0;  //Frequency maximum
29  //Max number of iterations
30  int t_max=1000;
31  //Dimension of the search variables
32  int d=5;
33  //Initialization of arrays
34  float Freq[n];     //Frequency
35  float v[n][d];     //Velocities
36  for(int i=0; i<n; i++)
37  {
38      for(int j=0; j<d; j++){v[i][j]=0;}
39  }
40  float Lb[d];  //Lower bounds
41  float Ub[d];  //Upper bounds
42  float Sol[n][d];    //Solutions
43  float Fitness[n];   //Fitness
44  float S[n][d];      //New solutions
45  float rnum;
46  float Fnew;
47  //Prepare arrays
48  for(int j=0; j<d; j++) {Lb[j]=-5.0; Ub[j]=5.0;}
49  //Initialize the population/solutions
50  for(int i=0; i<n;i++)
51  {
52      for(int j=0; j<d; j++)
53      {
54          Sol[i][j]=Lb[j]+(Ub[j]-Lb[j])*ra();
55      }
56      Fitness[i]=Fun(Sol[i],d);
57  }
58  //Find the best solution of the initial population
59  int I=0; float fmin;
60  fmin=Fitness[I];
61  for(int i=1; i<n; i++)
62  {
63      if(Fitness[i]<=fmin)
64      {
65          fmin=Fitness[i];
66          I=i;
67      }
68  }
69  float best[d];
70  for(int i=0; i<d; i++)
71      {
72          best[i]=Sol[I][i];
73      }
74  //Start the iterations --- Bat Algorithm
75  for(int t=1; t<=t_max; t++)
76  {
77      //Varying parameters
78      r=r*(1-exp(-gamma*t));
79      A=alpha*A;
80      //Loop over all bats/solutions
81      for(int i=0; i<n; i++)
82      {
```

```
83      Freq[i]=Freq_min+(Freq_max-Freq_min)*ra();
84      rnum=ra();
85      for(int j=0; j<d; j++)
86      {
87          v[i][j]=v[i][j]+(Sol[i][j]-best[j])*Freq[i];
88          S[i][j]=Sol[i][j]+v[i][j];
89          if (rnum>r)
90          {
91              S[i][j]=best[j]+0.1*A*distribution(generator);
92          }
93      }
94      //Evaluate new solutions
95      Fnew=Fun(S[i],d);
96      //If the solution improves or not to loud
97      if ((Fnew<=Fitness[i]) && (ra()>A))
98      {
99          Fitness[i]=Fnew;
100         for(int j=0; j<d; j++)
101         {
102             Sol[i][j]=S[i][j];
103         }
104     }
105     //Update the current best solution
106     if (Fnew<=fmin)
107     {
108         fmin=Fnew;
109         for(int j=0; j<d; j++)
110         {
111             best[j]=S[i][j];
112         }
113     }
114     }
115 }
116 //Output the best solution
117 cout<<"#### fmin = "<<fmin<<endl;
118 cout<<"#### Best = [ ";
119 for(int i=0; i<d; i++) {cout<<best[i]<<" ";}
120 cout<<"]"<<endl;
121 getchar();
122 return 0;
123 }
```

Listing 4.4
Bat algorithm in C++.

4.5   A worked example

Let us use the bat algorithm to find the minimum of

$$f(x) = (x_1 - 1)^2 + (x_2 - 1)^2 + (x_3 - 1)^2, \qquad (4.10)$$

in the simple ranges of $-5 \leq x_i \leq 5$. This simple 3-variable problem has the global minimum solution $x_{\text{best}} = [1, 1, 1]$.

For the purpose of algorithm demonstration, we only use $n = 5$ bats. Suppose the initial population is randomly initialized and their objective (fitness) values are as follows:

$$\begin{cases} \boldsymbol{x}_1 = (\ 5.00 \quad 0.00 \quad 5.00 \ ), & f_1 = f(\boldsymbol{x}_1) = 33.00, \\ \boldsymbol{x}_2 = (\ 2.00 \quad 2.00 \quad 3.00 \ ), & f_2 = f(\boldsymbol{x}_2) = 6.00, \\ \boldsymbol{x}_3 = (\ -3.00 \quad -2.00 \quad 0.00 \ ), & f_3 = f(\boldsymbol{x}_3) = 26.00, \\ \boldsymbol{x}_4 = (\ -5.00 \quad 0.00 \quad 5.00 \ ), & f_4 = f(\boldsymbol{x}_4) = 53.00, \\ \boldsymbol{x}_5 = (\ 3.00 \quad 4.00 \quad 5.00 \ ), & f_5 = f(\boldsymbol{x}_5) = 29.00, \end{cases} \qquad (4.11)$$

where we have only used two decimal places for simplicity. Though the actual random numbers generated for the initial population can be any numbers in the simple bounds, here we have used the numbers that are rounded up as the initialization; this is purely for simplicity and clarity.

Obviously, the best solution with the lowest (or best) value of the objective function is $\boldsymbol{x}_2$ with $f_2 = 6.00$ for this population at $t = 0$. Thus, we have

$$\boldsymbol{x}_{\text{best}} = [2.00, \ 2.00, \ 3.00]. \qquad (4.12)$$

In principle, each bat in the bat population can have its own pulse emission rate r_i and loudness A_i. However, for simplicity of implementation and discussion here, we can use the same r and A for all bats.

At the first iteration $t = 1$, starting with $r^{(0)} = 1$, $A^{(0)} = 1$, $\alpha = 0.97$ and $\gamma = 0.1$, we have

$$r = r^{(0)}[1 - \exp(-\gamma t)] = 1 \times [1 - \exp(-0.1 \times 1)] = 0.0952, \qquad (4.13)$$

$$A = \alpha A^{(0)} = 0.97 \times 1 = 0.97. \qquad (4.14)$$

When carrying out the loop over all the 5 bats, if we draw a random number $\beta = 0.2$, then the frequency becomes $f_i = 0 + 0.2 \times (2 - 0) = 0.4$, so the velocity $\boldsymbol{v}_1$ can be calculated by

$$\boldsymbol{v}_1^1 = \boldsymbol{v}_1^0 + (\boldsymbol{x}_1 - \boldsymbol{x}_{\text{best}}) \times 0.4 = [4.20, -2.80, 2.80], \qquad (4.15)$$

where we have used $f_{\min} = 0$ and $f_{\max} = 2$. This gives the new solution

$$\boldsymbol{x}_1^1(\text{new}) = \boldsymbol{x}_1 + \boldsymbol{v}_1 = [9.20, -2.80, 7.8], \qquad (4.16)$$

which in turn gives $f(\boldsymbol{x}_1^1) = 127.92$. This is worse than the previous solutions. However, the next step is to generate a uniformly distributed random number. Suppose we get $\beta = 0.5$, which is greater than $r = 0.0952$; we carry out a local search by

$$\boldsymbol{x}_1^1(\text{new}) = \boldsymbol{x}_{\text{best}} + 0.1\epsilon A. \qquad (4.17)$$

If we draw a random number vector, we get

$$\epsilon = [-2, 0.5, -1.5], \qquad (4.18)$$

and we have

$$\boldsymbol{x}_1^1(\text{new}) = \boldsymbol{x}_{\text{best}} + 0.1\epsilon \times 0.97 = [1.8060, 2.0485, 2.8545]. \qquad (4.19)$$

This gives $f(\boldsymbol{x}_1^1) = 5.1882$, which is better than $\boldsymbol{x}_{\text{best}}$. So we update it as the new best solution

$$\boldsymbol{x}_{\text{best}} = [1.8060, 2.0485, 2.8545]. \tag{4.20}$$

Similarly, we do this for the rest of the other four solutions in the population. Once the first iteration is done, we vary r and A again, and then carry out another loop over the whole population.

The bat algorithm can be efficient. For the objective function, the optimal solution is $\boldsymbol{x}_* = [1, 1, 1]$. Using the above codes, we can typically get $f_{\text{best}} = 4.9 \times 10^{-16}$ after 500 iterations.

4.6 Conclusion

The bat algorithm is a simple, flexible and yet efficient algorithm for solving optimization problems. In addition, this algorithm is relatively simple to implement with a minimum use of memory, and the computation cost is also low. The bat algorithm has been extended to other forms with many variants and applications [2, 10, 11].

References

1. X.S. Yang, "A new metaheuristic bat-inspired algorithm", in: *Nature Inspired Cooperative Strategies for Optimization (NICSO 2010)*, pp. 65-74, Springer, Berlin, 2010.

2. X.S. Yang, X.-S. He, "Bat algorithm: literature review and applications", *Int. J. Bio-Inspired Computation*, 5(3): 141-149, 2013.

3. X.S. Yang, "Bat algorithm for multi-objective optimisation", *Int. J. Bio-Inspired Computation*, 3(5):267-274, 2011.

4. A.H. Gandomi and X.-S. Yang, "Chaotic bat algorithm", *Journal of Computational Science*, 5(2): 224–232, 2014.

5. X.S. Yang and A. H. Gandomi, "Bat algorithm: a novel approach for global engineering optimization", *Engineering Computation*, 29(5): 464-483, 2012.

6. A.H. Gandomi, X.-S. Yang, A.H. Alavi, S. Talatahari, "Bat algorithm for constrained optimization tasks", *Neural Computing and Applications*, 22(6): 1239-1255, 2013.

7. I. Fister Jr., D. Fister, X.-S. Yang, "A hybrid bat algorithm", *Elekrotehnivški Vestn.*, 80(1-2): 1-7, 2013.

8. A. Chakri, R. Khelif, M. Benouaret, X.-S. Yang, "New directional bat algorithm for continuous optimization problems", *Expert Systems with Applications*, 69(1): 159-175, 2017.

9. A. Chakri, X.-S. Yang, R. Khelif, M. Benouaret, "Reliability-based design optimization using the directional bat algorithm", *Neural Computing and Applications*, 30(8): 2381-2402, 2018.

10. E. Osaba, X.-S. Yang, I. Fister Jr., J. Del Ser, P. López-García, A.J. Vazquez-Pardavila, "A discrete and improved bat algorithm for solving a medical good distribution problem with pharmacological waste collection", *Swarm and Evolutionary Computation*, 44(1): 273-286, 2019.

11. S. Chen, G.-H. Peng, X.-S. He, X.-S. Yang, "Global convergence analysis of the bat algorithm using a Markovian framework and dynamical system theory", *Expert Systems with Applications*, 114(1): 173-182, 2018.

5

Cat Swarm Optimization

Dorin Moldovan
Department of Computer Science
Technical University of Cluj-Napoca, Romania

Viorica Chifu
Department of Computer Science
Technical University of Cluj-Napoca, Romania

Ioan Salomie
Department of Computer Science
Technical University of Cluj-Napoca, Romania

Adam Slowik
Department of Electronics and Computer Science
Koszalin University of Technology, Koszalin, Poland

CONTENTS

5.1 Introduction

CSO is introduced in [1] by Chu et al. where the first version of the CSO algorithm is proposed. The main inspiration of the CSO algorithm is represented by the behavior of the cats and the algorithm optimizes the searching of a solution in an M-dimensional space according to a fitness function. The solution sets are portrayed by cats and each cat is characterized by a number of dimensions, velocities for each dimension, a flag which describes whether the cat is in seeking mode or in tracing mode and a fitness value. The best solution is represented by the cat that has the best fitness value. The seeking mode models the situation in which the cat in a resting position looks around and seeks the next position to move to while the tracing mode models the situation in which the cat traces some targets. In literature there are many modifications of the CSO algorithm: CSO for clustering [2], parallel CSO [3], a parallel version of CSO that is based on the Taguchi Method [4], a modified version of the CSO algorithm in which the concept of craziness is introduced which is called Crazy-CSO [5], Multi-Objective Binary Cat Swarm Optimization (MOBCSO) [6], a gray image segmentation algorithm based on CSO [7], Harmonious Cat Swarm Optimization (HCSO) [8], Discrete Binary Cat Swarm Optimization (DBCSO) [9] and Quantum Cat Swarm Optimization Clustering (QCSOC) algorithm [10]. The CSO algorithm is generally applied for solving various engineering problems and we applied this algorithm successfully in [11] for generating diets for elders. The rest of the paper is organized as follows: Section 5.2 presents and discusses the global version and the local version of the CSO algorithm, Section 5.3 presents the source-code of the global version of the CSO algorithm in Matlab, Section 5.4 presents the source-code of the global version of the CSO algorithm in C++, in Section 5.5 is presented a detailed numerical example of the global version of the CSO algorithm and in Section 5.6 are presented the main conclusions.

5.2   Original CSO algorithm

5.2.1   Pseudo-code of global version of CSO algorithm

The pseudo-code for the global version of the CSO algorithm is presented in Algorithm 4 and is adapted after the one that is described in [12]. The algorithm takes as input the following parameters: M - the number of dimensions of the search space, $\left[P_{i,j}^{min}, P_{i,j}^{max}\right]$ - the range of variability for the positions of the cats, MR - the mixture ratio which is a percent that describes how many cats are in seeking mode and how many cats are in tracing mode, SMP - the seeking memory pool, SRD - the seeking range of the selected dimension, CDC - the count of dimension to change, SPC - the self-position consideration, N - the total number of cats, max - the maximum number of iterations and c_1 - a constant which is used for updating the velocity of the cats in tracing mode. The output is represented by the global best, which is the best position achieved by a cat. In this chapter we present a simplified version of the global version of the CSO algorithm in which we suppose that the flag SPC is equal to 0 and the point to move to in the seeking mode is chosen randomly from the SMP copies, each copy having the same probability of being chosen. In step 1 is created the initial population of N cats randomly and each cat is represented using an M-dimensional vector of values. Then the $Gbest$, the global best, is updated considering the best cat (step 3). While the stopping criterion is not satisfied or the number of iterations is less than the maximum number of iterations, a set of steps is repeated. According to MR some cats will be in tracing mode, while the rest will be in seeking mode (step 5). The global best cat is updated after the fitness values of all cats are computed (step 6) and then for each cat the values of the new positions are determined considering whether the cat is in seeking mode or in tracing mode. If the cat is in seeking mode then SMP copies are created, the positions of the copies are updated considering the value of SRD and the new position to move is selected randomly. In the tracing mode, the velocity is updated first (step 15) and the new position is determined using the updated velocity (step 17). Finally, in step 21 the global best $Gbest$ is returned as the final result of the algorithm.

Algorithm 4 Pseudo-code of the global version of CSO.

Input M - the number of dimensions
$\left[P_{i,j}^{min}, P_{i,j}^{max}\right]$ - range of variability for i-th cat and j-th dimension
MR - mixture ratio
SMP - seeking memory pool
SRD - seeking range of the selected dimension
CDC - count of dimensions to change
SPC - self-position consideration
N - the total number of cats
max - maximum number of iterations
c_1 - a constant for updating the velocity of the cats in tracing mode
Output $Gbest$ - the best position achieved by a cat
1: randomly create the population of cats $X_i(i = 1, 2, ..., N)$ (each cat is a M-dimensional vector)
2: create the M-dimensional $Gbest$ vector
3: assign the best cat X_i to the $Gbest$
4: **while** stopping criterion not satisfied or $I < I_{max}$ **do**
5:     Set the cats in tracing mode or seeking mode according to MR
6:     Calculate the fitness values of all cats and update $Gbest$
7:     **for** $i = 1 : N$ **do**
8:         **if** X_i is in seeking mode **then**
9:             create SMP copies
10:             update the position of each copy using the formula
11:             $X_{cn} = X_c \times (1 \pm SRD \times R)$
12:             pick randomly a position to move to from the set of SMP copies
13:         **else**

14:          update the velocity of the cat using the formula
15:          $v_{i,d} = v_{i,d} + R \times c_1 \times (X_{best,d} - X_{i,d})$
16:          update the position of the cat using the formula
17:          $X_{i,d,new} = X_{i,d,old} + v_{i,d}$
18:       **end if**
19:    **end for**
20: **end while**
21: return the *Gbest* as a result

5.2.2   Description of global version of CSO algorithm

In this section is described the global version of the CSO algorithm. The algorithm starts by defining the algorithm's parameters and then the initial positions and velocities of the cats are generated randomly. The cats are distributed into tracing mode or seeking mode according to the value of a parameter that is called MR (Mutation Ratio). The fitness values of the cats are computed using an objective function and the best solution so far is the position of the best cat (X_{best}). Next are applied the seeking mode or the tracing mode process steps based on the values of the flags. If the termination criteria are met then the process is terminated otherwise the steps which follow after initialization are repeated. Some common termination criteria for the CSO are the number of the iterations, the running time and the amount of improvement. The seeking mode and the tracing mode are described in more detail next.

5.2.2.1   Seeking mode (resting)

The seeking mode describes the mode when the cat is resting. If the cat senses a danger then it moves cautiously and slowly. In the seeking mode the cat observes the M-dimensional space of solutions in order to decide what is the next move. The cat is aware of (a) its environment, (b) its own situation and (c) the choices it can make. In the CSO algorithm these facts are represented by the following parameters: (a) seeking memory pool (SMP), (b) seeking range of selected dimension (SRD), (c) count of dimensions to change (CDC) - the number of dimensions that will be mutated and (d) self-position consideration (SPC). The seeking mode process is described by the next steps: (1) for each cat X_i, SMP copies are created, and if the flag SPC is true then the cat's current position is considered as one of those SMP copies; (2) according to CDC the new position for each cat is computed using the following equation:

$$X_{cn} = X_c \times (1 \pm SRD \times R) \tag{5.1}$$

where X_{cn} is the new position, X_c is the current position and R is a random number from the interval $[0, 1]$; (3) for each position compute the fitness values of the new positions and if all the fitness values are equal then set the probability for all of the candidates' points to 1, otherwise use the equation described in the next step; (4) pick the point to move to randomly from the set of candidate points and replace the position of the cat X_i using the following equation:

$$P_i = \frac{FS_i - FS_b}{FS_{max} - FS_{min}} \tag{5.2}$$

where $0 < i < j$, P_i is the probability of the candidate cat X_i, FS_i is the fitness value of the cat X_i, FS_{max} is the maximum value of the fitness function, FS_{min} is the minimum value of the fitness function, $FS_{max} = FS_b$ for maximization problems and $FS_b = FS_{min}$ for minimization problems.

5.2.2.2 Tracing mode (movement)

The tracing mode simulates the chasing of a prey by a cat. The equations for updating the velocities and the positions of the cats are:

$$v_{k,d} = v_{k,d} + R \times c_1 \times (X_{best,d} - X_{k,d}) \tag{5.3}$$

and

$$X_{k,d,new} = X_{k,d,old} + v_{k,d} \tag{5.4}$$

where c_1 is a constant whose optimal value is determined considering the particularities of the optimization problem that will be solved and it does not have a specific range of variability, its value being determined in most of the cases using a trial and error procedure, R is a random value from the interval $[0, 1]$, $v_{k,d}$ is the velocity of the cat k for the d dimension, $X_{k,d}$ is the position of the cat k for the d dimension, $X_{best,d}$ is the position of the cat that has the best solution for the d dimension, $X_{k,d,new}$ is the new value of the position of cat k for the d dimension and $X_{k,d,old}$ is the current position of the cat k for the d dimension.

5.2.3 Description of local version of CSO algorithm

In the global version of the CSO algorithm each cat goes towards the best cat from the whole swarm. On the other hand in the local version of the CSO algorithm each cat goes towards the best cat from its neighborhood. Thus the formula:

$$v_{k,d} = v_{k,d} + R \times c_1 \times (X_{best,d} - X_{k,d}) \tag{5.5}$$

is replaced with the formula:

$$v_{k,d} = v_{k,d} + R \times c_1 \times (X_{Lbest,d} - X_{k,d}) \tag{5.6}$$

where X_{Lbest} describes the local best and X_{best} describes the global best. Another variant is the one that includes ω which is the inertia factor.

$$v_{k,d} = \omega \times v_{k,d} + R \times c_1 \times (X_{Lbest,d} - X_{k,d}) \qquad (5.7)$$

The inertia factor ω is a numerical value which is used for controlling the rate at which the value of the velocity decreases. In literature its initial value is a constant from the interval $[0, 1]$, usually greater than 0.9, and its value decreases at each iteration of the algorithm with a constant numerical value such as 0.01, 0.001 or 0.0001. The neighborhood can be geometrical or social. Geometrical neighborhoods can be computed using several distances such as the Euclidean distance, the Manhattan distance or the Chebyshev distance. The social neighborhood depends on the index of the cat. Some common topologies are the ring topology in which the left neighbor and the right neighbor are considered or the star topology in which there is a central element connected to all other nodes. In the ring topology the neighbors of the i-th cat are determined using the following formulas:

$$Left_i = \begin{cases} i+1 & \text{if } i < N \\ 1 & \text{if } i = N \end{cases} \qquad (5.8)$$

$$Right_i = \begin{cases} i-1 & \text{if } i > 1 \\ N & \text{if } i = 1 \end{cases} \qquad (5.9)$$

The local best for cat i is updated considering the best value among its value, $Left_i$ and $Right_i$.

The star topology considers the relation between the cat mother and its kittens and the initial swarm of cats is divided into several smaller swarms that consist of one mother cat and its offspring. In this case the local best for each cat from the swarm is updated considering the best value obtained by a kitten or the mother cat. Figure 5.1 presents illustrative examples for these two topologies.

Some advantages are the fact that the algorithm might explore different areas and it can detect different local optima. However a disadvantage might be the fact that the local groups will have a smaller number of individuals and thus this modified version of the algorithm might not detect the global optimum.

5.3   Source-code of global version of CSO algorithm in Matlab

In Listing 5.1 is presented the source-code for the objective function which will be optimized by the CSO algorithm. In the function $OF(P)$, the input parameter is the position of a cat C. The objective function is given by formula 5.10, where $P_{i,j}$, $1 \leq j \leq M$, is the position of cat C_i for dimension j.

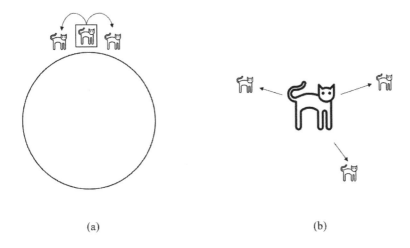

(a)   (b)

FIGURE 5.1
(a) Ring topology that considers the left and the right neighbor cats, (b) Star topology in which the cat mother is in the center.

$$OF(P) = \sum_{j=1}^{M} P_{i,j}^2 \qquad \text{where } -5.12 \leqslant P_{i,j} \leqslant 5.12 \qquad (5.10)$$

```matlab
function [out]=OF(P)
[x,y]=size(P);
out=zeros(x,1);
for i=1:x
  for j=1:y
    out(i,1)=out(i,1)+P(i,j)^2;
  end
end
```

Listing 5.1
Definition of objective function $OF(P)$ in Matlab.

In Listing 5.2 is presented the definition of the shuffle function in Matlab, a function that is used to permute randomly the initial values from the array given as input.

```matlab
function [out]=shuffle(x)
out=x(randperm(length(x)));
end
```

Listing 5.2
Definition of shuffle function $shuffle(.)$ in Matlab.

```matlab
%--- definition of the parameters of CSO algorithm
M=5; Pmin=zeros(1,M); Pmax=zeros(1,M); N=6; MAX=100; MR=0.2;
SMP=5; SRD=0.2; CDC=3; c1=0.5;
```

```
 4  %--- definition of the constraints value
 5  Pmin(1,:)=-5.12; Pmax(1,:)=5.12;
 6  VMAX=10; VMIN=-10;
 7  %--- definition of arrays required by CSO algorithm
 8  copies=zeros(SMP,M); numberOfCopies=SMP;
 9  P=zeros(N,M); V=zeros(N,M);
10  EvalPbest=zeros(N,1); Gbest=zeros(1,M);
11  flag=zeros(N,1); dimensionsToChange=zeros(1,M); Iter=0;
12  %--- population of cats is randomly created
13  P=P+(Pmax(1,:)-Pmin(1,:)).*rand(N,M)+Pmin(1,:);
14  Pbest=P;
15  %--- population of cats is evaluated
16  Eval=OF(P);
17  EvalPbest(:,1)=Eval(:,1);
18  %--- seeking the best cat
19  [Y,I]=min(Eval(:,1));
20  TheBest=I;
21  EvalGbest=Y;
22  %--- the main loop of the CSO algorithm
23  while(Iter<MAX)
24    Iter=Iter+1;
25    catsInSeekingMode=(1-MR)*N;
26    for i=1:N
27      if i<catsInSeekingMode
28        flag(i,1)=0;
29      else
30        flag(i,1)=1;
31      end
32    end
33    flag=shuffle(flag);
34    for i=1:N
35      if flag(i,1)==0
36        dimensionsToChange(1,:)=0;
37        dimensionsToChange(1,1:CDC)=1;
38        for c=1:numberOfCopies
39          copies(c,:)=P(i,:);
40        end
41        for c=1:numberOfCopies
42          dimensionsToChange=shuffle(dimensionsToChange);
43          for j=1:M
44            if dimensionsToChange(1,j)==1
45              if rand()>0.5
46                copies(c,j)=copies(c,j)+SRD*copies(c,j)*rand();
47              else
48                copies(c,j)=copies(c,j)-SRD*copies(c,j)*rand();
49              end
50            end
51          end
52        end
53        selectedCopy=randi(SMP);
54        for j=1:M
55          P(i,j)=copies(selectedCopy,j);
56          if P(i,j)>Pmax(j) P(i,j)=Pmax(j); end
57          if P(i,j)<Pmin(j) P(i,j)=Pmin(j); end
58        end
59      else
60        for j=1:M
61          V(i,j)=V(i,j)+rand()*c1*(P(TheBest,j)-P(i,j));
62          if V(i,j)>VMAX V(i,j)=VMAX; end
63          if V(i,j)<VMAX V(i,j)=VMIN; end
64          P(i,j)=P(i,j)+V(i,j);
65          if P(i,j)>Pmax(j) P(i,j)=Pmax(j); end
66          if P(i,j)<Pmin(j) P(i,j)=Pmin(j); end
67        end
68      end
69    end
70    TheBest=1;
```

```
71    for  i=1:N
72      Eval(i,1)=OF(P(i,:));
73      if Eval(i,1)<EvalPbest(i,1)
74        Pbest(i,:)=P(i,:);
75        EvalPbest(i,1)=Eval(i,1);
76      end
77      if Eval(i,1)<Eval(TheBest,1)
78        TheBest=i;
79      end
80    end
81    if Eval(TheBest,1)<EvalGbest
82      Gbest(1,:)=P(TheBest,:);
83      EvalGbest=Eval(TheBest,1);
84    end
85  end
86  disp(EvalGbest);
```

Listing 5.3
Source-code of the global version of the CSO in Matlab.

5.4 Source-code of global version of CSO algorithm in C++

```
1  #include <iostream>
2  #include <cstdlib>
3  #include <ctime>
4  #include <string>
5  using namespace std;
6  float OF(float x[], int size_array) {
7    float t = 0;
8    for(int i = 0; i < size_array; i++) {
9      t = t + x[i] * x[i];
10   }
11   return t;
12 }
13 float r() {
14   return (float)(rand()%1000)/1000;
15 }
16 void shuffle(int x[], int size_array) {
17   int temporaryValue = 0; int randomIndex = 0;
18   for(int i = 0; i < size_array; i++) {
19     randomIndex = rand() % size_array; temporaryValue = x[i];
20     x[i] = x[randomIndex]; x[randomIndex] = temporaryValue;
21   }
22 }
23 int main() {
24   srand(time(NULL));
25   int M = 5; float Pmin[M]; float Pmax[M]; int N = 6;
26   int TheBest = 0; int MAX = 100; int Iter = 0; float MR = 0.2;
27   for(int i = 0; i < M; i++) {
28     Pmin[i] = -5.12; Pmax[i] = 5.12;
29   }
30   int SMP = 5; float SRD = 0.2; int CDC = 3; float c1 = 0.5;
31   float VMAX = 10.00; float VMIN = -10.00;
32   float P[N][M]; float Pbest[N][M]; float V[N][M];
33   float Eval[N]; float EvalPbest[N]; float Gbest[M]; float EvalGbest;
34   int flag[N]; int dimensionsToChange[M];
35   for(int i = 0; i < N; i++) {
36     flag[i] = 0;
```

```
37      for(int  j = 0;  j < M;  j++) {
38        P[i][j] = (Pmax[j] - Pmin[j]) * r() + Pmin[j];
39        Pbest[i][j] = P[i][j];  V[i][j] = 0;
40      }
41      Eval[i] = OF(P[i], M);  EvalPbest[i] = Eval[i];
42      if(Eval[i] < Eval[TheBest]) TheBest = i;
43    }
44    EvalGbest = Eval[TheBest];
45    while(Iter < MAX) {
46      Iter++;  int  catsInSeekingMode = (1 - MR) * N;
47      for(int  i = 0;  i < N;  i++) {
48        flag[i] = (i < catsInSeekingMode) ? 0 : 1;
49      }
50      shuffle(flag, N);
51      for(int  i = 0;  i < N;  i++) {
52        if(flag[i] == 0) {
53          int  numberOfCopies = SMP;  float  copies[numberOfCopies][M];
54          for(int  j = 0;  j < M;  j++) dimensionsToChange[j] = 0;
55          for(int  j = 0;  j < CDC;  j++) {
56            dimensionsToChange[j] = (j < CDC) ? 1 : 0;
57          }
58          for(int  c = 0;  c < numberOfCopies;  c++) {
59            for(int  j = 0;  j < M;  j++) {
60              copies[c][j] = P[i][j];
61            }
62          }
63          for(int  c = 0;  c < numberOfCopies;  c++) {
64            shuffle(dimensionsToChange, M);
65            for(int  j = 0;  j < M;  j++) {
66              if(dimensionsToChange[j] == 1) {
67                copies[c][j] = (r() > 0.5)
68                  ? copies[c][j] + SRD * copies[c][j] * r()
69                  : copies[c][j] - SRD * copies[c][j] * r();
70              }
71            }
72          }
73          int  selectedCopy = std::rand() % SMP;
74          for(int  j = 0;  j < M;  j++) {
75            P[i][j] = copies[selectedCopy][j];
76            P[i][j] = P[i][j] > Pmax[j] ? Pmax[j] :
77              (P[i][j] < Pmin[j] ? Pmin[j] : P[i][j]);
78          }
79        } else {
80          for(int  j = 0;  j < M;  j++) {
81            V[i][j] += r() * c1 * (P[TheBest][j] - P[i][j]);
82            if(V[i][j] < VMIN || V[i][j] > VMAX) {
83              V[i][j] = (V[i][j] < VMIN) ? VMIN : VMAX;
84            }
85            P[i][j] = P[i][j] + V[i][j];
86            P[i][j] = P[i][j] > Pmax[j] ? Pmax[j] :
87              (P[i][j] < Pmin[j] ? Pmin[j] : P[i][j]);
88          }
89        }
90      }
91      TheBest = 0;
92      for(int  i = 0;  i < N;  i++) {
93        Eval[i] = OF(P[i], M);
94        if(Eval[i] < EvalPbest[i]) {
95          for(int  j = 0;  j < M;  j++) {
96            Pbest[i][j] = P[i][j];
97          }
98          EvalPbest[i] = Eval[i];
99        }
100       TheBest = (Eval[i] < Eval[TheBest]) ? i : TheBest;
101     }
102     if(Eval[TheBest] < EvalGbest) {
103       for(int  j = 0;  j < M;  j++) {
```

```
104        Gbest[j] = P[TheBest][j];
105      }
106      EvalGbest = Eval[TheBest];
107    }
108  }
109  cout << "EvalGBest = " << EvalGbest << endl;
110  return 0;
111 }
```

Listing 5.4
Source-code of the global version of the CSO in C++.

5.5 Step-by-step numerical example of global version of CSO algorithm

In the first step let us assume that we want to minimize an objective function $OF(.)$ given by equation 5.10, where M, the number of dimensions, is equal to 5.

In the second step we determine the parameter values of the CSO algorithm such as $N = 6$, $MR = 0.2$, $SMP = 5$, $SRD = 0.2$, $CDC = 3$, $MAX = 100$ and $c_1 = 0.5$.

In the third step the swarm which consists of 6 cats is created. Each cat is a 5-dimensional vector.
$Cat_1 = \{0.849, -0.501, 4.761, 1.597, 3.983\}$
$Cat_2 = \{0.460, -3.553, 0.061, -3.338, -1.095\}$
$Cat_3 = \{-1.802, -2.856, -2.662, 3.061, -1.464\}$
$Cat_4 = \{2.385, 2.467, 0.215, -3.061, 0.849\}$
$Cat_5 = \{-1.720, 2.682, -2.314, 4.300, -1.249\}$
$Cat_6 = \{-0.399, 4.874, -2.457, 2.785, -0.296\}$

In the fourth step, the 5-dimensional *Gbest* vector is created.
$Gbest = \{0, 0, 0, 0, 0\}$

In the fifth step, the $Catbest_i$ cat is created for each i-th cat Cat_i. In the beginning the cats $Catbest_i$ have the same values as the cats Cat_i.
$Catbest_1 = \{0.849, -0.501, 4.761, 1.597, 3.983\}$
$Catbest_2 = \{0.460, -3.553, 0.061, -3.338, -1.095\}$
$Catbest_3 = \{-1.802, -2.856, -2.662, 3.061, -1.464\}$
$Catbest_4 = \{2.385, 2.467, 0.215, -3.061, 0.849\}$
$Catbest_5 = \{-1.720, 2.682, -2.314, 4.300, -1.249\}$
$Catbest_6 = \{-0.399, 4.874, -2.457, 2.785, -0.296\}$

In the sixth step, for each cat Cat_i the 5-dimensional velocity vector V_i is created. At the start each vector V_i consists of zeros.

$V_1 = \{0, 0, 0, 0, 0\}$, $V_2 = \{0, 0, 0, 0, 0\}$, $V_3 = \{0, 0, 0, 0, 0\}$
$V_4 = \{0, 0, 0, 0, 0\}$, $V_5 = \{0, 0, 0, 0, 0\}$, $V_6 = \{0, 0, 0, 0, 0\}$

In the seventh step we evaluate each cat Cat_i using the objective function $OF(.)$ (please see equation 5.10).

$Eval\_Catbest_1 = OF(Cat_1) = 42.065$, $Eval\_Catbest_2 = OF(Cat_2) = 25.186$

$Eval\_Catbest_3 = OF(Cat_3) = 30.017$, $Eval\_Catbest_4 = OF(Cat_4) = 21.925$

$Eval\_Catbest_5 = OF(Cat_5) = 35.570$, $Eval\_Catbest_6 = OF(Cat_6) = 37.803$

In the eighth step we take the best cat Cat_i that has the smallest value for the objective function as our goal is to minimize the value of $OF(.)$ and assign that value to *Gbest*. In our case the best cat is Cat_4 and therefore $Gbest = Cat_4 = \{2.385, 2.467, 0.215, -3.061, 0.849\}$ and $OF(Gbest) = 21.925$.

In the ninth step, the main loop of the algorithm starts. We check whether the algorithm termination condition is fulfilled and in the example presented here we check whether the value of the current iteration is less than $MAX = 100$. If the termination condition is met then we jump to the final step; otherwise we continue with the tenth step.

In the tenth step we determine the number of cats in seeking mode and the number of cats in tracing mode. The number of cats in seeking mode is given by the formula $(1 - MR) \times N = 0.8 \times 6 = 4.8$ in which we consider only the integer part; thus the actual number is 4. The number of cats in tracing mode is equal to $6 - 4 = 2$. The cats are selected randomly for the seeking mode and for the tracing mode and after this selection the cats in seeking mode are $\{cat_1, cat_2, cat_3, cat_6\}$ and the cats is tracing mode are $\{cat_4, cat_5\}$.

Seeking mode

In the first substep of seeking mode we start the creation of $SMP = 5$ copies for each cat.

In the second substep of seeking mode we consider $CDC = 3$ dimensions to change for each copy. In other words for each copy we change 3 out of the 5 dimensions that characterize the current position of the copy and the 3 dimensions to change are selected randomly from those 5 dimensions. The dimensions to change for each copy of each cat are the following:

$D_{cat_1} = \{(d_1, d_2, d_4), (d_1, d_3, d_4), (d_1, d_2, d_5), (d_1, d_2, d_3), (d_2, d_3, d_4)\}$
$D_{cat_2} = \{(d_1, d_3, d_4), (d_1, d_2, d_5), (d_1, d_3, d_4), (d_1, d_3, d_5), (d_3, d_4, d_5)\}$

$D_{cat_3} = \{(d_1, d_3, d_4), (d_1, d_4, d_5), (d_1, d_3, d_5), (d_2, d_4, d_5), (d_2, d_3, d_5)\}$
$D_{cat_6} = \{(d_2, d_4, d_5), (d_1, d_2, d_4), (d_1, d_2, d_4), (d_2, d_3, d_4), (d_1, d_2, d_3)\}$

The dimensions to change will take new values using the formula from the next substep while the other dimensions will have the same values as the original. For example, in the case of the first copy of cat_1, the dimensions d_1, d_2 and d_4 will be changed. That means that this copy will have the value $(x_1, x_2, x_3 = 4.761, x_4, x_5 = 3.983)$ where x_1, x_2 and x_4 are new values that will computed in the next substep.

In the third substep of seeking mode we create the 5 copies for each cat, $5 \times 4 = 20$ total copies, using the following formula:
$X_{cn} = X_c \times (1 \pm SRD \times R)$
where X_{cn} is the new position, X_c is the current position, R is a random number from $[0, 1]$ and SRD has the value 0.2 and describes the percentage of how much the current dimension is modified. We consider that in the formula the signs plus and minus are chosen randomly with a probability of 50%.

In the fourth substep of seeking mode for each cat we select randomly one copy from the 5 copies and we check whether the values are in the acceptable range of variability. In the original version of the algorithm for each copy the probability of being selected is proportional with the value returned by $OF(.)$. In the version presented in this article we consider that each cat has the same probability of being chosen and that is why we do not necessarily choose the cat with the lowest value for $OF(.)$. The positions of the cats after this step are:
$P_1 = Copy(Cat_1) = \{0.766, -0.461, 4.761, 1.569, 3.983\}$
$P_2 = Copy(Cat_2) = \{0.465, -3.553, 0.056, -3.338, -1.013\}$
$P_3 = Copy(Cat_3) = \{-1.802, -2.890, -2.190, 3.061, -1.390\}$
$P_6 = Copy(Cat_6) = \{-0.448, 5.12, -2.422, -2.785, -0.296\}$

Tracing mode

In the first substep of tracing mode we update the velocities of the cats using the formula:
$v_{k,d} = v_{k,d} + R \times c_1 \times (X_{best,d} - X_{k,d})$
where $v_{k,d}$ is the velocity value for cat k and dimension d, R is a random number from the interval $[0, 1]$, c_1 is a constant that has value 0.5, $X_{best,d}$ is the value of the position of the best cat for the dimension d and $X_{k,d}$ is the current position of the cat k.

In the second substep of tracing mode we check whether the values of velocity are in the accepted range of variability and if they are not we update those values that are outside the range of variability. After this step the velocities of the two cats in tracing mode are:
$V_4 = \{0, 0, 0, 0, 0\}$
$V_5 = \{1.554, -0.030, 0.838, -3.055, 0.044\}$

The velocity of cat_4 is zero because cat_4 was the best cat in the first step of the algorithm and the equation that is used for updating the velocities of the cats considers the position of the best cat.

In the third substep of tracing mode we update the current position of the cat using the formula:

$X_{k,d,new} = X_{k,d,old} + v_{k,d}$

where $X_{k,d,new}$ is the new value for the k-th cat for dimension d, $X_{k,d,old}$ is the old value for the k-th cat for dimension d and $v_{k,d}$ is the value of velocity of the k-th cat for dimension d.

In the fourth substep of tracing mode we update the values of the positions in order to be in the accepted range of variability. The values of the positions for the two cats in tracing mode are:

$P_4 = \{2.385, 2.467, 0.215, -3.061, 0.849\}$

$P_5 = \{-0.166, 2.652, -1.475, 1.245, -1.205\}$

In the eleventh step we compute the fitness values of the cats, we update the values of their local bests and we update the value of *Gbest* if a better value is found. Then we return to the tenth step.

$OF(Cat_1) = 41.803$, $OF(Cat_2) = 25.016$, $OF(Cat_3) = 27.705$

$OF(Cat_4) = 21.925$, $OF(Cat_5) = 12.243$, $OF(Cat_6) = 40.128$

$Eval\_Catbest_1 = 41.803$, $Eval\_Catbest_2 = 25.016$, $Eval\_Catbest_3 = 27.705$

$Eval\_Catbest_4 = 21.925$, $Eval\_Catbest_5 = 12.243$, $Eval\_Catbest_6 = 37.803$

$Gbest = \{-0.166, 2.652, -1.475, 1.245, -1.205\}$

$OF(Gbest) = 12.243$

In the twelfth step we return *Gbest* as the result of the algorithm operation and we stop the algorithm. After 100 iterations the value of *Gbest* is $\{0.371, -0.054, -1.085, -0.024, -0.136\}$ and $OF(Gbest)$ is equal with 1.337

5.6   Conclusions

In this chapter we showed the main principles of the CSO algorithm. We showed how the algorithm works in the global version and we provided the source codes both in Matlab and in C++. These source codes could help others for a better understanding of the CSO algorithm and we believe that this chapter will make the implementations of others of their own versions of the CSO algorithm in any programming language easier to achieve.

References

1. S.-C. Chu, P.-w. Tsai, J.-S. Pan. "Cat Swarm Optimization" in *Lecture Notes in Artificial Intelligence*, vol. 4099, 2006, pp. 854-858.

2. B. Santosa, M. K. Ningrum. "Cat Swarm Optimization for Clustering" in *Proc. of IEEE 2009 International Conference of Soft Computing and Pattern Recognition*, 2009, pp. 54-59.

3. P.-W. Tsai, J.-S. Pan, S.-M. Chen, B.-Y. Liao. "Parallel Cat Swarm Optimization" in *Proc. of the 7th International Conference on Machine Learning and Cybernetics*, 2008, pp. 3328-3333.

4. P.-W. Tsai, J.-S. Pan, S.-M. Chen, B.-Y. Liao. "Enhanced Parallel Cat Swarm Optimization Based on the Taguchi Method" in *Expert Systems with Applications*, vol. 39, 2012, pp. 6309-6319.

5. A. Sarangi, S. K. Sarangi, M. Mukherjee, S. P. Panigrahi. "System Identification by Crazy-cat Swarm Optimization" in *Proc. of the 2015 International Conference on Microwave, Optical and Communication Engineering (ICMOCE)*, 2015, pp. 439-442.

6. L. Pappula, D. Ghosh. "Planar Thinned Antenna Array Synthesis using Multi-objective Binary Cat Swarm Optimization" in *Proc. of the 2015 IEEE International Symposium on Antennas and Propagation & USNC/URSI National Radio Science Meeting*, 2015, pp. 2463-2464.

7. W. Ansar, T. Bhattacharya. "A New Gray Image Segmentation Algorithm Using Cat Swarm Optimization" in *Proc. of the 2016 International Conference on Communication and Signal Processing (ICCSP)*, 2016, pp. 1004-1008.

8. K. C. Lin, K. Y. Zhang, J. C. Hung. "Feature Selection of Support Vector Machine Based on Harmonious Cat Swarm Optimization" in *Proc. of the 2014 7th International Conference on Ubi-Media Computing and Workshops*, 2014, pp. 205-208.

9. Y. Sharafi, M. A. Khanesar, M. Teshnehlab. "Discrete Binary Cat Swarm Optimization Algorithm" in *Proc. of the 2013 3rd IEEE International Conference on Computer, Control and Communication (IC4)*, 2013, pp. 1-6.

10. D. Yan, Y. Wang, P. Zhang, H. Cao, X. Yu. "Working Conditions Classification of Ball Mill Pulverizing System Based on Quantum Cat Swarm Optimization Clustering Algorithm" in *Proc. of the 2016 31st Youth Academic Annual Conference on Chinese Association of Automation (YAC)*, 2016, pp. 348-352.

11. D. Moldovan, P. Stefan, C. Vuscan, V. R. Chifu, I. Anghel, T. Cioara, I. Salomie. "Diet Generator for Elders using Cat Swarm

Optimization and Wolf Search" in *Proc. of the International Conference on Advancements of Medicine and Health Care through Technology*, 2016, pp. 238-243.

12. M. Bahrami, O. Bozorg-Haddad, X. Chu. "Cat Swarm Optimization (CSO) Algorithm" in *O. Bozorg-Haddad (ed.), Advanced Optimization by Nature-Inspired Algorithms, Studies in Computational Intelligence*, vol. 720, 2018, pp. 9-18.

6

Chicken Swarm Optimization

Dorin Moldovan
Department of Computer Science
Technical University of Cluj-Napoca, Romania

Adam Slowik
Department of Electronics and Computer Science
Koszalin University of Technology, Koszalin, Poland

CONTENTS

6.1   Introduction

Chicken Swarm Optimization (CSO) was introduced in [1] and its main inspiration is represented by the behavior of chicken swarms, in particular the behavior of hens, chicks and roosters, and by the hierarchical order of the chicken swarms. The CSO algorithm can be applied in various optimization problems in which the solutions have a fitness function and they can be represented in a D-dimensional space. The solutions are represented by chickens and each chicken has a position and a fitness value. The chicken with the best fitness value represents the best solution. CSO identifies three types of chickens: roosters, hens and chicks. The mathematical version of CSO respects four rules which are briefly presented next and described in more detail in the next section of the chapter: (1) a chicken swarm has several groups, (2) the types of the chickens depend on the fitness values of the chickens, (3) the

71

statuses in each group are updated every G time steps and (4) the chickens update their positions in each time step respecting rules which are inspired from the searching for food behavior. Several modifications of the CSO algorithm are Binary Chicken Swarm Optimization (BCSO) [2], Mutation Chicken Swarm Optimization (MCSO) [3] and Chaotic Chicken Swarm Optimization (CCSO) [4]. CSO was also used in combination with other algorithms and two illustrative examples are Bat-Chicken Swarm Optimization (B-CSO) [5] and Cuckoo Search-Chicken Swarm Optimization (CS-CSO) [6]. In literature it was applied for solving various types of optimization problems such as the 0-1 knapsack problem [7], features selection [8], classification problems [9], attitude determination [10], tuning the number of nodes for a deep learning model [11] and constrained optimization problems [12]. The paper is organized as follows: Section 6.2 presents the original CSO algorithm, Section 6.3 presents the source-code of the global version of the CSO algorithm in Matlab, Section 6.4 presents the source-code of the global version of the CSO algorithm in C++, Section 6.5 presents a numerical example of the global version of the CSO algorithm and Section 6.6 presents the main conclusions.

6.2   Original CSO algorithm

6.2.1   Pseudo-code of global version of CSO algorithm

The global version of the CSO algorithm is presented in Algorithm 5 and it is adapted after the one from [1]. The input parameters of the algorithm are: D - the number of dimensions of the search space, N - the number of chickens, RN - the number of roosters, HN - the number of hens, CN - the number of chicks, MN - the number of mother hens, FL - a random number from the interval $[0.5, 0.9]$, G - a number which indicates how often the hierarchy of the swarm is changed, I_{max} - the maximum number of iterations and $[C_{min}, C_{max}]$ - an interval that describes the minimum and the maximum possible values of the positions of the chickens. The output of the algorithm is represented by the global best which is the best position achieved by a chicken so far. The initial population of N chickens C_i with $i = 1, ..., N$ is created randomly in *step 1* such that each chicken is represented by a D-dimensional vector. In *step 2* of the algorithm are computed the fitness values of all chickens, in *step 3* the *Gbest* D-dimensional vector is created and in *step 4* the best chicken according to the fitness value is assigned to *Gbest*. Initially the value of t is equal to 0 (step 5). While t is less than a maximum number of iterations I_{max} repeat the following sequence of steps. In *step 7* the condition $(t \bmod G) == 0$ is verified and if the outcome is positive then in *step 8* the fitness values of the chickens are ranked establishing the hierarchical order of the swarm and in *step 9* the swarm of chickens is divided into several groups and for each group the relations between the chicks and the associated mother are established. Next for each chicken C_i from the swarm of N chickens the new positions are updated using formulas that are different for each type of chicken (steps 12-20). In *step 21* the value of $C_{i,j}^{t+1}$ is updated if it is not in the interval $[C_{min}, C_{max}]$ as follows: if the value is less than C_{min} then it takes the value C_{min} and if the value is greater than C_{max} then it takes the value C_{max}. The fitness value of the new solution is evaluated in *step 22* and if the new solution is better than the current solution *Gbest* then the value of *Gbest* is updated (step 23). In *step 25* the current iteration t is updated using the formula $t = t + 1$ and finally in *step 27 Gbest* is returned as the final result of the algorithm.

Algorithm 5 Pseudo-code of the global version of CSO.

1: create the initial population of chickens $C_i (i = 1, 2, ..., N)$ randomly such that each chicken is represented by a D-dimensional vector

2: evaluate the fitness values of all N chickens

3: create the D-dimensional vector *Gbest*

4: assign the best chicken C_i to *Gbest*

5: $t = 0$

6: **while** $t < I_{max}$ **do**

7:     **if** t modulo $G == 0$ **then**

8:         rank the fitness values of the chickens and establish the hierarchical order of the swarm

9:         divide the swarm of chickens into several groups and determine the relations between the chicks and the associated mother hens in each group

10:     **end if**

11:     **for** $i = 1 : N$ **do**

12:         **if** $C_i ==$ rooster **then**

13:             $C_{i,j}^{t+1} = C_{i,j}^t \times \left(1 + \mathcal{N}(0, \sigma^2)\right)$

14:         **end if**

15:         **if** $C_i ==$ hen **then**

16:             $C_{i,j}^{t+1} = C_{i,j}^t + S_1 \times R_1 \times \left(C_{r_1,j}^t - C_{i,j}^t\right) + S_2 \times R_2 \times \left(C_{r_2,j}^t - C_{i,j}^t\right)$

17:         **end if**

18:         **if** $C_i ==$ chick **then**

19:             $C_{i,j}^{t+1} = C_{i,j}^t + FL \times (C_{m,j}^t - C_{i,j}^t)$

20:         **end if**

21:         update the value $C_{i,j}^{t+1}$ if it is not in the interval $[C_{min}, C_{max}]$

22:         evaluate the fitness value of the new solution

23:         if the new solution is better than *Gbest* then update *Gbest*

24:     **end for**

25:     $t = t + 1$

26: **end while**

27: return the *Gbest* as a result

6.2.2 Description of global version of CSO algorithm

This section presents the global version of the CSO algorithm. Initially the following parameters of the algorithm are defined: D, N, RN, HN, CN, MN, FL, G, I_{max} and $[C_{min}, C_{max}]$. The value of FL is usually in the interval $[0.5, 0.9]$ and the value of G is in the interval $[2, 20]$. In [1] the original authors of the algorithm use the following values for the initial parameters of the algorithm: $RN = 0.2 \times N$, $HN = 0.6 \times N$, $CN = N - RN - HN$, $MN = 0.1 \times N$ and $G = 10$. The values of C_{min} and C_{max} are problem dependent. The initial population of chickens C_i with $i = 1, 2, ..., N$ is created randomly and each chicken is described by a D-dimensional vector. For each chicken from the set of N chickens the fitness value is evaluated and the best chicken is assigned to the $Gbest$ D-dimensional vector. For a number of iterations I_{max} that is given as input a sequence of steps is repeated. If the current iteration is divisible without rest by G then the hierarchy of the swarm of chickens is established considering the fitness values of the chickens and the swarm is then divided into several groups that have a dominant rooster, several hens and chicks. The chickens that have the best RN fitness values are roosters, the chickens with the next best HN fitness values are hens and the remaining CN chickens are chicks. The relations between the chicks and the mother hens are determined for each group considering the fitness values of the chickens. For each chicken C_i from the group of N chickens where $i = 1, 2, ..., N$ the positions in the next iteration of the algorithm are updated considering the type of the chicken.

If the chicken is a rooster then the formula that is used for updating the position is:

$$C_{i,j}^{t+1} = C_{i,j}^t \times \left(1 + \mathcal{N}(0, \sigma^2)\right) \tag{6.1}$$

where $t + 1$ is the next iteration, t is the current iteration, j takes values from the set $\{1, ..., D\}$ and $\mathcal{N}(0, \sigma^2)$ is a Gaussian with standard deviation σ^2 and mean 0. The value of σ^2 is given by the following formula:

$$\sigma^2 = \begin{cases} 1 & \text{if } F_i \leq F_k \\ e^{\frac{F_k - F_i}{|F_i| + \epsilon}} & \text{otherwise} \end{cases} \tag{6.2}$$

where ϵ is a small positive constant that is used in order to avoid division by 0, F_i is the fitness value of the i-th chicken and F_k is the fitness value of a rooster that is selected randomly from the group of roosters.

If the chicken is a hen then the formula which is used for updating the position is:

$$C_{i,j}^{t+1} = C_{i,j}^t + S_1 \times R_1 \times \left(C_{r_1,j}^t - C_{i,j}^t\right) + S_2 \times R_2 \times \left(C_{r_2,j}^t - C_{i,j}^t\right) \tag{6.3}$$

where R_1 is a random number from $[0, 1]$, R_2 is a random number from $[0, 1]$, r_1 is the index of the rooster which is the hen's mate and r_2 is the index of a rooster or of a hen which is randomly chosen from the swarm such that $r_1 \neq r_2$. The rooster that is a given hen's mate is selected randomly from the set of roosters when the hierarchical order of the swarm is updated.

The formulas for S_1 and S_2 are the following:

$$S_1 = e^{\frac{F_i - F_{r_1}}{|F_i| + \epsilon}} \tag{6.4}$$

$$S_2 = e^{(F_{r_2} - F_i)} \tag{6.5}$$

where F_i is the fitness value of the chicken C_i, F_{r_1} is the fitness value of the rooster r_1 and F_{r_2} is the fitness value of the rooster r_2.

If the chicken is a chick then the formula that is used for updating the position is:

$$C_{i,j}^{t+1} = C_{i,j}^t + FL \times (C_{m,j}^t - C_{i,j}^t) \tag{6.6}$$

where m is the mother of the i-th chicken. We decided that m is the mother of the i-th chicken when we updated the hierarchy of the swarm.

6.3 Source-code of global version of CSO algorithm in Matlab

Listing 6.1 presents the source-code of the objective function that is optimized by the CSO algorithm. In the function $OF(x, D)$ the input parameter x is the vector of decision variables, and D is the number of dimensions. The objective function is given by formula 6.7.

$$OF(x, D) = \sum_{i=1}^{D} x_i^2 \qquad \text{where } -5.12 \leqslant x_i \leqslant 5.12 \tag{6.7}$$

```
1 function [y]=OF(x,D)
2 y=0;
3 for i=1:D
4 y=y+x(1,i)*x(1,i);
5 end
6 end
```

Listing 6.1
Definition of objective function $OF(x,D)$ in Matlab.

```
1 D=5; N=10; G=10; FL_MIN=0.5; FL_MAX=0.9; e=10^(-9); RP=20;
2 HP=60; MAX=30; TheBest=1; Cmin=zeros(1,D); Cmax=zeros(1,D);
3 cType=zeros(1,N); cPositionInSwarm=zeros(1,N); C=zeros(N,D);
4 E=zeros(1,N); Gbest=zeros(1,D); Cmin(1,:)=-5.12; Cmax(1,:)=5.12;
5 nRoosters=round((N*RP)/100); nHens=round((N*HP)/100);
6 hens=zeros(1,nHens); nChicks=N-nRoosters-nHens;
7 chicks=zeros(1,nChicks); roosters=zeros(1,nRoosters);
8 cPositionInSwarm(1,:)=linspace(1,N,N);
9 hRoosterRelation=zeros(1,N); cHenRelation=zeros(1,N);
10 for i=1:N
11   for j=1:D
12     C(i,j)=(Cmax(1,j)-Cmin(1,j))*rand()+Cmin(1,j);
13   end
14   E(1,i)=OF(C(i,:),D);
```

```
15    if E(1,i)<E(1,TheBest)
16       TheBest=i;
17    end
18 end
19 Gbest(1,:)=C(TheBest,:);  EGbest=E(1,TheBest);
20 t=0;
21 while(t<=MAX)
22    if mod(t,G)==0
23    f_values=E(1,:);
24    cType(1,:)=0;
25    f_values=sort(f_values);
26    t_roosters=f_values(1,nRoosters);
27    t_hens=f_values(1,nRoosters+nHens);
28    rIndex=1;  hIndex=1;  cIndex=1;
29    for i=1:N
30       if E(1,i)<=t_roosters
31          cType(1,i)=2;  roosters(1,rIndex)=i;
32          if rIndex>nRoosters
33             break;
34          end
35          rIndex=rIndex+1;
36       end
37    end
38    for i=1:N
39       if E(1,i)<=t_hens && cType(1,i)~=2
40          cType(1,i)=1;  hens(1,hIndex)=i;
41          if hIndex>nHens
42             break;
43          end
44          hIndex=hIndex+1;
45       end
46    end
47    for i=1:N
48       if cType(1,i)==0
49          chicks(1,cIndex)=i;
50          cIndex=cIndex+1;
51       end
52    end
53    for i=1:nHens
54       hIndex=hens(1,i);
55       r=randi(nRoosters);  rIndex=roosters(1,r);
56       hRoosterRelation(1,hIndex)=rIndex;
57    end
58    for i=1:nChicks
59       cIndex=chicks(1,i);
60       hen=randi(nHens);  hIndex=hens(1,hen);
61       cHenRelation(1,cIndex)=hIndex;
62    end
63    end
64    for i=1:N
65       for j=1:D
66          if cType(1,i)==2
67             sigma_squared=1;  k=randi(nRoosters);
68             if E(1,k)<E(1,i)
69                sigma_squared=exp((E(1,k) - E(1,i))/(abs(E(1,i)) + e));
70             end
71             C(i,j)=C(i,j)*(1+sigma_squared*randn());
72          elseif cType(1,i)==1
73             r1=hRoosterRelation(1,i);  r2=r1;
74             while (r2~=r1)
75                type=randi(2);
76                if type==0
77                   r2=roosters(1,randi(nRoosters));
78                else
79                   r2=hens(randi(nHens));
80                end
81             end
```

```
82    s1=exp((E(1,i) − E(1,r1))/(abs(E(1,i)) + e));
83    s2=exp(E(1,r2) − E(1,i));
84    C(i,j)=C(i,j)+s1*rand()*(C(r1,j)−C(i,j))
85        +rand()*s2*(C(r2,j)−C(i,j));
86    else
87      m=cHenRelation(1,i);
88      C(i,j)=C(i,j)+(FL_MIN+rand()
89          *(FL_MAX−FL_MIN))*(C(m,j)−C(i,j));
90    end
91    if C(i,j) < Cmin(1,j)
92      C(i,j) = Cmin(1,j);
93    end
94    if C(i,j) > Cmax(1,j)
95      C(i,j) = Cmax(1,j);
96    end
97    end
98  end
99  TheBest=1;
100 for i=1:N
101   E(1,i)=OF(C(i,:),D);
102   if E(1,i)<E(1,TheBest)
103     TheBest=i;
104   end
105 end
106 if E(1,TheBest)<EGbest
107   EGbest=E(1,TheBest);
108   Gbest(1,:)=C(TheBest,:);
109 end
110 t=t+1;
111 end
112 disp('FINAL RESULT'); disp(Gbest(1,:)); disp(EGbest);
```

Listing 6.2
Source-code of the global version of CSO in Matlab.

6.4 Source-code of global version of CSO algorithm in C++

```
1  #include <iostream>
2  #include <cstdlib>
3  #include <ctime>
4  #include <string>
5  #include <bits/stdc++.h>
6  #include <math.h>
7  using namespace std;
8  float OF(float x[], int size_array) {
9    float t = 0;
10   for(int i = 0; i < size_array; i++) t = t + x[i] * x[i];
11   return t;
12 }
13 float r() {return (float)(rand()%1000)/1000;}
14 double rand(double min, double max) {
15   return min + (double) rand() / RAND_MAX * (max − min);
16 }
17 int main() {
18   srand(time(NULL));
19   int D = 5; int N = 10; int G = 10; double FL_MIN = 0.5;
20   double FL_MAX = 0.9; double e = 0.000000001; int RP = 20;
21   int HP = 60; float Cmin[D]; float Cmax[D]; int TheBest = 0;
```

```
22  int MAX = 30; int cType[N]; int cPositionInSwarm[N];
23  for(int i = 0; i < D; i++) {Cmin[i] = -5.12; Cmax[i] = 5.12;}
24  float C[N][D]; float E[N]; float Gbest[D]; float EGbest;
25  int nRoosters = (N * RP) / 100; int nHens = (N * HP) / 100;
26  int hens[nHens]; int nChicks = N - nRoosters - nHens;
27  int chicks[nChicks]; int roosters[nRoosters];
28  unordered_map<int, int> hRoosterRelation;
29  unordered_map<int, int> cHenRelation;
30  for(int i = 0; i < N; i++) cPositionInSwarm[i] = i;
31  for(int i = 0; i < N; i++) {
32    for(int j = 0; j < D; j++)
33      C[i][j] = (Cmax[j] - Cmin[j]) * r() + Cmin[j];
34    E[i] = OF(C[i], D); if(E[i] < E[TheBest]) TheBest = i;
35  }
36  for(int j = 0; j < D; j++) Gbest[j] = C[TheBest][j];
37  EGbest = E[TheBest]; int t = 0;
38  while(t < MAX) {
39    if(t % G == 0) {
40      vector<double> f_values;
41      for(int i = 0; i < N; i++) f_values.push_back(E[i]);
42      for(int i = 0; i < N; i++) cType[i] = 0;
43      sort(f_values.begin(), f_values.end());
44      double t_roosters = f_values[nRoosters - 1];
45      double t_hens = f_values[nRoosters + nHens - 1];
46      int rIndex = 0; int hIndex = 0; int cIndex = 0;
47      for(int i = 0; i < N; i++) {
48        if(E[i] <= t_roosters) {
49          cType[i] = 2; roosters[rIndex++] = i;
50          if(rIndex >= nRoosters) break;
51        }
52      }
53      for(int i = 0; i < N; i++) {
54        if(E[i] <= t_hens && cType[i] != 2) {
55          cType[i] = 1; hens[hIndex++] = i;
56          if(hIndex >= nHens) break;
57        }
58      }
59      for(int i = 0; i < N; i++)
60        if(cType[i] == 0) chicks[cIndex++] = i;
61      for(int i = 0; i < nHens; i++) {
62        int hIndex = hens[i]; int r = rand() % nRoosters;
63        int rIndex = roosters[r];
64        hRoosterRelation[hIndex] = rIndex;
65      }
66      for(int i = 0; i < nChicks; i++) {
67        int cIndex = chicks[i]; int hen = rand() % nHens;
68        int hIndex = hens[hen]; cHenRelation[cIndex] = hIndex;
69      }
70    }
71    for(int i = 0; i < N; i++) {
72      for(int j = 0; j < D; j++) {
73        if (cType[i] == 2) {
74          double sigma_squared = 1; int k = rand() % nRoosters;
75          if(E[k] < E[i])
76            sigma_squared = exp((E[k] - E[i])/(abs(E[i]) + e));
77          default_random_engine generator;
78          normal_distribution<double> d(0,sigma_squared);
79          C[i][j] = C[i][j] * (1 + d(generator));
80        } else if (cType[i] == 1) {
81          int r1 = hRoosterRelation[i]; int r2 = r1;
82          while(r2 != r1) {
83            int type = rand() % 2;
84            if(type == 0) {r2 = roosters[rand() % nRoosters];}
85            else {r2 = hens[rand() % nHens];}
86          }
87          double s1 = exp((E[i] - E[r1])/(abs(E[i]) + e));
88          double s2 = exp(E[r2] - E[i]);
```

```
89      C[i][j] = C[i][j] + s1 * rand(0, 1) * (C[r1][j]
90          - C[i][j]) + rand(0, 1) * s2 * (C[r2][j] - C[i][j]);
91      } else {
92          int m = cHenRelation[i];
93          C[i][j] = C[i][j] + rand(FL_MIN, FL_MAX)
94              * (C[m][j] - C[i][j]);
95      }
96      if(C[i][j] < Cmin[j]) C[i][j] = Cmin[j];
97      if(C[i][j] > Cmax[j]) C[i][j] = Cmax[j];
98      }
99  }
100 TheBest = 0;
101 for(int i = 0; i < N; i++) {
102     E[i] = OF(C[i], D);
103     TheBest = (E[i] < E[TheBest]) ? i : TheBest;
104 }
105 if(E[TheBest] < EGbest) {
106     EGbest = E[TheBest];
107     for(int j = 0; j < D; j++) Gbest[j] = C[TheBest][j];
108 }
109 t = t + 1;
110 }
111 cout << EGbest << endl; return 0;
112 }
```

Listing 6.3
Source-code of the global version of CSO in C++.

6.5 Step-by-step numerical example of global version of CSO algorithm

In the first step, we assume that we want to minimize the objective function $OF(x, D)$ given by equation 6.7 where the number of dimensions D is equal to 5. For simplicity in this step-by-step numerical example we consider that $OF(C) = OF(C, D) = OF(C, 5)$ where C is an array that describes the position of a chicken.

In the second step, we initialize the parameters of CSO such that the number of chickens N is equal to 10, the swarm of chickens updates the hierarchy every $G = 10$ iterations, $FL = [0.5, 0.9]$, $\epsilon = 10^{-9}$, the roosters percent is 20%, the hens percent is 60%, the maximum number of iterations is 30, $C_{min} = -5.12$ and $C_{max} = 5.12$.

In the third step, the chicken swarm that consists of 10 chickens is randomly created and each chicken is represented by a 5-dimensional vector that describes the position of the chicken.
$C_1 = \{-4.700, -0.337, -1.699, 0, -3.389\}$
$C_2 = \{2.293, -0.225, -1.454, 4.730, -0.368\}$
$C_3 = \{2.099, -3.635, -2.242, 3.348, 4.720\}$
$C_4 = \{-0.092, 5.068, 4.526, 3.348, -0.655\}$
$C_5 = \{-1.116, 1.064, 4.116, -3.553, -2.129\}$
$C_6 = \{-1.208, -0.808, 2.211, 2.232, 4.044\}$

$C_7 = \{-0.542, 2.314, 2.775, 0.389, 3.778\}$
$C_8 = \{4.218, 1.710, -2.058, -4.761, 4.034\}$
$C_9 = \{2.078, 3.184, -1.822, -1.710, 1.771\}$
$C_{10} = \{1.679, -3.676, 2.160, -2.529, 3.768\}$

In the fourth step, the 5-dimensional *Gbest* vector is created.
$Gbest = \{0, 0, 0, 0, 0\}$

In the fifth step, we evaluate each chicken C_i using the objective function $OF(.)$.
$E_1 = OF(C_1) = 36.583, E_2 = OF(C_2) = 29.943,$
$E_3 = OF(C_3) = 56.147, E_4 = OF(C_4) = 57.828,$
$E_5 = OF(C_5) = 36.487, E_6 = OF(C_6) = 28.350,$
$E_7 = OF(C_7) = 27.780, E_8 = OF(C_8) = 63.910,$
$E_9 = OF(C_9) = 23.848, E_{10} = OF(C_{10}) = 41.600$

In the sixth step, we consider the best chicken C_i which has the smallest value for the objective function because the goal of the algorithm is to minimize the value of $OF(.)$ and we assign the vector that represents the chicken with the smallest $OF(.)$ value to *Gbest*. The best chicken is C_9 and thus $Gbest = \{2.078, 3.184, -1.822, -1.710, 1.771\}$ and $OF(Gbest) = 23.848$.

In the seventh step of the algorithm, the main loop is started. We check whether the current iteration is less than $I_{max} = 30$ and if that condition is true the algorithm jumps to the final step, otherwise the algorithm continues with the eighth step.

In the eighth step of the algorithm, we check whether the value of the current iteration is divisible without rest by $G = 10$ and if that condition is true then the algorithm updates the hierarchy of the chicken swarm, otherwise the algorithm continues with the ninth step.

Chicken swarm hierarchy update

In the first substep, the chickens are sorted according to their fitness value and the best $20\% \times 10 = 2$ chickens are considered roosters, the next best $60\% \times 10 = 6$ chickens are considered hens and the rest of the chickens are considered chicks.
$TypeOf(C_1) = Hen, TypeOf(C_2) = Hen,$
$TypeOf(C_3) = Hen, TypeOf(C_4) = Chick,$
$TypeOf(C_5) = Hen, TypeOf(C_6) = Hen,$
$TypeOf(C_7) = Rooster, TypeOf(C_8) = Chick,$
$TypeOf(C_9) = Rooster, TypeOf(C_{10}) = Hen$

In the second substep, each hen is associated with a rooster randomly and that rooster is considered the head of the group the hen belongs to.
$RoosterOf(C_1) = C_9, RoosterOf(C_2) = C_7,$
$RoosterOf(C_3) = C_7, RoosterOf(C_5) = C_9,$
$RoosterOf(C_6) = C_9, RoosterOf(C_{10}) = C_9$

In the third substep, each chick is associated with a hen randomly and that hen is considered the mother of the chick.
$HenMotherOf(C_4) = C_{10}$
$HenMotherOf(C_8) = C_5$

In the ninth step of the algorithm, each chicken updates its current position considering one of the equations 6.1, 6.3 or 6.6 such that the roosters use the equation 6.1, the hens use the equation 6.3 and the chicks use the equation 6.6. Then we check whether the value of each decision variable is within the range $[C_{min}, C_{max}]$. The new positions of the chickens are:
$C_1 = \{3.366, 1.537, -1.817, -0.021, -1.372\}$
$C_2 = \{0.275, 2.273, 0.850, 1.026, 2.125\}$
$C_3 = \{-1.160, 2.271, 3.869, 0.540, 4.484\}$
$C_4 = \{1.094, -2.112, 2.854, -1.917, 2.886\}$
$C_5 = \{0.444, 3.035, 3.582, -2.237, 3.108\}$
$C_6 = \{2.304, 0.628, 1.835, -0.931, 1.686\}$
$C_7 = \{-0.476, 2.031, 2.436, 0.341, 3.317\}$
$C_8 = \{2.097, 2.639, 2.414, -3.090, 3.467\}$
$C_9 = \{1.825, 2.796, -1.600, -1.501, 1.555\}$
$C_{10} = \{1.706, 4.601, -0.818, -1.858, 0.549\}$

In the tenth step of the algorithm, we compute the new fitness values of the chickens.
$E_1 = OF(C_1) = 18.886, E_2 = OF(C_2) = 11.536,$
$E_3 = OF(C_3) = 41.880, E_4 = OF(C_4) = 25.815,$
$E_5 = OF(C_5) = 36.916, E_6 = OF(C_6) = 12.786,$
$E_7 = OF(C_7) = 21.416, E_8 = OF(C_8) = 38.768,$
$E_9 = OF(C_9) = 18.385, E_{10} = OF(C_{10}) = 28.513$

In the eleventh step of the algorithm, we update the value of *Gbest* if the algorithm finds a better value. Then we return to the seventh step of the algorithm.
$Gbest = \{0.275, 2.273, 0.850, 1.026, 2.125\}$
$OF(Gbest) = 11.536$

In the twelfth step the algorithm returns *Gbest* as the final result. After 30 iterations the value of *Gbest* is $\{0.043, 0.064, -0.038, -0.036, 0.035\}$ and the value of $OF(Gbest)$ is 0.010.

6.6 Conclusions

This chapter presented the main principles of the Chicken Swarm Optimization algorithm. We presented the pseudo code of the algorithm, the source code in Matlab and the source code in C++. In addition we showed how this algorithm works in the global version providing a step-by-step numerical example. We believe that this chapter will facilitate the development of other versions of the algorithm in other programming languages.

References

1. X. Meng, Y. Liu, X. Gao, H. Zhang. "A New Bio-inspired Algorithm: Chicken Swarm Optimization" in *Lecture Notes in Computer Science*, vol. 8794, 2014, pp. 86-94.

2. M. Han, S. Liu. "An Improved Binary Chicken Swarm Optimization Algorithm for Solving 0-1 Knapsack Problem" in *Proc. of the 2017 13th International Conference on Computational Intelligence and Security (CIS)*, 2017, pp. 207-210.

3. K. Wang, Z. Li, H. Cheng, K. Zhang. "Mutation Chicken Swarm Optimization Based on Nonlinear Inertia Weight" in *Proc. of the 2017 3rd IEEE International Conference on Computer and Communications (ICCC)*, 2017, pp. 2206-2211.

4. K. Ahmed, A. E. Hassanien, S. Bhattacharyya. "A Novel Chaotic Chicken Swarm Optimization Algorithm for Feature Selection" in *Proc. of the 2017 Third International Conference on Research in Computational Intelligence and Communication Networks (ICR-CICN)*, 2017, pp. 259-264.

5. S. Liang, T. Feng, G. Sun, J. Zhang, H. Zhang. "Transmission Power Optimization for Reducing Sidelobe via Bat-chicken Swarm Optimization in Distributed Collaborative Beamforming" in *Proc. of the 2016 2nd IEEE International Conference on Computer and Communications (ICCC)*, 2016, pp. 2164-2168.

6. S. Liang, T. Feng, G. Sun. "Sidelobe-level suppression for linear and circular antenna arrays via the cuckoo search-chicken swarm optimisation algorithm" in *IET Microwaves, Antennas & Propagation*, vol. 11, 2017, pp. 209-218.

7. M. Han, S. Liu. "An Improved Binary Chicken Swarm Optimization Algorithm for Solving 0-1 Knapsack Problem" in *Proc. of the 2017*

13th International Conference on Computational Intelligence and Security (CIS), 2017, pp. 207-210.

8. A. I. Hafez, H. M. Zawbaa, E. Emary, H. A. Mahmoud, A. E. Hassanien. "An Innovative Approach for Feature Selection Based on Chicken Swarm Optimization" in *Proc. of the 2015 7th International Conference of Soft Computing and Pattern Recognition (SoCPaR)*, 2015, pp. 19-24.

9. Roslina, M. Zarlis, I. T. R. Yanto, D. Hartama. "A Framework of Training ANFIS using Chicken Swarm Optimization for Solving Classification Problems" in *Proc. of the 2016 International Conference on Informatics and Computing (ICIC)*, 2016, pp. 437-441.

10. Y. Ji, Y. Wu, X. Sun, S. Yan, Q. Chen, B. Du. "A New Algorithm for Attitude Determination Based on Chicken Swarm Optimization" in *Proc. of the 2018 7th International Conference on Digital Home (ICDH)*, 2018, pp. 121-126.

11. D. Moldovan, V. Chifu, C. Pop, T. Cioara, I. Anghel, I. Salomie. "Chicken Swarm Optimization and Deep Learning for Manufacturing Processes" in *Proc. of the 2018 17th RoEduNet Conference: Networking in Education and Research (RoEduNet)*, 2018, pp. 1-6.

12. J. Wang, Z. Cheng, O. K. Ersoy, M. Zhang, K. Sun, Y. Bi. "Improvement and Application of Chicken Swarm Optimization for Constrained Optimization" in *IEEE Access*, vol. 7, 2019, pp. 58053-58072.

7

Cockroach Swarm Optimization

Joanna Kwiecien

AGH University of Science and Technology
Department of Automatics and Robotics, Krakow, Poland

CONTENTS

7.1 Introduction

The cockroach swarm optimization algorithm is one of the algorithms belonging to the group of swarm intelligence algorithms. It should be mentioned that swarm intelligence has focused on how social insects solve various problems by their mutual cooperation. CSO draws inspiration from the social behavior of cockroaches looking for food. The most common behaviors of cockroaches such as chasing, swarming, escaping from light and being ruthless are the basis of this approach. In 2010, Chen and Tang in [1] proposed the basic version of the CSO algorithm. Typically, each iteration of the CSO algorithm consists of three procedures, namely, chase-swarming, dispersion, and ruthless. In CSO, each individual represents a solution vector, and initialization is randomly over the entire search space. Changes to the position of cockroaches within the search space are based on using varying combinations of movements.

It should be mentioned that a number of basic modifications to CSO have been developed to improve the quality of solutions obtained with this algorithm. One of the first proposals was a modified CSO algorithm, where an inertia weight was introduced [2]. Furthermore, Ogagbuwa and Adewumi follow a somewhat different approach by introducing the hunger behavior after chase-swarming operation [3]. Some papers illustrated the ability of CSO to solve optimization problems. With some additional assumptions, the cockroach swarm optimization algorithm was used to solve discrete optimization problems, including route planning [4-6]. Moreover, cockroach-inspired algorithms for robot path planning were used in [7].

The rest of this paper is organized as follows: in Section 7.2 the basic principles of CSO, and its pseudo-code are presented and described in detail in order to allow for easy implementation by future users, and, consequently, practical applications of this algorithm. In Sections 7.3 and 7.4 the source-codes of the CSO algorithm are shown in Matlab and C++ programming language, respectively. In Section 7.5, we illustrate in detail the numerical example and simulation results of the CSO algorithm, and some conclusions are revealed in Section 7.6.

7.2   Original cockroach swarm optimization algorithm

In general, the CSO algorithm can search through possible movement choices to find a movement sequence that will guide the cockroach individuals from initial solutions to the global optimum. During each cycle, the CSO algorithm looks around possible solutions in order to find an even better one. The solutions are modified through some procedures, including chase-swarming, dispersion, and ruthlessness.

7.2.1   Pseudo-code of CSO algorithm

The pseudo-code for the basic version of the CSO algorithm is presented in Algorithm 6.

Algorithm 6 Pseudo-code of CSO.

1: determine the D-dimensional objective function $OF(.)$
2: initialize the CSO algorithm parameter values such as N – number of cockroaches in the swarm, $visual$ – the visibility parameter, $Stop$ – termination condition
3: randomly create swarm, the $i - th$ individual represents a vector $X_i = (x_{i1}, x_{i2}, ..., x_{iD})$
4: **for** each $i - th$ cockroach from swarm **do**
5:    evaluate quality of the cockroach X_i using $OF(.)$ function
6: **end for**
7: select the best cockroach P_g in initial swarm
8: **while** termination condition not met **do**

9:     **for** $i = 1$ to N **do**
10:       **for** $j = 1$ to N **do**
11:         **if** $OF(X_i)$ is local optimum **then**
12:           move cockroach i towards P_g using formula
13:           $X_i = X_i + step \cdot rand \cdot (P_g - X_i)$
14:         **else**
15:           move cockroach i towards P_i (within visual scope) using
        formula
16:           $X_i = X_i + step \cdot rand \cdot (P_i - X_i)$
17:         **end if**
18:       **end for** j
19:     **end for** i
20:     **if** $OF(X_i)$ is better than $OF(P_g)$ **then**
21:       $P_g = X_i$
22:     **end if**
23:     **for** $i = 1$ to N **do**
24:       move cockroach randomly using formula
25:       $X_i = X_i + rand(1, D)$
26:       **if** $OF(X_i)$ is better than $OF(P_g)$ **then**
27:         $P_g = X_i$
28:       **end if**
29:     **end for**
30:     select cockroach h randomly
31:     $X_h = P_g$
32: **end while**
33: return the best one as a result

7.2.2 Description of the CSO algorithm

In Algorithm 6 the pseudo-code of the first version of the CSO algorithm
was presented. In order to better understand CSO, it is necessary to conduct
its detailed description. At the start, we should have defined an objective
function (step 1) and the CSO algorithm parameters (step 2) such as number of
cockroaches (N), visual range (*visual*), and the stopping criterion (*Stop*). The
visibility parameter denotes the visual distance of cockroaches. In the third
step, we have to initialize the swarm with random solutions. A D-dimensional
vector $X_i = (x_{i1}, x_{i2}, ..., x_{iD})$ represents the *ith* cockroach, $i = 1, 2, ..., N$. The
position of each individual is a potential solution, and the objective function
value is calculated for all entities (step 5). Given such evaluation, the global
best position is kept and denoted as global optimum P_g (step 7). At the core
of CSO lies a loop starting from the 8*th* step, until a stopping condition is

satisfied. The simplest criterion is a maximum number of iterations that the CSO algorithm executes, or a limited number of fitness function evaluations.

In the chase-swarming procedure (steps 9 to 19), in the new cycle, the strongest cockroaches carry the local best solutions P_i, form small swarms, and move forward to the global optimum P_g according to the formula (step 13): $X_i = X_i + step \cdot rand \cdot (P_g - X_i)$. Within this procedure, each individual X_i moves to its local optimum P_i in the range of its visibility (step 16) using the formula: $X_i = X_i + step \cdot rand \cdot (P_i - X_i)$.

There can occur a situation when a cockroach moving in a small group becomes the strongest by finding a better solution, because individuals follow in other ways than their local optimum. When the global best position of cockroaches found so far is improved (with respect to the objective function), this new cockroach position will become the new P_g (step 21).

In addition, a dispersion procedure is incorporated into the running process (steps 23 to 29). In order to improve the ability of local searching and to avoid getting stuck in local minima, each cockroach is randomly dispersed (step 25) using the formula: $X_i = X_i + rand(1, D)$, where $rand(1, D)$ is a D-dimensional random vector (D is the space dimension).

In step 27, the position of the best cockroach is updated. In step 31, the phenomenon of replacing a randomly chosen individual by the current best individual (step 31) is given. This is a behavior that corresponds the situation, when the stronger cockroach eats the weaker. All aforementioned procedures are applied repeatedly until the stopping criterion is satisfied. When the stopping criterion is met, the global best position of individuals (a local or global optimum) is returned.

7.3   Source-code of CSO algorithm in Matlab

In Listing 7.1 the source-code for the objective function which will be optimized by the CSO algorithm is shown. The result of $OF(.)$ function is an D-dimensional column vector with the objective function values for each cockroach from the swarm. We used the well-known Sum Squares function as the objective function. It is convex and unimodal, and has the global optimum (0) at x* = (0, 0, ..., 0). Therefore, we assume that the CSO algorithm minimizes the objective function given by formula 7.1 and evaluated on the hypercube $x_i \in [-5.12, 5.12]$. To simplify, the cockroach form is augmented to include the solution quality.

$$OF(X_i) = \sum_{j=1}^{D} j X_{i,j}^2 \qquad \text{where } -5.12 \leqslant X_{i,j} \leqslant 5.12 \qquad (7.1)$$

```
1  function [X]=OF(X, i, D)
2  X(i,D+1)=0
3  for j=1:D
4    X(i,D+1)=X(i,D+1)+j*X(i,j)^2;;
5  end
```

Listing 7.1
Definition of objective function $OF(.)$ in Matlab.

```
1   % declaration of the parameters of the CSO algorithm
2   N=5; visual=10; step=2; iter=50; D=3;
3   % constraints definition for all decision variables
4   Xmin=-5.12; Xmax=5.12;
5   % the initial swarm is created randomly
6   X=zeros(N,D+1);
7   for i=1:N;
8     for j=1:(D);
9       X(i,j)=Xmin+(Xmax-Xmin)*rand;
10    end
11   [X]=OF(X,i,D+1);
12  end
13  % find Pg
14  [M Nr]=min(X(:,D+1));
15  Pg=X(Nr,:); % vector of Pg
16  vPg=M; % value of the objective function for Pg
17  %main program loop starts
18  for it=1:iter
19  % chase-swarming procedure
20  for i=1:N
21    if X(i,D+1)~=vPg
22      s1=X(i,:); % s1 is the vector of the i-th cockroach
23      vs1=X(i,D+1); % the position of the i-th cockroach
24      Pi=s1; % assign Pi
25      vPi=vs1; % assign value of the objective function for Pi
26      flag=0; % if flag = 0 the i-th cockroach does not see better
            individual
27      for l=1:N
28        if (i ~= l)
29          vl=X(l,D+1); % value of the objective function for the l-th
            cockroach
30          dist=abs(vl-vs1); % distance between two cockroaches
31          if (dist <= visual) & (vl < vPi) % the l-th individual is
            better than Pi within visual scope
32            Pi=X(l,:);
33            vPi=vl;
34            flag=1; % the i-th cockroach sees better individual
35          end
36        end
37      end
38  % if the i-th cockroach sees better individual, it moves toward
        local optimum Pi
39      if flag==1
40        for j=1:D
41          X(i,j)=X(i,j)+step*rand*(Pi(1,j)-X(i,j));
42          if X(i,j)<Xmin
43            X(i,j)=Xmin;
44          end
45          if X(i,j)>Xmax
46            X(i,j)=Xmax;
47          end
48        end
49        [X]=OF(X,i,D);
50      end
51      if flag==0;
52        for j=1:D
53          X(i,j)=X(i,j)+step*rand*(Pg(1,j)-X(i,j));
```

```
54        if X(i,j)<Xmin
55          X(i,j)=Xmin;
56        end
57        if X(i,j)>Xmax
58          X(i,j)=Xmax;
59        end
60      end
61 % evaluation of all solutions
62      [X]=OF(X,i,D);
63    end
64  end
65 end
66 %update Pg
67 [M Nr]=min(X(:,D+1));
68 newPg=X(Nr,:);
69 newvPg=M;
70 if newvPg<vPg
71   Pg=newPg;
72   vPg=newvPg;
73 end
74 % dispersion procedure
75  for i=1:N
76    for j=1:D
77      X(i,j)=X(i,j)+rand;
78      if X(i,j)<Xmin
79          X(i,j)=Xmin;
80      end
81      if X(i,j)>Xmax
82          X(i,j)=Xmax;
83      end
84    end
85 % evaluation of all solutions
86      [X]=OF(X,i,D);
87  end
88  % update Pg
89  [M Nr]=min(X(:,D+1));
90  newPg=X(Nr,:);
91  newvPg=M;
92  if newvPg<vPg
93    Pg=newPg;
94    vPg=newvPg;
95  end
96  % ruthless behavior
97  r=randi(N);
98  X(r,:)=Pg;
99  end
100   % the result of the CSO algorithm is returned
101  disp(Pg);
102
```

Listing 7.2
Source-code of CSO in Matlab.

7.4   Source-code of CSO algorithm in C++

```
1 #include <iostream>
2 #include <cmath>
3 using namespace std;
4 // definition of the objective function OF
5 double OF(double x[], int size_array)
6 {
7     double f=0;
```

```
8      for(int j=0; j<size_array; j++){
9         f=f+j*x[j]*x[j];
10     }
11     return f;
12  }
13  // main program function
14  int main(){
15  // initialization of the parameters
16     int N=5; int iter=50; int D=3;
17     double visual = 10;
18     double step = 2;
19     int i,j,it;
20     double r;
21     bool flag;
22     double X[N][D]; //cockroach position
23     double Pg[D]; //the best solution found so far
24     double Pi[N][D]; //the local optimum
25     double v[N]; //the fitness value
26     double vPg; //value of the objective function for Pg
27     int Best=0;
28     double Xmin[D]; double Xmax[D];
29  // initialization of the constraints
30    for (int j=0; j<D; j++){
31      Xmin[j]=-5.12; Xmax[j]=5.12;
32    }
33  //initialize the cockroach swarm
34     for(int i=0; i<N; i++) {
35        for(int j=0; j<D; j++) {
36          r=((double)rand()/((double)(RAND_MAX)+(double(1))));
37          X[i][j]=(Xmax[j]-Xmin[j])*r()+Xmin[j];
38          Pi[i][j]=X[i][j];
39        }
40  // evaluate all the cockroaches in the swarm
41        v[i]=OF(X[i],D);
42        vPi[i]=v[i];
43  // find the best individual in the swarm
44        if (v[i]<v[Best]) Best=i;
45     }
46  // assign the best cockroach to Pg
47     for(j=0; j<D; j++) {
48        Pg[j]=X[Best][j];
49        vPg=v[Best];
50     }
51  //main program loop
52    while(it<iter)
53    {
54     for (i=0; i<N, i++){
55       for (k=0; k<N, k++){
56         if(fabs(v[i]-v[k])<=visual && (v[k])<v[i]))
57         {
58         Pi[i][j]=X[k][j];
59         flag=true;
60         }
61          if(flag==false){
62          Pi[i][j]=X[i][j];
63          }
64          flag=false;
65          if(Pi[i][j]==X[i][j]){
66          r=((double)rand()/((double)(RAND_MAX)+(double(1))));
67          X[i][j]=X[i][j]+((step*r)*(Pg[j]-X[i][j]));
68          }
69          else {
70          r=((double)rand()/((double)(RAND_MAX)+(double(1))));
71            X[i][j]=X[i][j]+(step*r*(Pi[i][j]-X[i][j]));
72          }
73       }
74       v[i]=OF(X[i],D);
```

```
75    }
76  //update Pg
77    if (v[i]<vPg) {Best = i;
78    for(int j=0; j<D; j++) Pg[j]=X[Best][j];
79    vPg=v[Best];
80    }
81  //dispersion
82    for (i=0; i<N; i++){
83      for (j=0; j<D;j++){
84        X[i][j]=X[i][j]+rand();
85        if (X[i][j]<Xmin[j]) X[i][j]=Xmin[j];
86        if (X[i][j]>Xmax[j]) X[i][j]=Xmax[j];
87      }
88      v[i]=OF(X[i],D);
89      if (v[i]<vPg) {Best = i;
90      for(int j=0; j<D; j++) Pg[j]=X[Best][j];
91      vPg=v[Best];
92      }
93    }
94  // ruthless behavior
95    int RANDOM = rand() %N;
96    for (j=0; j<D; j++){
97    if(X[RANDOM][j]!=Pg[j]{
98      X[RANDOM][j]=Pg[j];}
99    }
100   }
101   cout<<"Value of the global best solution =" << vPg << endl;
102 }
```

Listing 7.3
Source-code of CSO in C++.

7.5   Step-by-step numerical example of CSO algorithm

In this section, we show step-by-step the CSO algorithm based on one mathematical function. First, we assume that the aim is to minimize the objective function given by equation 7.1, where D is equal to 3. In the second step, we initialize the following fixed parameters: the population size (N) equalling 5 individuals, $visual = 10$, $step = 2$. The stopping criterion is taken to be a predefined maximum number of iterations ($iter$) which equals 50. X_{min} and X_{max} are used to represent the minimum and maximum limits of variables. Next, CSO creates a swarm of 5 cockroaches randomly. To simplify numerical illustrations, consider the case that the cockroach i is expressed in the D-dimensional vector. Therefore, the swarm consists of 5 cockroaches:

$X_1 = \{-0.0635, 0.9296, -4.6603\}$
$X_2 = \{0.6601, 1.3675, -0.0195\}$
$X_3 = \{-4.5095, 0.1553, -4.5753\}$
$X_4 = \{3.5188, 0.9694, 1.9560\}$
$X_5 = \{1.5076, 1.2640, 4.6475\}$.

As mentioned in the previous section, each cockroach has a fitness value computed by optimization function $OF(.)$ and represents a complete solution. Hence, for all individuals we have the following values:

$OF(X_1) = \sum_{j=1}^{D} j X_{1,j}^2 = 66.888$

$OF(X_2) = \sum_{j=1}^{D} j X_{2,j}^2 = 4.1769$

$OF(X_3) = \sum_{j=1}^{D} j X_{3,j}^2 = 83.1831$

$OF(X_4) = \sum_{j=1}^{D} j X_{4,j}^2 = 25.7397$

$OF(X_5) = \sum_{j=1}^{D} j X_{5,j}^2 = 70.2664.$

In the next step, the best cockroach from the swarm is selected. In our case, the best cockroach is X_2 with the value vP_g (see the $16-th$ line in Listing 7.2) equalling 4.1769. Then, the main loop of the CSO algorithm starts, until the stopping criterion has been satisfied, usually a maximum number of iterations.

For all cockroaches different from P_g we calculate distance ($dist$) between their values of objective function. Therefore, the chase-swarming procedure starts on the $1-st$ cockroach. The following distances are obtained:

$dist(OF(X_1), OF(X_2)) = 62.7111$
$dist(OF(X_1), OF(X_3)) = 16.2951$
$dist(OF(X_1), OF(X_4)) = 41.1483$
$dist(OF(X_1), OF(X_5)) = 3.3784.$

Next, during the run of CSO, the flag of the cockroaches is setting to make them move towards P_i or P_g according to the equations of the chase-swarming procedure. Once a cockroach goes into the chase-swarming mode, it moves according to the better one for every dimension.

Only $dist(OF(X_1), OF(X_5))$ is within the range of visibility ($visual = 10$), but $OF(X_5)$ is worse. In that way, the first individual cannot see a better one ($flag = 0$), and it moves towards the swarm leader P_g. Here, we assume that after movement of the first cockroach, other individuals see only its new position and new value of $OF(X_1)$.

Taking account $rand = 0.2197$, the $1-st$ cockroach has the following form: $X_1=\{0.2544, 1.1220, -2.6213\}$ with its OF equalling 23.1962. The new solution should satisfy the constraint of the range, and in this case it is done. The same steps are done for all cockroaches (apart from P_g). In other words, the distance between individuals is calculated and checked, and after setting the flag their specific movement toward P_g or P_i is applied.

Therefore, for the $3-rd$ cockroach, we have the following distances:

$dist(OF(X_3), OF(X_1)) = 59.9869$
$dist(OF(X_3), OF(X_2)) = 79.0062$
$dist(OF(X_3), OF(X_4)) = 57.4434$
$dist(OF(X_3), OF(X_5)) = 12.9167.$

This individual cannot see a better one ($flag = 0$), hence it moves towards P_g. With $rand = 0.4596$, we obtain new solution $X_3=\{0.2428, 1.2697, -0.3872\}$ with its OF equalling 3.7329.

For the $4-th$ cockroach, we have the following distances:

$dist(OF(X_4), OF(X_1)) = 2.5435$
$dist(OF(X_4), OF(X_2)) = 21.5628$

$dist(OF(X_4), OF(X_3)) = 22.0068$
$dist(OF(X_4), OF(X_5)) = 44.5267.$

Note that X_1 is the locally best individual for X_4 (its $flag = 1$), so the $4-th$ cockroach can move toward X_1. Assume $rand = 0.9585$, the new cockroach X_4 has the following vector: $\{-2.7392, 1.2619, -5.1200, OF = 89.3316\}$.

For the last $(5 - th)$ cockroach, we have the following distances:

$dist(OF(X_5), OF(X_1)) = 47.0702$
$dist(OF(X_5), OF(X_2)) = 66.0895$
$dist(OF(X_5), OF(X_3)) = 66.5335$
$dist(OF(X_5), OF(X_4)) = 19.0652.$

If $rand= 0.79$, then the positions of all cockroaches are as follows:

$X_1 = \{0.25441.1220 - 2.6213\}$
$X_2 = \{0.6601, 1.3675, -0.0195\}$
$X_3 = \{0.24281.2697 - 0.3872\}$
$X_4 = \{-2.73921.2619 - 5.1200\}$
$X_5 = \{0.16851.4275 - 2.7268\}$

Their objective functions have the following values:

$OF(X_1) = 23.1962,$ $OF(X_2) = 4.1769,$ $OF(X_3) = 3.7329,$ $OF(X_4) = 89.3316,$
$OF(X_5) = 26.4102.$

After calculating the objective function one by one, the update of the position and the value of the cockroach are considered, to determine which owns the best fitness value we find so far. It should be noted, that the third cockroach becomes new P_g.

For calculations in the dispersion procedure, assume that:

$rand(1, D) = [0.4519, 0.3334, 0.0591]$ for the $1 - st$ cockroach,
$rand(1, D) = [0.7409, 0.5068, 0.1999]$ for the $2 - nd$ cockroach,
$rand(1, D) = [0.4272, 0.1687, 0.7517]$ for the $3 - th$ cockroach,
$rand(1, D) = [0.3684, 0.9418, 0.0172]$ for the $4 - th$ cockroach,
$rand(1, D) = [0.8291, 0.6266, 0.5387]$ for the $5 - th$ cockroach.

For such values, it is easy to determine values for the cockroaches:

$X_1 = \{0.7063, 1.4554, -2.5622\}$
$X_2 = \{1.4010, 1.8743, 0.1804\}$
$X_3 = \{0.6700, 1.4383, 0.3645\}$
$X_4 = \{-2.3709, 2.2038, -5.1028\}$
$X_5 = \{0.9975, 2.0541, -2.1880\}$

The fitness of individuals has the following values:

$OF(X_1) = 24.4302,$ $OF(X_2) = 9.0864,$ $OF(X_3) = 4.9851,$ $OF(X_4) = 93.4507,$
$OF(X_5) = 23.7965.$

It may, of course, happen that a new best position is found, but in this procedure P_g is not updated.

To explain the ruthless procedure, assume that the $4-th$ individual is replaced by updated P_g. Then:

$X_1=\{0.7063, 1.4554, -2.5622\}$
$X_2=\{1.4010, 1.8743, 0.1804\}$
$X_3=\{0.6700, 1.4383, 0.3645\}$
$X_4=\{0.2428, 1.2697, -0.3872\}$
$X_5=\{0.9975, 2.0541, -2.1880\}$,

and their objective functions have the following values: $OF(X_1)=24.4302$, $OF(X_2)=9.0864$, $OF(X_3)=4.9851$, $OF(X_4)=3.7329$, $OF(X_5)=23.7965$.

It should be noted at this point that after the first iteration the better global solution is obtained with $OF = 3.7329$ ($X_4 = [0.2428, 1.2697, -0.3872]$).

7.6   Conclusions

In this chapter, we have described the cockroach swarm optimization algorithm that has a flexible architecture and can be used for various optimization problems. We presented the basic principles to boost the interest of readers by providing a comprehensive tutorial to entry into this method. We show how CSO can be implemented using Matlab and C++ languages. One example of function optimization was given to illustrate the application of CSO. Although the procedures of the CSO algorithm were presented for a certain task, one can design CSO for other similar problems. To apply the CSO algorithm to solve the discrete problem, we therefore need defining movement in the search space and to employ it in the structure of the considered algorithm. We believe that there still is a great potential for solving problems in various domains.

References

1. Z.Chen, H. Tang. "Cockroach swarm optimization" in *Proc. of 2nd International Conference on Computer Engineering and Technology (ICCET)*, 2010, pp. 652-655.

2. Z. Chen. "A modified cockroach swarm optimization" in *Energy Procedia*, vol. 11, pp. 4-9, 2011.

3. I.C. Obagbuwa, A.O. Adewumi. "An Improved Cockroach Swarm Optimization" in *The Scientific World Journal*, Article ID 375358, 2014.

4. J. Kwiecien. "Use of different movement mechanisms in cockroach swarm optimization algorithm for traveling salesman problem" in *Artificial Intelligence and Soft Computing*, vol. 9693, pp. 484-493, 2016.

5. J. Kwiecien, M. Pasieka. "Cockroach swarm optimization algorithm for travel planning" in *Entropy*, 19, 213, 2017.

6. L. Cheng, Z. Wang, S. Yanhong, A. Guo. "Cockroach swarm optimization algorithm for TSP" in *Adv. Eng. Forum*, vol. 1, pp. 226-229, 2011.

7. C. Le, H. Lixin, Z. Xiaoqin, B. Yuetang, Y. Hong. "Adaptive cockroach colony optimization for rod-like robot navigation" in *Journal of Bionic Engineering*, vol. 12, pp. 324-337, 2015.

8

Crow Search Algorithm

Adam Slowik
Department of Electronics and Computer Science
Koszalin University of Technology, Koszalin, Poland

Dorin Moldovan
Department of Computer Science
Technical University of Cluj-Napoca, Romania

CONTENTS

8.1 Introduction

Crow Search Algorithm CSA is a novel metaheuristic that was introduced in [1] and its main application is the solving of the constrained engineering optimization problems. Its main source of inspiration is the behavior of crows, birds which are considered to be among the most intelligent animals in the whole world, and the main principles of the algorithm are the organization of the crows in the form of flocks, the memorization of the hiding places that are used for storing excess food, the following of each other when they do a theft and the protection of their caches from being stolen. These principles lead to the development of a unique algorithm that is much different from other algorithms that have as main inspiration the behavior of the birds in nature such as: Chicken Swarm Optimization (CSO) [2], Cuckoo Search (CS) [3], Bird Swarm Algorithm (BSA) [4], Bird Mating Optimizer (BMO) [5] and Peacock Algorithm (PA) [6]. The algorithms that might be considered as the main source of inspiration for CSA are Particle Swarm Optimization (PSO) [7], Genetic Algorithms (GA) [8] and Harmony Search (HS) [9], but compared to those algorithms CSA has fewer configurable parameters and thus it reduces the effort of parameter setting which is time-consuming work. CSA can be applied for various engineering optimization problems and in the original article that introduces the algorithm [1], several applications are presented such as the three-bar truss, the welded beam and the gear train design problems. The solutions of CSA are represented by crows, and each crow has a position in the D-dimensional space, a memory and a fitness value. Even though CSA is a novel bio-inspired algorithm, in literature there are already a lot of variations of it such as: Multi-Objective Crow Search Algorithm (MOCSA) [10], Binary Crow Search Algorithm (BCSA) [11] and Chaotic Crow Search Algorithm (CCSA) [12]. Moreover CSA was also used in combination with other algorithms and several examples are Hybrid Cat Swarm Optimization - Crow Search Algorithm (HCSO-CSA) [13] and hybrid Grey Wolf Optimizer (GWO) with CSA (GWOCSA) [14]. CSA has already been used in literature for solving a diversity of optimization problems such as data clustering [15], electromagnetic optimization [16], parameter estimation of Software Reliability Growth Models (SRGMs) [17], photovoltaic model parameters identification [18], economic environmental dispatch [19], performance improvement for inverter-based distributed generation systems [20] and enhancement of the performance of medium-voltage distribution systems [21]. The chapter has the following structure: Section 8.2 presents the original version of CSA, Section 8.3 presents the source-code of CSA in Matlab, Section 8.4 presents the source-code of CSA in C++, Section 8.5 presents a numerical example of CSA and Section 8.6 presents conclusions.

8.2   Original CSA

The pseudo-code of CSA is presented in Algorithm 7 and it is adapted after the original version that is presented in [1].

Algorithm 7 Pseudo-code of CSA.

1: **for** $i = 1 : N$ **do**
2:     initialize crow C_i in the D-dimensional search space
3:     initialize memory M_i to C_i
4:     evaluate the fitness value of C_i
5: **end for**
6: $t = 0$
7: **while** $t < Iter_{max}$ **do**
8:     **for** $i = 1 : N$ **do**
9:         select a random value k from $\{1, ..., N\}$
10:         **if** $r \geq AP$ **then**
11:             **for** $j = 1 : D$ **do**

12: $\qquad C_{i,j} = C_{i,j} + r_i \times fl \times (M_{k,j} - C_{i,j})$

13: $\qquad$ **end for**

14: $\qquad$ **else**

15: $\qquad$ **for** $j = 1 : D$ **do**

16: $\qquad\qquad C_{i,j} = r_j \times (C_{max} - C_{min}) + C_{min}$

17: $\qquad$ **end for**

18: $\qquad$ **end if**

19: $\qquad$ **end for**

20: $\qquad$ **for** $i = 1 : N$ **do**

21: $\qquad$ **for** $j = 1 : D$ **do**

22: $\qquad\qquad$ update the value of $C_{i,j}$ to be in the interval $[C_{min}, C_{max}]$

23: $\qquad$ **end for**

24: $\qquad$ **end for**

25: $\qquad$ **for** $i = 1 : N$ **do**

26: $\qquad\qquad$ evaluate the new position of C_i

27: $\qquad\qquad$ evaluate the new value of memory M_i

28: $\qquad\qquad$ **if** $OF(C_i) < OF(M_i)$ **then**

29: $\qquad\qquad\qquad M_i$ is updated to C_i

30: $\qquad\qquad$ **end if**

31: $\qquad$ **end for**

32: $\qquad t = t + 1$

33: **end while**

34: return M_i from memory for which $OF(M_i)$ is minimal

The **inputs** of the algorithm are: N - the number of crows, D - the number of dimensions of the search space, $Iter_{max}$ - the maximum number of iterations, $[C_{min}, C_{max}]$ - the range of variability of the positions of the crows, AP - the awareness probability and fl - the flight length. The **output** of CSA is represented by that i-th value from memory M for which the value of $OF(M_i)$ is minimal in the minimization case or maximal in the maximization case.

In steps 1-5 of the algorithm the initial population of N crows is initialized as follows: for each crow the D-dimensional vector C_i that describes the position is initialized with random numbers from the interval $[C_{min}, C_{max}]$ and the initial value of the memory M_i is initialized with the value of C_i. Initially $Memory(M) = Flock(C)$.

$$Flock = \begin{bmatrix} C_{1,1} & ... & C_{1,D} \\ ... & ... & ... \\ C_{N,1} & ... & C_{N,D} \end{bmatrix} \qquad (8.1)$$

$$Memory = \begin{bmatrix} M_{1,1} & ... & M_{1,D} \\ ... & ... & ... \\ M_{N,1} & ... & M_{N,D} \end{bmatrix} \qquad (8.2)$$

The fitness of C_i is evaluated using the objective function that is described by formula 8.5. In *step 6* the value of the current iteration is initialized to 0.

The next steps, from *step 8* to *step 32*, are repeated for a number of iterations equal to $Iter_{max}$. In steps 8-19 the initial population of N crows is initialized using formula 8.3 if the value of a random variable r from $[0, 1]$ is greater than or equal to AP or using the formula 8.4 otherwise. The first case corresponds to the situation in which the crow C_i follows another crow C_j from the flock having as main objective the discovery of the memory M_j of that crow, and the second case corresponds to the situation in which the new position is initialized randomly in the D-dimensional search space.

$$C_{i,j} = C_{i,j} + r_i \times fl \times (M_{k,j} - C_{i,j}) \tag{8.3}$$

$$C_{i,j} = r_j \times (C_{max} - C_{min}) + C_{min} \tag{8.4}$$

In formula 8.3 the value of r_i is a random value from $[0, 1]$ and k is a number from $\{1, ..., N\}$ selected randomly prior to the updating of the crow position. In formula 8.4 the value of r_j is a random number from $[0, 1]$ for each dimension j such that $j \in \{1, ..., D\}$.

The feasibility of C_i is checked for each crow. In this chapter a position C_i is considered feasible if all values of the D-dimensional vector C_i are in the interval $[C_{min}, C_{max}]$. In steps 20-24 the positions of the crows are updated to take values from the interval $[C_{min}, C_{max}]$ as follows: if $C_{i,j} < C_{min}$ then $C_{i,j} = C_{min}$ and if $C_{i,j} > C_{max}$ then $C_{i,j} = C_{max}$.

In steps 25-31 the memory of each crow C_i is updated as follows: in *step 26* the value of the position C_i is evaluated using the formula 8.5, in *step 27* the value of the memory M_i is evaluated using the same formula 8.5 and if the value of $OF(C_i)$ is less than the value of $OF(M_i)$ (or greater than the value of $OF(M_i)$ for maximization problems) then M_i is updated to the value of C_i (*step 29*). In *step 32* the value of the current iteration t is increased with 1.

Finally, in *step 34* the algorithm returns the memory M_i from the entire set of memory values for which the value of $OF(M_i)$ is minimal in the minimization case or maximal in the maximization case.

8.3   Source-code of CSA in Matlab

The objective function that is optimized by CSA is given by formula 8.5. In the function $OF(x, D)$ the input parameter x is the vector of decision variables, and D represents the number of dimensions of the search space.

$$OF(x, D) = \sum_{i=1}^{D} x_i^2 \quad \text{where } -5.12 \leqslant x_i \leqslant 5.12 \tag{8.5}$$

Listing 8.1 presents the source-code of the objective function.

```
1 function [y]=OF(x,D)
2   y=0;
3   for i=1:D
4     y=y+x(1,i)*x(1,i);
5   end
6 end
```

Listing 8.1
Definition of objective function $OF(x,D)$ in Matlab.

```
1 N=10; d=5; Pmax=5.12; Pmin=-5.12;
2 IterMax=30;
3 fl=0.9;
4 AP=0.5;
5 Crows=zeros(N,d);
6 Memory=zeros(N,d);
7 EvalCrows=zeros(N,1);
8 EvalMemory=zeros(N,1);
9 for i=1:N
10   for j=1:d
11     Crows(i,j)=rand()*(Pmax-Pmin)+Pmin;
12   end
13 end
14 Memory=Crows;
15 EvalCrows=OF(Crows);
16 Iter=0;
17 while Iter<IterMax
18   for i=1:N
19     ri=rand();
20     k=randi(N);
21     if rand()>=AP
22       for j=1:d
23         Crows(i,j)=Crows(i,j)+ri*fl*(Memory(k,j)-Crows(i,j));
24       end
25     else
26       for j=1:d
27         Crows(i,j)=rand()*(Pmax-Pmin)+Pmin;
28       end
29     end
30   end
31   for i=1:N
32     for j=1:d
33       if Crows(i,j)<Pmin
34         Crows(i,j)=Pmin;
35       end
36       if Crows(i,j)>Pmax
37         Crows(i,j)=Pmax;
38       end
39     end
40   end
41   EvalCrows=OF(Crows);
42   EvalMemory=OF(Memory);
43   for i=1:N
44     if EvalCrows(i,1)<EvalMemory(i,1)
45       Memory(i,:)=Crows(i,:);
46       EvalMemory(i,1)=EvalCrows(i,1);
47     end
48   end
49   Iter=Iter+1;
50 end
51 [x,y]=min(EvalMemory);
52 disp('FINAL RESULT: ');
53 disp(Memory(y,:));
54 disp(x);
```

Listing 8.2
Source-code of CSA in Matlab.

8.4   Source-code of CSA in C++

```
 1  #include <iostream>
 2  #include <cstdlib>
 3  #include <ctime>
 4
 5  using namespace std;
 6  float OF(float x[], int size_array) {
 7    float t = 0;
 8    for(int i = 0; i < size_array; i++) t = t + x[i] * x[i];
 9    return t;
10  }
11  float r() {return (float)(rand()%1000)/1000;}
12
13  int main()
14  {
15    srand(time(NULL));
16    int N=10; int d=5; int IterMax=30; int Iter=0;
17    float Crowmax=5.12; float Crowmin=-5.12;
18    float fl=0.9; float AP=0.5;
19    float Crows[N][d]; float Memory [N][d];
20    float EvalCrows[N]; float EvalMemory[N];
21    for(int i=0; i<N; i++) {
22      for(int j=0; j<d; j++) {
23        Crows[i][j]=r()*(Crowmax-Crowmin)+Crowmin;
24        Memory[i][j]=Crows[i][j];
25      }
26      EvalCrows[i]=OF(Crows[i],d);
27    }
28    while(Iter<IterMax) {
29      for(int i=0; i<N; i++) {
30        float ri=r();
31        int k=rand()%N;
32        if (r()>=AP) {
33          for(int j=0; j<d; j++) {
34            Crows[i][j]+=ri*fl*(Memory[k][j]-Crows[i][j]);
35          }
36        } else {
37          for(int j=0; j<d; j++) {
38            Crows[i][j]=r()*(Crowmax-Crowmin)+Crowmin;
39          }
40        }
41      }
42      for(int i=0; i<N; i++) {
43        for(int j=0; j<d; j++) {
44          if (Crows[i][j]<Crowmin) {
45            Crows[i][j]=Crowmin;
46          }
47          if (Crows[i][j]>Crowmax) {
48            Crows[i][j]=Crowmax;
49          }
50        }
51      }
52      for(int i=0; i<N; i++) {
53        EvalCrows[i]=OF(Crows[i], d);
54        EvalMemory[i]=OF(Memory[i], d);
55        if(EvalCrows[i]<EvalMemory[i]) {
56          for(int j=0; j<d; j++) {
57            Memory[i][j]=Crows[i][j];
58          }
59          EvalMemory[i]=EvalCrows[i];
60        }
61      }
62      Iter=Iter+1;
```

```
63    }
64    float minimum=EvalMemory[0]; int minimumIndex=0;
65    for(int i=1; i<N; i++) {
66       if(EvalMemory[i]<minimum) {
67          minimum=EvalMemory[i];
68          minimumIndex=i;
69       }
70    }
71    cout<<"FINAL RESULT: ["<<Memory[minimumIndex][0];
72    for(int j=1; j<d; j++) {
73       cout<<", "<<Memory[minimumIndex][j];
74    }
75    cout<<"]"<<endl;
76    cout<<"OF="<<EvalMemory[minimumIndex]<<endl;
77    getchar();
78    return 0;
79 }
```

Listing 8.3
Source-code of CSA in C++.

8.5   Step-by-step numerical example of CSA

First step
The objective of the algorithm is to minimize the value of the function $OF(x, D)$ which is given by equation 8.5 where the number of dimensions D is equal to 5.

Second step
The values of the parameters of CSA are initialized as follows:
N - the number of crows is equal to 10
D - the number of dimensions is equal to 5
$Iter_{max}$ - the maximum number of iterations is equal to 30
$[C_{min}, C_{max}]$ - the positions of the crows are in the interval $[-5.12, 5.12]$
AP - the awareness probability is 0.5
fl - the flight length is equal to 0.9

Third step
The initial population of crows that consists of 10 crows is created randomly and each crow is represented by one 5-dimensional vector that describes the position and one 5-dimensional vector that describes the memory. Initially for each crow the value of the position is equal to the value of the memory.
$C_1 = M_1 = \{1.300, -5.027, 1.904, -0.696, -2.314\}$
$C_2 = M_2 = \{1.525, -5.017, -1.894, -1.290, 3.307\}$
$C_3 = M_3 = \{-2.969, -3.962, -1.914, -4.515, -0.993\}$
$C_4 = M_4 = \{0.389, 1.730, 1.464, 2.836, -0.778\}$
$C_5 = M_5 = \{-2.222, 0.460, 2.703, 1.884, 0.716\}$
$C_6 = M_6 = \{0.378, 0.491, -0.870, 0.153, 0.143\}$
$C_7 = M_7 = \{-1.484, -2.232, -0.993, 2.928, 1.689\}$
$C_8 = M_8 = \{2.519, 0.286, 3.307, -4.075, 0.757\}$

$C_9 = M_9 = \{-1.280, 4.751, 1.710, 3.440, 0.092\}$
$C_{10} = M_{10} = \{-4.546, -0.184, -0.983, -4.423, 2.058\}$

Fourth step

The position of each crow is evaluated using the objective function $OF(x, D)$.
$EvalCrow_1 = OF(C_1) = 36.438, EvalCrow_2 = OF(C_2) = 43.697,$
$EvalCrow_3 = OF(C_3) = 49.569, EvalCrow_4 = OF(C_4) = 13.941,$
$EvalCrow_5 = OF(C_5) = 16.522, EvalCrow_6 = OF(C_6) = 1.186,$
$EvalCrow_7 = OF(C_7) = 19.606, EvalCrow_8 = OF(C_8) = 34.551,$
$EvalCrow_9 = OF(C_9) = 38.984, EvalCrow_{10} = OF(C_{10}) = 45.476$

For a number of iterations equal to $Iter_{max} = 30$ repeat the **Fifth step**, the **Sixth step** and the **Seventh step**.

Fifth step

The position of each crow is updated considering the following two possible cases: (case 1) if the value of a random numerical value is greater than the value of $AP = 0.5$ then the new position considers both the current value of the position and the current value of the memory of the crow, (case 2) otherwise the new position is initialized randomly. The new positions of the crows are:

(case 2) $C_1 = \{-4.792, 3.143, -1.505, -4.997, -4.751\}$
(case 1) $C_2 = \{1.363, -4.780, -1.695, -1.153, 3.195\}$
(case 1) $C_3 = \{-1.679, -2.247, -1.512, -2.717, -0.555\}$
(case 1) $C_4 = \{0.389, 1.730, 1.464, 2.836, -0.778\}$
(case 1) $C_5 = \{-2.222, 0.460, 2.703, 1.884, 0.716\}$
(case 2) $C_6 = \{4.710, -4.741, 3.502, 1.392, 2.682\}$
(case 1) $C_7 = \{-1.400, 0.646, 0.121, 3.139, 1.031\}$
(case 1) $C_8 = \{2.519, 0.286, 3.307, -4.075, 0.757\}$
(case 1) $C_9 = \{-2.117, 3.485, 1.019, 1.423, 0.596\}$
(case 2) $C_{10} = \{-4.116, -2.283, -3.604, 5.027, -0.798\}$

Sixth step

The values of the positions of the crows are updated so that they are in the interval $[C_{min}, C_{max}] = [-5.12, 5.12]$. After this step the positions have the same values because all values are already in the interval $[-5.12, 5.12]$.

Seventh step

For each crow the fitness value of the position and the fitness value of the memory are calculated. In addition if the fitness value of the position of the crow is less than the fitness value of the memory of the crow then the value of the memory is initialized with the value of the crow position and the fitness value of the memory is also updated with the fitness value of the crow position in order to reflect the change.

$EvalCrow_1 = OF(C_1) = 82.661, EvalMemory_1 = OF(M_1) = 36.438$
$EvalCrow_2 = OF(C_2) = 39.134, EvalMemory_2 = OF(M_2) = 43.697$
$OF(C_2) < OF(M_2) \Rightarrow M_2 = C_2, EvalMemory_2 = OF(C_2) = 39.134$
$EvalCrow_3 = OF(C_3) = 17.850, EvalMemory_3 = OF(M_3) = 49.569$
$OF(C_3) < OF(M_3) \Rightarrow M_3 = C_3, EvalMemory_3 = OF(C_3) = 17.850$
$EvalCrow_4 = OF(C_4) = 13.941, EvalMemory_4 = OF(M_4) = 13.941$
$EvalCrow_5 = OF(C_5) = 16.522, EvalMemory_5 = OF(M_5) = 16.522$
$EvalCrow_6 = OF(C_6) = 66.067, EvalMemory_6 = OF(M_6) = 1.186$
$EvalCrow_7 = OF(C_7) = 13.314, EvalMemory_7 = OF(M_7) = 19.606$
$OF(C_7) < OF(M_7) \Rightarrow M_7 = C_7, EvalMemory_7 = OF(C_7) = 13.314$
$EvalCrow_8 = OF(C_8) = 34.551, EvalMemory_8 = OF(M_8) = 34.551$
$EvalCrow_9 = OF(C_9) = 20.054, EvalMemory_9 = OF(M_9) = 38.984$
$OF(C_9) < OF(M_9) \Rightarrow M_9 = C_9, EvalMemory_9 = OF(C_9) = 20.054$
$EvalCrow_{10} = OF(C_{10}) = 61.069, EvalMemory_{10} = OF(M_{10}) = 45.476$
$OF(C_{10}) < OF(M_{10}) \Rightarrow M_{10} = C_{10}, EvalMemory_{10} = OF(C_{10}) = 45.476$

Finally after $Iter_{max} = 30$ iterations, the best solution from memory is returned. As the best solution we mean the solution for which the value of the objective function is the lowest (in the minimization case). The content of the memory after 30 iterations is:
$M_1 = \{-0.323, 0.516, -0.406, -0.576, -0.676\}$
$EvalMemory_1 = OF(M_1) = 1.327$
$M_2 = \{-0.060, -0.166, 0.284, -0.199, -0.361\}$
$EvalMemory_2 = OF(M_2) = 0.282$
$M_3 = \{1.150, 0.437, -0.162, 0.052, 0.188\}$
$EvalMemory_3 = OF(M_3) = 1.578$
$M_4 = \{-0.242, -0.003, -0.025, -0.203, 0.249\}$
$EvalMemory_4 = OF(M_4) = 0.162$
$M_5 = \{0.041, -0.071, -0.011, -0.231, -0.368\}$
$EvalMemory_5 = OF(M_5) = 0.196$
$M_6 = \{0.378, 0.491, -0.870, 0.153, 0.143\}$
$EvalMemory_6 = OF(M_6) = 1.186$
$M_7 = \{0.068, 0.247, -0.618, -0.464, -0.362\}$
$EvalMemory_7 = OF(M_7) = 0.795$
$M_8 = \{-0.050, -0.365, 0.067, -0.299, -0.071\}$
$EvalMemory_8 = OF(M_8) = 0.235$
$M_9 = \{0.131, 0.096, 0.144, -0.142, 0.184\}$
$EvalMemory_9 = OF(M_9) = 0.102$
$M_{10} = \{0.230, 0.022, -0.035, -0.166, -0.240\}$
$EvalMemory_{10} = OF(M_{10}) = 0.140$

Eighth step
The final result corresponds to the memory of the ninth crow.
$Result = M_9 = \{0.131, 0.096, 0.144, -0.142, 0.184\}$
$EvalResult = EvalMemory_9 = OF(M_9) = 0.102$

8.6   Conclusions

This chapter presented the main principles of CSA. It presented the pseudo-code of CSA and the corresponding source-code both in Matlab and in C++. In addition it showed how this algorithm works providing a step-by-step numerical example. This chapter will facilitate the development of other versions of the algorithm in other programming languages and it might be the source of inspiration for new modifications that can be applied in complex engineering optimization problem solving.

References

1. A. Askarzadeh. "A novel metaheuristic method for solving constrained engineering optimization problems: Crow search algorithm" in *Computers and Structures*, vol. 169, 2016, pp. 1-12.

2. X. Meng, Y. Liu, X. Gao, H. Zhang. "A New Bio-inspired Algorithm: Chicken Swarm Optimization" in *Lecture Notes in Computer Science*, vol. 8794, Springer, 2014, pp. 86-94.

3. C. Zefan, Y. Xiaodong. "Cuckoo search algorithm with deep search" in *Proc. of the 2017 3rd IEEE International Conference on Computer and Communications (ICCC)*, 2017, pp. 2241-2246.

4. X.-B. Meng, X. Z. Gao, L. Lu, Y. Liu, H. Zhang. "A new bio-inspired optimisation algorithm: Bird Swarm Algorithm" in *Journal of Experimental & Theoretical Artificial Intelligence*, vol. 28, no. 4, 2016, pp. 673-687.

5. A. Arram, M. Z. A. Nazri, M. Ayob, A. Abunadi. "Bird mating optimizer for discrete berth allocation problem" in *Proc. of the 2015 International Conference on Electrical Engineering and Informatics (ICEEI)*, 2015, pp. 450-455.

6. R. Chaudhary, H. Banati. "Peacock Algorithm" in *Proc. of the 2019 IEEE Congress on Evolutionary Computation (CEC)*, 2019, pp. 2331-2338.

7. R. Eberhart, J. Kennedy. "A new optimizer using particle swarm theory" in *MHS'95. Proceedings of the Sixth International Symposium on Micro Machine and Human Science*, 1995, pp. 39-43.

8. J. Stender. "Introduction to genetic algorithms" in *IEE Colloquium on Applications of Genetic Algorithms*, 1994.

9. J. Zhang, P. Zhang. "A study on harmony search algorithm and applications" in *Proc. of 2018 Chinese Control And Decision Conference (CCDC)*, 2018, pp. 736-739.

10. H. Nobahari, A. Bighashdel. "MOCSA: A multi-objective crow search algorithm for multi-objective optimization" in *Proc. of the 2017 2nd Conference on Swarm Intelligence and Evolutionary Computation (CSIEC)*, 2017, pp. 60-65.

11. R. C. T. De Souza, L. d. S. Coelho, C. A. De Macedo, J. Pierezan. "A V-shaped binary crow search algorithm for feature selection" in *Proc. of the 2018 IEEE Congress on Evolutionary Computation (CEC)*, 2018, pp. 1-8.

12. G. I. Sayed, A. Darwish, A. E. Hassanien. "Chaotic crow search algorithm for engineering and constrained problems" in *Proc. of the 2017 12th International Conference on Computer Engineering and Systems (ICCES)*, 2017, pp. 676-681.

13. A. B. Pratiwi. "A hybrid cat swarm optimization - crow search algorithm for vehicle routing problem with time windows" in *Proc. of the 2017 2nd International conferences on Information Technology, Information Systems and Electrical Engineering (ICITISEE)*, 2017, pp. 364-368.

14. S. Arora, H. Singh, M. Sharma, S. Sharma, P. Anand. "A new hybrid algorithm based on grey wolf optimization and crow search algorithm for unconstrained function optimization and feature selection" in *IEEE Access*, vol. 7, 2019, pp. 26343-26361.

15. Z.-X. Wu, K.-W. Huang, A. S. Girsang. "A whole crow search algorithm for solving data clustering" in *Proc. of the 2018 Conference on Technologies and Applications of Artificial Intelligence (TAAI)*, 2018, pp. 152-155.

16. L. dos Santos Coelho, C. Richter, V. C. Mariani, A. Askarzadeh. "Modified crow search approach applied to electromagnetic optimization" in *Proc. of the 2016 IEEE Conference on Electromagnetic Field Computation (CEFC)*, 2016, pp. 1-1.

17. A. Chaudhary, A. P. Agarwal, A. Rana, V. Kumar. "Crow search optimization based approach for parameter estimation of SRGMs" in *Proc. of the 2019 Amity International Conference on Artificial Intelligence (AICAI)*, 2019, pp. 583-587.

18. A. Omar, H. M. Hasanien, M. A. Elgendy, M. A. L. Badr. "Identification of the photovoltaic model parameters using the crow search algorithm" in *The Journal of Engineering*, vol. 2017, 2017, pp. 1570-1575.

19. A. A. A. E. Ela, R. A. El-Sehiemy, A. M. Shaheen, A. S. Shalaby. "Application of the crow search algorithm for economic environmen-

tal dispatch" in *Proc. of the 2017 Nineteenth International Middle East Power Systems Conference (MEPCON)*, 2017, pp. 78-83.

20. D. A. Zaki, H. M. Hasanien, N. H. El-Amary and A. Y. Abdelaziz. "Crow search algorithm for improving the performance of an inverter-based distributed generation system" in *Proc. of the 2017 Nineteenth International Middle East Power Systems Conference (MEPCON)*, 2017, pp. 656-663.

21. A. M. Shaheen, R. A. El-Sehiemy. "Optimal allocation of capacitor devices on MV distribution networks using crow search algorithm" in *CIRED - Open Access Proceedings Journal*, vol. 2017, 2017, pp. 2453-2457.

9

Cuckoo Search Algorithm

Xin-She Yang

School of Science and Technology
Middlesex University, London, United Kingdom

Adam Slowik

Department of Electronics and Computer Science
Koszalin University of Technology, Koszalin, Poland

CONTENTS

9.1 Introduction

The original cuckoo search (CS) was first developed by Xin-She Yang and Suash Deb in 2009 [1], which was inspired by the brooding parasitism of some cuckoo species. CS has been used for engineering optimization [2], and has been extended to multiobjective optimization [3], and structural optimization problems [4]. In fact, the literature is rapidly expanding with a diverse range of applications [5]. One of the main advantages of cuckoo search is that it uses Lévy flights [6] that consist of a fraction of long-distance, non-local moves in addition to many local moves, which makes the algorithm much more efficient in exploring the sparse search space [5, 6]. Other variants have been developed, including the quantum-inspired CS [7], modified cuckoo search [8], random-key discrete cuckoo search [9] and others [10]. Recently, CS has been extended to a new variant of multi-species cuckoo search [11]. For a more comprehensive review of the cuckoo search, its variants and applications, please refer to more advanced literature [5, 10].

The rest of this chapter will first introduce the main steps of the original cuckoo search, and explain in detail how to carry out Lévy flights and implementations, followed by the introduction of the demo codes in both Matlab and C++.

9.2   Original cuckoo search

Many cuckoo species engage in aggressive reproduction strategies by laying eggs in the nests of host species so as to let host species to raise their young chicks. From the evolutionary point of view, this maximizes the reproductivity probability of the cuckoo species. However, host species can also fight back by abandoning such eggs or nests with contaminated alien eggs. Thus, there is an ongoing arms race between cuckoo species and host species. Cuckoo search has been developed, based on the inspiration of such cuckoo-host characteristics.

9.2.1   Description of the cuckoo search

In essence, the main characteristics for the cuckoo search with a population of n cuckoos can be summarized as

- Each cuckoo lays one egg and dumps it into a randomly chosen host nest. An egg can be considered as a solution vector x to an optimization problem.

- The best eggs in nests will be passed on to the next generation. This means the best solutions are retained.

- The number of available nests is fixed, and each egg laid by a cuckoo may be discovered and abandoned with a probability p_a. This is equivalent to that a fraction p_a of the total population will be modified at each iteration t.

In the real-world, each nest of a host bird typically has 3 or 4 or more eggs, and each cuckoo can attack quite a few nests by laying its eggs in such nests. For simplicity, we assume that each cuckoo can only lay one egg and affect one host nest, which means that the number of eggs is equal to the number of nests and cuckoos. Thus, we can encode the location of an egg as a solution vector to an optimization problem. This way, there is no need to distinguish eggs, cuckoos and nests. Thus, we have 'egg=cuckoo=nest'.

In the original CS, there are two main equations for two different actions. One equation is

$$x_i^{t+1} = x_i^t + \alpha s \otimes H(p_a - \epsilon) \otimes (x_j^t - x_k^t), \qquad (9.1)$$

where x_j^t and x_k^t are two different solution vectors, randomly selected from the population for $i = 1, 2, ..., n$. The switching condition with a probability p_a is realized by a Heaviside function (or a step function) by comparing p_a with a random number ϵ drawn from a uniform distribution in $[0, 1]$. In addition, s is a step size by random permutation of solution differences, and this is scaled by a scaling factor α. Furthermore, the product of two vectors is an entry-wise multiplication using $\otimes$.

The other equation is mainly for non-local random walks by Lévy flights

$$x_i^{t+1} = x_i^t + \alpha L(s, \beta), \tag{9.2}$$

where the step sizes are drawn from the Lévy distribution

$$L(s, \beta) \sim \frac{\beta \Gamma(\beta) \sin(\pi\beta/2)}{\pi} \frac{1}{s^{1+\beta}}, \tag{9.3}$$

which is an approximation to a Lévy distribution with an exponent $0 \le \beta \le 2$. The gamma function is defined by

$$\Gamma(q) = \int_0^\infty z^{q-1} e^{-u} du. \tag{9.4}$$

The notation '$\sim$' highlights the fact that this is a pseudo-random number to be drawn from the probability distribution on the right-hand side of the above equation. Strictly speaking, the Lévy step sizes should be given by the inverse integral form of the following Fourier transform [6]

$$L(s, \beta) = \frac{1}{\pi} \int_0^\infty \cos(ks) \exp[-w|k|^\beta] dk, \tag{9.5}$$

where w is a scaling parameter.

However, the realization of this integral is difficult, so we will use an efficient but simplified algorithm, called Mantegna's algorithm, in the implementation below.

9.2.2 Pseudo-code of CS

Based on the above descriptions, the main steps of the CS consist of initialization, the iteration loop, modifications of solutions by two different ways. The pseudo-code is shown in Algorithm 8.

9.2.3 Parameters in the cuckoo search

Comparing with other swarm optimization algorithms, CS has a relatively low number of parameters, which makes it easier to tune the algorithm. The three parameters are population size n, switching probability p_a and the Lévy exponent β. In our implementation here, we have used $n = 25$, $p_a = 0.25$ and $\beta = 1.5$.

Algorithm 8 Pseudo-code of the cuckoo search algorithm.

1: Define the objective function $f(\boldsymbol{x})$
2: Initialize all the parameters
3: Generate an initial population of n nests $\boldsymbol{x}_i$ $(i = 1, 2, ..., n)$
4: **for** t=1:MaxGeneration **do**
5:     Get a cuckoo/solution randomly
6:     Generate a solution by Lévy flights [Eq. (9.3)]
7:     Evaluate the new solution $\boldsymbol{x}_i$ and its fitness f_i
8:     Choose a nest (say, j) among n nests randomly
9:     **if** $f_i < f_j$ **then**
10:        Replace $\boldsymbol{x}_j$ by $\boldsymbol{x}_i$
11:    **end if**
12:    A fraction (p_a) of the worse nests are abandoned
13:    New nests/solutions are generated by Eq. (9.1)
14:    Update the best solution if a better solution is found
15: **end for**
16: Output results

9.3   Source code of the cuckoo search in Matlab

The implementation of the cuckoo search algorithm is straightforward using Matlab, though care should be taken when generating Lévy flights in terms of drawing random numbers.

For simplicity and to be consistent with other tutorial chapters such as the tutorial on the firefly algorithm, we use the same function benchmark $f(\boldsymbol{x})$

$$\text{minimize}\quad f(\boldsymbol{x}) = (x_1 - 1)^2 + (x_2 - 1)^2 + ... + (x_D - 1)^2, \quad x_i \in \mathbb{R}, \quad (9.6)$$

which has a global minimum of $\boldsymbol{x}_* = (1, 1, ..., 1)$. We also apply the following lower bound (Lb) and upper bound (Ub):

$$Lb = [-5, \ -5, \ ..., \ -5], \quad Ub = [+5, \ +5, \ ..., \ +5]. \quad (9.7)$$

The Matlab code for the cuckoo search algorithm here consists of five parts: initialization, the main loop, Lévy flights, simple bounds, and the objective function. The whole codes should consist of all the lines of codes in a sequential order. For ease of description, we discuss each part in sequential order.

The first part is mainly initialization of parameter values such as the population size $n = 25$ and the switching probability $pa = 0.25$. All the solutions of 25 nests or cuckoos in the initial population are evaluated so as to obtain the initial best solution. As mentioned earlier, we do not distinguish among an egg, a nest or a cuckoo and we have used here 'an egg=a cuckoo=a nest=a solution vector'.

```
1 function [bestnest ,fmin]=cuckoo_search_demo
2 %% Initialization of parameters
3 n=25;          % Number of nests (or population size)
4 pa=0.25;       % Discovery rate of alien eggs/solutions
5 nd=15;         % Number of dimensions
6 Lb=-5*ones(1,nd);   % Lower bounds
7 Ub=5*ones(1,nd);    % Upper bounds
8 N_IterTotal=1000;   % Increase it to get better results
9
10 % Random initial solutions for the population with n cuckoos
11 for i=1:n,
12 nest(i,:)=Lb+(Ub-Lb).*rand(size(Lb));
13 end
14 % Get the best solution among the initial population
15 fitness=10^10*ones(n,1);
16 [fmin,bestnest,nest,fitness]=get_best_nest(nest,nest,fitness);
```

Listing 9.1

CS demo initialization.

The second part is the main part, consisting of a loop over the whole cuckoo population at each iteration. New cuckoos or solutions are generated by `get_cuckoos()` or `empty_nests()` subroutines, which will be explained below.

```
1 %% Starting iterations
2 for iter=1:N_IterTotal,
3      % Generate new solutions (but keep the current best)
4      new_nest=get_cuckoos(nest,bestnest,Lb,Ub);
5      [fnew,best,nest,fitness]=get_best_nest(nest,new_nest,fitness);
6      % Discovery and randomization
7      new_nest=empty_nests(nest,Lb,Ub,pa) ;
8
9      % Evaluate this set of solutions
10      [fnew,best,nest,fitness]=get_best_nest(nest,new_nest,fitness);
11      % Find the best objective so far
12      if fnew<fmin,
13          fmin=fnew;
14          bestnest=best;
15      end
16 end %% End of iterations
17 %% Display all the nests
18 disp(strcat('Best solution=', num2str(bestnest)));
19 disp(strcat('Best objective=',num2str(fmin)));
```

Listing 9.2

Main part for the cuckoo search algorithm.

The third part is about the Lévy flights and the selection of the best solution. The generation of pseudo-random numbers obeying the Lévy distribution is carried out by Mantegna's algorithm [12]

$$s = \frac{u}{|v|^{1/\beta}}, \tag{9.8}$$

where u and v are drawn from normal distributions

$$u \sim N(0, \sigma_u^2), \quad v \sim N(0, 1). \tag{9.9}$$

Here, σ^2 should be calculated by

$$\sigma = \left\{ \frac{\Gamma(1+\beta) \sin(\pi\beta/2)}{\beta \, \Gamma[(1+\beta)/2] \, 2^{(\beta-1)/2}} \right\}^{1/\beta}, \tag{9.10}$$

where Γ is the standard gamma function. In practice, it is easier to use a fixed σ value, together with scaling factor such as 0.01. Thus, the Lévy flight step is replaced by

$$u \sim N(0, \sigma^2), \quad v \sim N(0, 1), \tag{9.11}$$

and

$$s = \frac{u}{|v|^{1/\beta}}, \quad \beta = 3/2, \tag{9.12}$$

which gives stepsizes for modifying $\boldsymbol{x}_i$ as

$$\text{stepsize} = 0.01(\boldsymbol{x}_i - \boldsymbol{x}_{\text{best}}) \cdot \frac{u}{|v|^{1/\beta}}. \tag{9.13}$$

```
1  %% Get cuckoos by ramdom walk
2  function nest=get_cuckoos(nest,best,Lb,Ub)
3  % Carry out Levy flights
4  n=size(nest,1);
5  beta=3/2;
6  sigma=(gamma(1+beta)*sin(pi*beta/2)/(gamma((1+beta)/2)*beta*2^((beta
       -1)/2)))^(1/beta);
7
8  for j=1:n,
9      s=nest(j,:);
10     %% Levy flights by Mantegna's algorithm
11     u=randn(size(s))*sigma;
12     v=randn(size(s));
13     step=u./abs(v).^(1/beta);
14     stepsize=0.01*step.*(s-best);
15     s=s+stepsize.*randn(size(s));
16     %% Apply simple bounds/limits
17     nest(j,:)=simplebounds(s,Lb,Ub);
18 end
19
20 %% Find the current best nest
21 function [fmin,best,nest,fitness]=get_best_nest(nest,newnest,fitness)
22 % Evaluating all new solutions
23 for j=1:size(nest,1),
24     fnew=fobj(newnest(j,:));
25     if fnew<=fitness(j),
26         fitness(j)=fnew;
27         nest(j,:)=newnest(j,:);
28     end
29 end
30 % Find the current best solution
31 [fmin,K]=min(fitness);
32 best=nest(K,:);
33
34 %% Replace some nests by constructing new solutions/nests
35 function new_nest=empty_nests(nest,Lb,Ub,pa)
36 % A fraction of worse nests are discovered with a probability pa
37 n=size(nest,1);
38 % Discovered or not -- a status vector
39 K=rand(size(nest))>pa;
40 %% Generate new solutions by biased/selective random walks
41 stepsize=rand*(nest(randperm(n),:)-nest(randperm(n),:));
42 new_nest=nest+stepsize.*K;
43 for j=1:size(new_nest,1)
44     s=new_nest(j,:);
45   new_nest(j,:)=simplebounds(s,Lb,Ub);
46 end
```

Listing 9.3
Lévy flights and selection of the best.

The fourth part applies the simple lower and upper bounds. This will ensure the new solution vector should be within the regular bounds.

```
1  % Application of simple bounds
2  function s=simplebounds(s,Lb,Ub)
3    % Apply the lower bound
4    ns_tmp=s;
5    I=ns_tmp<Lb;
6    ns_tmp(I)=Lb(I);
7
8    % Apply the upper bounds
9    J=ns_tmp>Ub;
10   ns_tmp(J)=Ub(J);
11   % Update this new move
12   s=ns_tmp;
```

Listing 9.4
Simple bounds are verified for new solutions.

The fifth part is the objective function. New solutions can be evaluated by calling the objective or cost function.

```
1  %% Cost or objective function
2  function z=cost(x)
3  z=sum((x-1).^2); % Solutions should be (1,1,...,1)
```

Listing 9.5
The objective function.

If we run this code with a maximum number of 1000 iterations, we can get the best minimum value as $f_{best} = 3.5 \times 10^{-12}$ for 15 decision variables. If we use $D = 5$ (five variables), we can easily get about 2.7×10^{-31}.

Since such algorithms contain randomness, multiple runs are needed to get some proper statistics and statistical measures.

9.4   Source code in C++

For the CS, we have also implemented it in C++. However, as there is no standard library for Lévy flights, we have also implemented the Lévy flights in C++ code below.

```
1  #include <iostream>
2  #include <time.h>
3  #include <cmath>
4  #include <random>
5  #include <algorithm>
6  using namespace std;
7  float simplebounds(float s, float Lb, float Ub)
8  {if (s>Ub) {s=Ub;} if (s<Lb) {s=Lb;} return s;}
9  //A very simple approximation of gamma function
10 float gamma(float x)
11 {return sqrt(2.0*M_PI/x)*pow(x/M_E, x);}
12 //Definition of the objective function OF(.)
13 float fobj(float x[], int size_array)
14 {float t=0;
15 for(int i=0; i<size_array; i++){t=t+(x[i]-1)*(x[i]-1);}
```

```
16 return t;}
17 //Generate pseudo random numbers in the range [0, 1)
18 float ra(){return (float)rand()/RAND_MAX;}
19 int main()
20 {//Normal distribution generator
21 std::default_random_engine generator;
22 std::normal_distribution<double> distribution(0,1.0);
23 srand(time(NULL));
24 int n=25;        //Number of nests (or population size)
25 float pa=0.25;   //Discovery rate of alien eggs/solutions
26 int nd=15;       //Number of dimensions
27 float Lb[nd];    //Lower bounds
28 float Ub[nd];    //Upper bounds
29 float fitness[n];      //Fitness values
30 float beta=3/2, sigma, fmin, best[nd], bestnest[nd];
31 int K, r1, r2, r, randperm1[n], randperm2[n];
32 for(int i=0; i<nd; i++){Lb[i]=-5; Ub[i]=5;}
33 int N_IterTotal=1000; //Increase it get better results
34 //Random initial solutions for the population with n cuckoos
35 float nest[n][nd], new_nest[n][nd];
36 for(int i=0; i<n; i++){
37     for(int j=0; j<nd; j++){
38         nest[i][j]=Lb[j]+(Ub[j]-Lb[j])*ra();}}
39 for(int i=0; i<n; i++){
40     fitness[i]=fobj(nest[i],nd);
41     randperm1[i]=i; randperm2[i]=i;}
42 K=0; fmin=fitness[K];
43 for(int i=1; i<n; i++){
44     if (fitness[i]<=fmin){
45         fmin=fitness[i];
46         K=i;}}
47 for(int j=0; j<nd; j++){
48     best[j]=nest[K][j];}
49 //Starting iterations
50 float s, u, v, step, stepsize, fnew;
51 for(int iter=0; iter<N_IterTotal; iter++)
52 {
53 //Generate new solutions (but keep the current best)
54 sigma=pow(gamma(1+beta)*sin(M_PI*beta/2)/(gamma((1+beta)/2)*beta*pow
    (2,(beta-1)/2)),(1/beta));
55 for(int i=0; i<n; i++)
56 {for(int j=0; j<nd; j++){
57     s=nest[i][j];
58     //Levy flights by Mantegna's algorithm
59     u=distribution(generator)*sigma;
60     v=distribution(generator);
61     step=u/pow(abs(v),(1/beta));
62     stepsize=0.01*step*(s-best[j]);
63     s=s+stepsize*distribution(generator);
64     //Apply simple bounds/limits
65     new_nest[i][j]=simplebounds(s, Lb[j], Ub[j]);}
66 fnew=fobj(new_nest[i], nd);
67 if (fnew<=fitness[i]){fitness[i]=fnew;
68     for(int j=0; j<nd; j++){
69         nest[i][j]=new_nest[i][j];}}}
70 //Get the best nest
71 K=min_element(fitness, fitness+nd)-fitness;
72 fmin=fitness[K];
73 for(int j=0; j<nd; j++){best[j]=nest[K][j];}
74 //Carry out random permutation
75 for(int i=0; i<n; i++){r1=rand()%n; r2=rand()%n;
76     if (r1!=r2){r=randperm1[r1]; randperm1[r1]=randperm1[r2];
        randperm1[r2]=r;}
77     r1=rand()%n; r2=rand()%n;
78     if (r1!=r2){r=randperm2[r1]; randperm2[r1]=randperm2[r2];
        randperm2[r2]=r;}
79     }
```

```
80 //Discovery  and  randomization
81 for(int i=0; i<n; i++)
82 {
83     for(int j=0; j<nd; j++)
84     {
85         stepsize=ra()*(nest[randperm1[i]][j]-nest[randperm2[i]][j]);
86         if(ra()>pa)
87         {
88             new_nest[i][j]=nest[i][j]+stepsize;
89         }
90         else
91         {
92             new_nest[i][j]=nest[i][j];
93         }
94         new_nest[i][j]=simplebounds(new_nest[i][j], Lb[j], Ub[j]);
95     }
96 }
97 //Evaluate  the  new  set  of  solutions
98 for(int i=0; i<n; i++)
99 {
100 fnew=fobj(new_nest[i], nd);
101     if (fnew<=fitness[i])
102     {fitness[i]=fnew;
103         for(int j=0; j<nd; j++)
104         {nest[i][j]=new_nest[i][j];}}}
105 //Get  the  best  nest
106 K=min_element(fitness, fitness+nd)-fitness;
107 fmin=fitness[K];
108 for(int j=0; j<nd; j++){best[j]=nest[K][j];}
109 //Find  the  best  objective  so  far
110 if (fnew<fmin)
111     {fmin=fnew;
112         for(int j=0; j<nd; j++)
113         {bestnest[j]=best[j];}}}
114 //Output  the  best  solution
115 cout<<"Best solution = [ ";
116 for(int i=0; i<nd; i++) {cout<<bestnest[i]<<" ";}
117 cout<<"]"<<endl;
118 cout<<"Best objective = "<<fmin<<endl;
119 getchar();
120 return 0;
121 }
```

Listing 9.6
Cuckoo search in C++.

9.5   A worked example

Let us use the cuckoo search algorithm to find the minimum of

$$f(\boldsymbol{x}) = (x_1 - 1)^2 + (x_2 - 1)^2 + (x_3 - 1)^2 + (x_4 - 1)^2 + (x_5 - 1)^2, \quad (9.14)$$

in the simple ranges of $-5 \leq x_i \leq 5$. This simple 5-variable problem has the global minimum solution $\boldsymbol{x}_{\text{best}} = [1, 1, 1, 1, 1]$.

For the purpose of algorithm demonstration, we only use $n = 3$ cuckoos. Suppose the initial population is randomly initialized and their objective (fitness) values are as follows:

$$
\begin{cases}
\boldsymbol{x}_1 = (\ 5.00 \quad -1.00 \quad 5.00 \quad 2.00 \quad 3.00\ ), & f_1 = f(\boldsymbol{x}_1) = 41.00, \\
\boldsymbol{x}_2 = (\ 3.00 \quad 2.00 \quad 4.00 \quad 5.00 \quad -5.00\ ), & f_2 = f(\boldsymbol{x}_2) = 66.00, \\
\boldsymbol{x}_3 = (\ 3.00 \quad -4.00 \quad 2.00 \quad 0.00 \quad -2.00\ ), & f_3 = f(\boldsymbol{x}_5) = 40.00, \\
\end{cases}
$$
$$(9.15)$$

where we have only used two decimal places for simplicity. Obviously, the actual random numbers generated for the initial population can be any numbers in the simple bounds, but here we have used the numbers that are rounded up as the initialization; this is purely for simplicity and clarity.

Obviously, the best solution with the lowest (or best) value of the objective function is $\boldsymbol{x}_3$ with $f_3 = 40.00$ for this population at $t = 0$. Thus, we have

$$\boldsymbol{x}_{\text{best}} = [3.00,\ -4.00,\ 2.00,\ 0.00,\ -2.00].\tag{9.16}$$

At the first iteration, let us suggest that we randomly pick solution $\boldsymbol{x}_2$ for modifying it by Lévy flights. If we generate two random number vectors u and v, then suppose we get

$$\text{stepsize} = \Delta\boldsymbol{x} = [0.50,\ -0.90,\ -1.40,\ -0.70,\ 2.30],\tag{9.17}$$

then the new modified solution will be

$$\boldsymbol{x}_2^1 = \boldsymbol{x}_2^0 + \Delta\boldsymbol{x} = [3.50,\ 1.10,\ 2.60,\ 4.30,\ -2.70],\tag{9.18}$$

which gives $f(\boldsymbol{x}_2^1) = 33.40$. This solution is better than the previous solution $\boldsymbol{x}_2$ and all the previous solutions. Thus, the next call to get the best solution will give the new best solution $f_{\min} = 33.40$ with

$$\boldsymbol{x}_{\text{best}} = [3.50,\ 1.10,\ 2.60,\ 4.30,\ -2.70].\tag{9.19}$$

After this Lévy flight modification, we randomly select another solution, say $\boldsymbol{x}_3$ via emptying a nest. We first generate a random number and compare it with p_a to decide if we should update this solution. Suppose the result is to update the solution. Now we have to do this by random permutation of the population, and also generate a random number (say $r = 0.42$). We can get the modification step size as

$$\Delta\boldsymbol{x}_3 = 0.42 \times (\boldsymbol{x}_2^1 - \boldsymbol{x}_1) = [-0.630,\ 0.882,\ -1.008,\ 0.966,\ -2.394],\tag{9.20}$$

which gives the new solution

$$\boldsymbol{x}_3^1 = \boldsymbol{x}_3 + \Delta\boldsymbol{x}_3 = [2.370,\ -3.118,\ 0.992,\ 0.966,\ -4.394].\tag{9.21}$$

Thus, the new objective value $f(\boldsymbol{x}_3^1) = 47.93$, which is no better than any of the solutions in the current population, so we will not update $f_{\min}$. However, we still replace $\boldsymbol{x}_3$ by the new solution for population diversity.

Once the above modifications over the whole population and updates are done, we move on to the next generation $t = 2$ of iterations.

Cuckoo search is very efficient, even with a small population. For example, if we use $n = 5$ (cuckoos), we can get $f_{\text{best}} = 1.34 \times 10^{-17}$ after 1000 iterations for the above example. Even for $D = 15$ and $n = 25$, we can typically get $f_{\text{best}} = 2.50 \times 10^{-12}$ after 1000 iterations.

Obviously, due to the stochastic nature, performance evaluations should be based on the statistical measures of results of multiple runs with proper parametric studies.

9.6   Conclusion

The cuckoo search algorithm can be very efficient and flexible. However, the implementation is slightly more complicated than the implementations of other algorithms due to its use of Lévy flights. The demo codes presented in this chapter provide a simple implementation to focus on the core ideas of the algorithm. It is hoped that this can form a basis for further improvements and modifications so as to solve a wide range of optimization problems in applications.

References

1. X.S. Yang, S. Deb, "Cuckoo search via Lévy flights", in *Proceedings of World Congress on Nature and Biologically Inspired Computing (NaBIC 2009)*, pp. 210-214, IEEE Publications, 2009.

2. X.S. Yang, S. Deb, "Engineering optimisation by cuckoo search", *Int. J. Mathematical Modelling and Numerical Optimisation*, 1(4): 330-343, 2010.

3. X.S. Yang, S. Deb, "Multiobjective cuckoo search for design optimization", *Computers & Operations Research*, 40(6): 1616-1624, 2013.

4. A.H. Gandomi, X.S. Yang, A. H. Alavi, "Cuckoo search algorithm: a metaheuristic approach to solve structural optimization problems", *Engineering with Computers*, 29(1): 17–35, 2013.

5. X.S. Yang, S. Deb, "Cuckoo search: recent advances and applications", *Neural Computing and Applications*, 24(1): 169-174, 2014.

6. I. Pavlyukevich, "Lévy flights, non-local search and simulated annealing", *J. Computational Physics*, 226(2): 1830-1844, 2007.

7. A. Layeb, "A novel quantum-inspired cuckoo search for knapsack problems", *Int. J. Bio-Inspired Computation*, 3(5): 297-305, 2011.

8. S. Walton, O. Hassan, K. Morgan, M.R. Brown, "Modified cuckoo search: a new gradient free optimization algorithm", *Chaos, Solitons & Fractals*, 44(9): 710-718, 2011.

9. A. Ouaarab, B. Ahiod, X.S. Yang, "Random-key cuckoo search for the travelling salesman problem", *Soft Computing*, 19(4): 1099-1106, 2015.

10. X.S. Yang, *Cuckoo Search and Firefly Algorithm: Theory and Applications*, Studies in Computational Intelligence, vol. 516, Springer, Heidelberg, 2013.

11. X.S. Yang, S. Deb, S.K. Mishra, "Multi-species cuckoo search algorithm for global optimization", *Cognitive Computation*, 10(6): 1085-1095, 2018.

12. R. N. Mantegna, "Fast, accurate algorithm for numerical simulation of Lévy stable stochastic process", *Physical Review E*, 49(5): 4677-4683, 1994.

10

Dynamic Virtual Bats Algorithm

Ali Osman Topal

Department of Computer Engineering
Epoka University, Tirana, Albania

CONTENTS

10.1 Introduction

Dynamic Virtual Bats Algorithm (DVBA) is proposed as another bat-inspired algorithm in 2014 by Topal and Altun [1, 2]. It is inspired by the bat's ability to manipulate frequency and wavelength of emitted sound waves. Although the algorithm is fundamentally inspired from Bat Algorithm (BA) [3], it is conceptually very different. In DVBA, a role based search is developed by using two bats: explorer and exploiter bat. Each bat has its own role in the algorithm and during the search they exchange these roles dynamically according to their positions. The bat which is in a better position becomes the exploiter; meanwhile the other one becomes the explorer. While the exploiter bat increases the intensification of the search around a preferable position, the explorer bat will keep looking for a better position. Until the explorer bat finds a better position, the exploiter bat will increase intensification of the search after each iteration to attain the optimal solution.

DVBA has been tested on well-known test functions (from CEC-2014 [4]) and a supply chain cost problem. The results show that DVBA, similar to other evolutionary algorithms, has some challenging problems, like the convergence speed and avoiding local optima traps. Recently, these problems in DVBA have been eradicated by introducing probabilistic selection restart techniques in an improved version of DVBA (IDVBA) [5]. The rest of this chapter is organized as follows. In Section 10.2 an overview of DVBA is given with its pseudo code, in Sections 10.3 and 10.4 source code of the algorithm are presented in Matlab and C++ programming language respectively, in Section 10.5 a numerical example of DVBA is shown in detail, and the conclusion is drawn in Section 10.6.

10.2 Dynamic virtual bats algorithm

In DVBA, there are two bats with two different roles. In Figure 10.1, these roles are simulated. Here, black triangles represent the bats and plus signs represent the prey. The black dots are the positions on the waves which are going to be checked for a better solution. As shown in Figure 10.1a during exploration the search points which are created by the explorer bat are distributed widely in the search space. However in Figure 10.1b the exploiter bat creates a very small search scope where search points have become closer to each other.

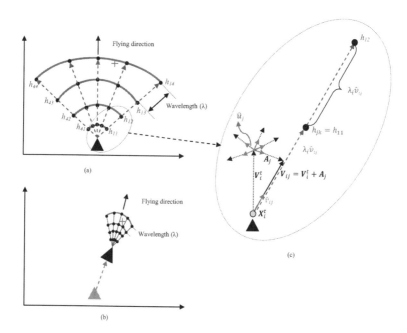

FIGURE 10.1

(a) Exploration: explorer bat searching a prey, (b) Exploitation: exploiter bat chasing a prey, (c) Search positions in the search scope of a bat.

10.2.1 Pseudo-code of DVBA

The pseudo-code of DVBA is represented in Algorithm 9. Step by step the algorithm is explained in the next section.

Algorithm 9 Description of DVBA with pseudo-code.

1: Objective function $f(x)$, $x = (x_1, ..., x_d)^T$
2: Initialize the bat population x_i and v_i, $(i = 1, 2)$
3: Initialize wavelength λ_i and frequency $freq_i$
4: Find the increment rate ρ
5: Define the increment divisor β, the scope width variable b
6: Initialize the number of the waves(j) and search points(k)
7: Find f_{gbest} based on the bats' starting position.
8: **while** $(t < $ Max number of FEs) **do**
9:     **for each** bat **do**
10:         Create a sound waves scope
11:         Evaluate the solutions on the waves
12:         Choose the best solution on the waves, h_*
13:         **if** $f(h_*)$ is better than $f(x_i)$ **then**
14:             Move to h_*, update x_i
15:             Change v_i towards to the better position
16:             Decrease λ_i and increase f_i
17:         **end if**
18:         **if** $f(x_i)$ is not better than the best solution f_{gbest} **then**
19:             Change the direction randomly
20:             Increase λ_i and decrease $freq_i$
21:         **end if**
22:         **if** $f(x_i)$ is the best found solution **then**
23:             Minimize λ_i and maximize $freq_i$
24:             Change the direction randomly
25:         **end if**
26:     **end for**
27:     Rank the bats and find the current best x_{gbest}
28: **end while**

---

where f_{gbest} is the global best solution and d is the number of dimensions.

10.2.2 Description of DVBA

In this section, we will explain how bats' hunting strategies became an optimization algorithm by presenting DVBA's pseudo-code in Algorithm 9. DVBA starts with defining a multidimensional objective function $f(x)$ in step 1. Similar to real bats, virtual bats are looking for superior solutions (prey) in their search scope $f(x)$. The solution (prey) in DVBA is the minimum value of the objective function $f(x)$.

Both bats start flying from random positions x_i with random velocities v_i using random wavelength and frequency (step 2 and 3). In step 4, 5, and 6 the parameters such as number of wave vectors, search points on the wave vectors, increment rate divisor β, and the scope width variable b are defined. Increment rate ρ is calculated by using the search space boundaries (10.8), so the size of the bat's search scope can be adapted with the size of the search space. In steps 16 and 20, ρ is used as the increment rate for frequency and wavelength (10.6-10.7). During the search, the range of the search scope changes dynamically: it expands when exploring (Figure 10.1a) or it shrinks if prey gets closer (Figure 10.1b). The length and the width of the search scope are controlled by wavelength λ_i and the frequency $freq_i$, respectively.

After defining all the parameters and initializing the bats' position and velocities, the main loop starts in step 8. Function of evaluation (FEs) is used as stopping criteria for the main loop. Generally, maximum number of FEs is set to $100,000$ for 10 dimensional functions and $300,000$ for 30 dimensional functions in optimization algorithms [4].

For each bat, the search scope $H_{j,k}$ (10.5) is generated in step 10. The search scope is simulated by using wave direction vectors V_j^i and search points h_{jk} on the vectors as shown in Figure 10.1a and Figure 10.1c. To distribute the wave direction vectors in the search scope the unit vectors $\hat{u}_j$ are generated which give direction to the scope width vectors $\mathbf{A}_j$ (Figure 10.1c). The unit vectors $\hat{u}_j$ are generated randomly; consequently, the wave vectors are distributed randomly in the search scope as well. The magnitude of A_j changes the width of the search scope, and is inversely proportional with the frequency of the waves $freq_i$ (10.1). When the frequency $freq$ increases reflects what is shown in Figure 10.1b. The search points h_{jk} are distributed by scaling the unit vector $\hat{v}_{ij}$ (10.3) with the same length as wavelength λ_i. The scope width vector $\mathbf{A}_j$, the wave vector $\mathbf{V}_{ij}$, and the positions of the search points h_{ij} in the bat's search scope at time step t are given by (10.1-10.5),

$$\mathbf{A}_j = \frac{\hat{u}_j b}{freq_i}, (j = 1, ..., w), (i = 1, ..., n) \tag{10.1}$$

$$\mathbf{V}_{ij} = \mathbf{V}_i^t + \mathbf{A}_j \tag{10.2}$$

$$\hat{\mathbf{v}}_{ij} = \frac{\mathbf{V}_{ij}}{\|\mathbf{V}_{ij}\|} \tag{10.3}$$

$$h_{j,k} = \mathbf{X}_i^t + \hat{\mathbf{v}}_{ij} k \lambda_i, (k = 1, ..., m) \tag{10.4}$$

$$\mathbf{H}_{j,k} = \begin{bmatrix} h_{1,1} & h_{1,2} & \cdots & h_{1,k} \\ h_{2,1} & h_{2,2} & \cdots & h_{2,k} \\ \vdots & \vdots & \ddots & \vdots \\ h_{j,1} & h_{j,2} & \cdots & h_{j,k} \end{bmatrix} \tag{10.5}$$

where w is the number of wave vectors, i is the index of bat (maximum 2), $b \in (20, 40)$ is the scope width variable, and m is the number of search positions on the wave vectors.

If the frequency increases, vectors will get closer and the wavelength (distance between search points) will get shorter (10.7). This is inspired by a bat's dynamic search ability for hunting. The changes of frequency $freq_i$, and wavelength λ_i at time step $t + 1$ are given by (10.6 - 10.8),

$$freq_i^{t+1} = freq_i^t \pm \rho \tag{10.6}$$

$$\lambda_i^{t+1} = \lambda_i^t \pm \rho \tag{10.7}$$

$$\rho = mean(\frac{\mathbf{U} - \mathbf{L}}{\beta}), \qquad \{\beta \in \Re : \beta > 0\} \tag{10.8}$$

where ρ and β are positive real constants, U is the upper bounds and L is the lower bounds of the search space, and ρ is used as the increment rate for frequency and wavelength. β is used as the increment rate divisor. Choosing the range of the wavelength is very important. If the wavelength is too short, in large search spaces, convergence could be very slow. Conversely, if the wavelength is too long, it might bypass the global best and fail to locate the optimum position. To overcome this problem the length of the wavelength is linked to the range of the problem. For simplicity the following approximations were used. The wavelength range λ, in a range $[\lambda_{min}, \lambda_{max}] = [0.1\rho, 10\rho]$ corresponds to a range of the frequency $[freq_{min}, freq_{max}] = [\rho, 10\rho]$.

If the best position of search scope h_* is better than the bat's current position x_i^t, the bat will fly to this position (10.9)(step 13 and 14). And also in step 15, its direction will be changed towards the next position (10.10). In step 16, the bat's wavelength will be shortened (10.7), and its frequency will be increased (10.6) so it will become the exploiter bat. As long as it has a better position in its sound waves scope it will increase intensification of the search. To avoid very small or big $\mathbf{V}_i^{t+1}$ it is normalized in equation 10.11.

$$x_i^{t+1} = h_* \tag{10.9}$$

$$\mathbf{V}_i^{t+1} = |\mathbf{x}_i^{t+1} - \mathbf{x}_i^t| \tag{10.10}$$

$$\hat{\mathbf{v}}_i = \frac{\mathbf{V}_i^{t+1}}{\|\mathbf{V}_i^{t+1}\|} \tag{10.11}$$

If h_* is worse than the bat's current position, the algorithm checks whether the current position x_i^t is the best one ever found (step 22) x_{gbest} or not (step 18).

If the current solution is not the best solution ever found then the bat becomes the explorer bat, changes its direction randomly, increases the wavelength, and decreases the frequency expanding the search scope (step 18, 19, and 20). These actions help the bat keep exploring the search space without getting trapped in a local optima and provides a random walk.

If the bat is already on the best found position (x_{gbest}) or a better one, the bat will become the exploiter bat. The wavelength will be minimized (λ_i^{t+1}) = (λ_{min}) , the frequency will be maximized ($freq_i^{t+1}$) = ($freq_{max}$), and the search direction will be changed randomly (step 22, 23, and 24). As a result, the bat can increase the intensification of the search around the best position.

After repeating same steps for the second bat, the first iteration of the while loop will be over. Until the number of the max function of evaluation is reached, the bats will dynamically change their roles and keep searching.

10.3   Source-code of DVBA in Matlab

The source code of an objective function is given in Listing 10.1. The input of the objective function "x" is the positions of the bats and the *sphere(x)* function returns the function values of the bats as a column vector. The aim of DVBA is to find the minimum value of this objective function. The objective function is given in (10.12). In Listing 10.2, the dimension of the function D is set to 10 and DVBA minimizes this function within 100.000 FEs.

$$f(x) = \sum_{i=1}^{D} x_i^2 \qquad \text{where } -5.12 \leqslant x_i \leqslant 5.12 \qquad (10.12)$$

```
1  % Objective function f
2  function f = sphere(x)
3      x = x.^2;
4      f = sum(x,2);
5  end
```

Listing 10.1
Definition of objective function $f(x)$ in Matlab.

```
1  f = @sphere;
2  ub = 5.12; lb = -ub; % [ub,lb]:upper and lower bounds
3  %  declaration of the parameters of DVBA
4  j=5; k=6; b=20; beta=100; D=10; P=2; maxAssesment= 100000;
5  % ro: increment rate of frequency and wavelength
6  ro = mean((ub-lb)/beta);
7  Lmax = ro * 10; Lmin = ro * 0.1; % wavelength range
8  freqMax = ro * 10.0; freqMin = ro * 1.0; % frequency range
9  freq = freqMin + (freqMax - freqMin) * rand(1,P);  % frequency
10 lambda = Lmin + (Lmax - Lmin) * rand(1,P);  % wavelength
11 % initialize the bat population and directions.
12 x = lb + (ub-lb)*rand(P,D);
```

```matlab
13  v = -1 + 2 * rand(P,D);
14  % find the best fitness value based on the bat's initial position.
15  fgbest = min(f(x));
16  % main program loop starts
17  while maxAssesment >= 1
18  % for each bat
19      for i=1:P
20  % wave vectors directions are generated
21          u = -1 + 2*rand(j,D);
22          A = u*b/freq(i);
23          V = bsxfun(@plus,v(i,:), A) ;
24          V = V./norm(V);
25  % search scope vector is generated for the bat
26          H = []; % search scope column vector
27          for r = 1:j     % wave vectors
28              for z = 1:k    % search points on the wave vector
29                  xtrans = x(i,:) + V(r,:)*z*lambda(i);
30  % keep the search points in the search scope [ub, lb]
31                  xtrans(xtrans>ub)=ub;
32                  xtrans(xtrans<lb)=lb;
33  % add each search points to the search scope matrix of the bat
34                  H = [H ; xtrans];
35              end
36          end
37  % compare the current position with the search scope points first
38          fnew = min(f(H));
39          maxAssesment = maxAssesment-(k*j + 5);
40  % in each loop, (k*j + 5) times FEs are done
41          idx = find(f(H)==fnew);
42          xnew = H(idx(1),:);
43          xtemp = x(i,:); % assign xtemp to use for the next steps
44  % the bat becomes the EXPLOITER bat,
45          if fnew < f(x(i,:))
46  % its direction will be changed towards the next position
47              v(i,:) = (xnew - x(i,:))./norm(xnew - x(i,:));
48              x(i,:) = xnew;
49  % shrink the search scope of the bat within the ranges
50              if lambda(i) > Lmin+ro
51                  lambda(i) = lambda(i) - ro;
52              else
53                  lambda(i) = Lmin;
54              end
55              if freq(i) < freqMax - ro
56                  freq(i) = freq(i) + ro;
57              else
58                  freq(i) = freqMax;
59              end
60          end
61  % the bat becomes the EXPLORER bat
62          if f(xtemp) > fgbest
63  % change its direction randomly
64              v(i,:) = -1 + 2 * rand(1,D);
65  % expand the search scope of the bat
66              if lambda(i) < Lmax - ro
67                  lambda(i) = lambda(i) + ro;
68              end
69              if freq(i) > freqMin + ro
70                  freq(i) = freq(i) - ro;
71              end
72          end
73  % the bat is already on the best found position
74          if f(xtemp) <= fgbest && f(xtemp) < fnew
75  % change its direction randomly and shrink the search scope
76              v(i,:) = -1 + 2 * rand(1,D);
77              lambda(i) = Lmin; freq(i) = freqMax;
78          end
79  % update best values: fgbest, xgbest
```

```
80            if f(x(i,:)) <= fgbest
81                fgbest = f(x(i,:)); xgbest = x(i,:);
82            end
83        end
84 end
85 % the result of DVBA is returned
86 disp(fgbest)
```

Listing 10.2
Source-code of DVBA in Matlab.

10.4   Source-code of DVBA in C++

```cpp
1  #include <iostream>
2  #include <cstdlib>
3  #include <ctime>
4  using namespace std;
5  // definition of the objective function f(x)
6  float ub=5.12; float lb=-5.12; //ub, lb: upper and lower bounds.
7  float f(float x[], int size_array){
8      float t = 0;
9      for(int i=0; i<size_array; i++){
10         t=t+x[i]*x[i];}
11     return t;}
12 // generate pseudo random values from the range [0, 1)
13 float r(){
14     return float(rand()%1000)/1000;}
15 // "checkBoundary" function to check the constraints:
16 float checkBoundary(float x, float min, float max){
17     if (x<min) x = min;
18     if (x>max) x = max;
19     return x;}
20 // main program function
21 int main(){
22 srand(time(0));
23 // initialization of the parameters
24 int P=2; int maxAssessment=100000; int assessment=0; int D=10;
25 int beta=100; int j=5; int k=6; int b=20;
26 int theBest=0;
27 float xmin[D]; float xmax[D]; float rho[D];
28 float freqMax[D], freqMin[D], freq[P][D];
29 float Lmax[D], Lmin[D], lambda[P][D];
30 // initialization of the constraints
31  for (int z=0; z<D; z++){
32     xmin[z] = -5.12; xmax[z] = 5.12;
33     rho[z] = (xmax[z] - xmin[z])/beta;
34     freqMax[z] = 10*rho[z]; freqMin[z] = rho[z];
35     Lmax[z] = 10*rho[z]; Lmin[z] = 0.01*rho[z];}
36 // initialization of all data structures which are needed by DVBA
37     float x[P][D]; float xtemp[D]; float xnew[D];
38     float v[P][D]; float xgbest[D];
39     float u[j][D], A[D][j], V[j][D], H[j*k][D];
40 // randomly create bat's positions, velocities, frequency(freq) and
        wavelength(lambda)
41     for(int i=0; i<P; i++){
42        for(int z=0; z<D; z++){
43           x[i][z]=(xmax[z]-xmin[z]) * r() + xmin[z];
44           v[i][z] = (-1 + 2 * r());
45           freq[i][z] = (freqMax[z]-freqMin[z]) * r() + freqMin[z];
46           lambda[i][z] = (Lmax[z]-Lmin[z]) * r() + Lmin[z];}
47 // evaluate the bats, find the best bat
48        if (f(x[i],D)<f(x[theBest],D)) theBest=i;}
```

```
49  // assign the best bat to xgbest
50      for (int i=0; i<D; i++) xgbest[i]=x[theBest][i];
51  // main program loop:
52      while(assessment<maxAssessment){
53  // for each bat
54          for(int i=0; i<P; i++){
55              assessment = assessment + (k*j);
56              int n=0, theHbest=0;
57  // generate the search scope matrix for the bat:
58              for(int z=0; z<j; z++){   // wave vectors
59                  for(int m=0; m<k; m++){  //search points
60                      for(int s=0; s<D; s++){ // for each dimension
61                          u[z][s]=-1+2*r();
62                          A[z][s]=b*u[z][s]/freq[i][s];
63                          V[z][s]=v[i][s]+A[z][s];
64                          V[z][s]=(V[z][s]-xmin[s])/(xmax[s]-xmin[s]);
65                          H[n][s]=x[i][s]+(z+1)*V[z][s]*lambda[i][s];
66                          H[n][s]=checkBoundary(H[n][s],xmin[s],xmax[s]);}
67  // find the best position on the bat's search scope xnew:
68                          if (f(H[n],D)<f(H[theHbest],D)) theHbest = n;
69                          n = n + 1;}}
70              for (int u=0; u<D; u++) xnew[u] = H[theHbest][u];
71              for (int t=0;t<D;t++) xtemp[t] = x[i][t];
72  // the bat becomes the EXPLOITER bat,
73              if (f(H[theHbest],D) < f(x[i],D)){
74                  for(int t=0; t<D; t++){
75  // move towards the next position, shrink the search scope
76                      v[i][t]=((xnew[t]-x[i][t]))/(xmax[t]-xmin[t]);
77                      x[i][t]=xnew[t];
78                      lambda[i][t]=checkBoundary(lambda[i][t]-rho[t],Lmin[t],
        Lmax[t]);
79                      freq[i][t]=checkBoundary(freq[i][t]+rho[t],freqMin[t],
        freqMax[t]);}}
80  // the bat becomes the EXPLORER bat
81              if (f(xtemp,D) > f(x[theBest],D)){
82  // change its direction randomly, expand the search scope
83                  for(int t=0; t<D; t++){
84                      v[i][t] = (-1 + 2 * r());
85                      lambda[i][t]=checkBoundary(lambda[i][t]+rho[t],Lmin[t],
        Lmax[t]);
86                      freq[i][t]=checkBoundary(freq[i][t]-rho[t],freqMin[t],
        freqMax[t]);}}
87  // the bat is already on the best found position
88              if (f(xtemp,D)<=f(x[theBest],D)&&f(xtemp,D)<f(xnew,D)){
89  // change its direction randomly, expand the search scope
90                  for(int t=0; t<D; t++){
91                      v[i][t] = (-1 + 2 * r());
92                      lambda[i][t] = Lmin[t];
93                      freq[i][t] = freqMax[t];}}
94  // update best values: fbest, xbest
95              if (f(x[i],D) <= f(x[theBest],D)) theBest = i;
96              for (int t=0; t<D; t++) xgbest[t] = x[theBest][t];}
97          cout<<"fbest = "<<f(x[theBest],D)<<endl;}
98      cout<<"xgbest = ";
99      for (int t=0;t<D;t++) cout<<xgbest[t]<<" ";}
```

Listing 10.3
Source-code of the DVBA in C++.

10.5   Step-by-step numerical example of DVBA

We will use DVBA to minimize the function given by equation 10.12. For simplification purposes, the dimension will be set to 3, $j = 2$, and $k = 3$. So there will be only two wave vectors, three search points on each of them, and in total $j \times k = 6$ points in the search scope of the bats. Search scope wideness variable b is set to 20, and increment rate divisor is set to 100. In the first step all these parameters will be set.

In the second step, increment rate ρ will be calculated by using equation 10.8, then the wavelength and the frequency will be initialized randomly for each bat within the ranges mentioned previously: $\lambda \in [0.1\rho, 10\rho]$ and the $freq \in [\rho, 10\rho]$.

$\rho = 0.1024, \quad stepsize$

$[freqMin, freqMax] = [0.1024 \quad 1.024], \quad frequency(freq) \quad range$

$[Lmin, Lmax] = [0.01024 \quad 1.0240], \quad wavelength(lambda) \quad range$

$freq1 = 0.5527$

$freq2 = 1.0065$

$lambda1 = 0.9682$

$lambda2 = 0.1649$

In the third step, two bats will start the search from random positions within the range of the search space $[-5.12 \quad 5.12]$ with random velocities(directions) in the range of (-1,1). And the best function value($fgbest$) will be updated based on the bats' current positions.

$x1 = [5.0652, \quad -4.2795, \quad -1.9859]$

$x2 = [1.3005, \quad -0.5550, \quad -0.5254]$

$v1 = [-0.4553, \quad 0.9136, \quad 0.7642]$

$v2 = [-0.5224, \quad 0.6805, \quad 0.8534]$

$$f(x1) = \sum_{i=1}^{3} x_i^2 = [5.0652^2 + (-4.2795)^2 + (-1.9859)^2] = 47.9140$$

$$f(x2) = \sum_{i=1}^{3} x_i^2 = [1.3005^2 + (-0.5550)^2 + (-0.5254)^2] = 2.2755$$

$fgbest = min(f) = 2.2755$

In the fourth step the main loop of the algorithm starts. If the algorithm termination condition is fulfilled we jump to the fourteenth step, if not we continue to step five.

In the fifth step, the first bat's search scope column vector(H) will be generated by using the equations from 10.1 to 10.4. Since there are just two wave vectors there will be only $A11$, $A12$, $V11$, and $V12$ vectors for the search scope.

$u = -1 + 2 * rand(j, d)$

$u1 = [0.0309, \quad 0.0278, \quad -0.0334]$

$u2 = [-0.0299, \quad 0.0252, \quad -0.0177]$

$A11 = u1 * b/freq1 = [1.1198, \quad 1.0074, \quad -1.2079]$
$A12 = u2 * b/freq1 = [-1.0808, \quad 0.9105, \quad -0.6400]$
$V11 = norm(v1 + A11) = [0.2420, \quad 0.6996, \quad -0.1616]$
$V12 = norm(v1 + A12) = [-0.5594, \quad 0.6644, \quad 0.0453]$

$H1 = x1 + V11 * 1 * lambda1 = [5.1200, \quad -3.6021, \quad -2.1424]$
$H2 = x1 + V11 * 2 * lambda1 = [5.1200, \quad -2.9247, \quad -2.2989]$
$H3 = x1 + V11 * 3 * lambda1 = [5.1200, \quad -2.2473, \quad -2.4553]$
$H4 = x1 + V12 * 1 * lambda1 = [4.5235, \quad -3.6362, \quad -1.9421]$
$H5 = x1 + V12 * 2 * lambda1 = [3.9819, \quad -2.9930, \quad -1.8983]$
$H6 = x1 + V12 * 3 * lambda1 = [3.4402, \quad -2.3497, \quad -1.8545]$

$H1$, $H2$, and $H3$ are the search points on the first wave vector, and the last three are the search points on the second wave vector.

In the sixth step, all the positions in the first bat's search scope (H) will be evaluated by using the objective function, the best value and the position will be assigned as $fnew$ and $xnew$, respectively.

$$f(H1) = \sum_{i=1}^{3} H_i^2 = [5.1200^2 + (-3.6021)^2 + (-2.1424)^2] = 43.7792$$

if we repeat the same competition for all the search points in the scope, we obtain f(H) values:

$f(H1) = 43.7792$
$f(H2) = 40.0529$
$f(H3) = 37.2932$
$f(H4) = 37.4563$
$f(H5) = 28.4168$
$f(H6) = 20.7954$
$fnew = min(f(H)) = f(H6) = 20.7954$
$xnew = H6 = [3.4402, \quad -2.3497, \quad -1.8545]$
$xtemp = x1$

$x1$ is assigned to $xtemp$, so we can use $xtemp$ for the next steps as $x1$'s not-updated value.

In the seventh step, $fnew$ and the bat's position $f(x1)$ will be compared and the bat's role will be decided. In this case, $fnew$ is better than the first bat's current position ($20.7954 < 47.9140$), so the first bat will become the exploiter bat. It will move to $xnew$, its direction will be changed towards the next position, and its search scope will be shrunk by decreasing its wavelength (*lambda*1) and increasing its frequency (*freq*1).

$$v1 = (xnew - x1)/norm(xnew - x1) = [-0.6432, \quad 0.7639, \quad 0.0520]$$
$$x1 = xnew = [3.4402, \quad -2.3497, \quad -1.8545]$$
$$lambda1 = lambda1 - \rho = 0.8658$$
$$freq1 = freq1 + \rho = 0.6551$$

Since *fnew* and $f(xtemp)$ are not better than *fgbest*, *fgbest* is not updated, only *x*1 is updated, and the first bat's evaluation is over here.

In the eighth step, we return to the fourth step and repeat the same competition for the second bat. Firstly, its search scope column vector (H) will be generated as follows:

$$u = -1 + 2 * rand(j, d)$$
$$u1 = [-0.0292, \quad -0.0081, \quad -0.0337]$$
$$u2 = [0.0002, \quad -0.0103, \quad 0.0184]$$
$$A21 = u1 * b/freq2 = [-0.5812, \quad -0.1617, \quad -0.6693]$$
$$A22 = u1 * b/freq2 = [0.0039, \quad -0.2046, \quad 0.3659]$$
$$V21 = norm(v2 + A21) = [-0.6572, \quad 0.3089, \quad 0.1096]$$
$$V22 = norm(v2 + A22) = [-0.3087, \quad 0.2834, \quad 0.7261]$$

$$H1 = x2 + V21 * 1 * lambda2 = [1.1922, \quad -0.5041, \quad -0.5073]$$
$$H2 = x2 + V21 * 2 * lambda2 = [1.0839, \quad -0.4531, \quad -0.4892]$$
$$H3 = x2 + V21 * 3 * lambda2 = [0.9755, \quad -0.4022, \quad -0.4712]$$
$$H4 = x2 + V22 * 1 * lambda2 = [1.2497, \quad -0.5083, \quad -0.4057]$$
$$H5 = x2 + V22 * 2 * lambda2 = [1.1988, \quad -0.4616, \quad -0.2860]$$
$$H6 = x2 + V22 * 3 * lambda2 = [1.1479, \quad -0.4149, \quad -0.1663]$$

In the ninth step we repeat the sixth step for the second bat's search scope H:

$$f(H1) = \sum_{i=1}^{3} H_i^2 = [1.1922^2 + (-0.5041)^2 + (-0.5073)^2] = 1.9328$$

if we repeat the same competition for all the search points in the scope we obtain f(H) values:

$$f(H1) = 1.9328$$
$$f(H2) = 1.6195$$
$$f(H3) = 1.3354$$
$$f(H4) = 1.9846$$

$f(H5) = 1.7318$

$f(H6) = 1.5173$

$fnew = min(f(H)) = f(H3) = 1.3354$

$xnew = H3 = [0.9755, \quad -0.4022, \quad -0.4712]$

$xtemp = x2$

In the tenth step, $fnew$ and the bat's position $f(x2)$ will be compared and the bat's role will be decided. In this case, $fnew(1.3354)$ is better than $f(x2)(2.2755)$. So the first bat will become the exploiter bat as well. It will move to $xnew$, its direction will be changed towards the next position, and its search scope will be shrunk by decreasing its wavelength ($lambda2$) and increasing its frequency ($freq2$).

$v2 = (xnew - x2)/norm(xnew - x2) = [-0.8949, \quad 0.4206, \quad 0.1493]$

$x2 = xnew = [0.9755, \quad -0.4022, \quad -0.4712]$

$lambda2 = lambda2 - \rho = 0.0625$

$freq2 = freq2 + \rho = 1.1089 = 1.0240$

$freq2$ was bigger than $freqMax$, so it is constrained by $freqMax = 1.0240$.

In the eleventh step, we compare $f(xtemp)$ with $fgbest$ whether the bat is already on the best found position or not. The bat's current position $f(x2)$ is better than $fgbest$, so steps 23 and 24 will be executed in Algorithm 9: The wavelength will be minimized, the frequency will be maximized, and the bat's search direction will be changed randomly. So its search scope will be at its minimum size.

$lambda2 = Lmin = 0.0102$

$freq2 = freqMax = 1.0240$

$v2 = -1 + 2 * rand(1, D) = [0.9415, \quad -0.2689, \quad -0.1303]$

In the twelfth step, since $fnew$ is better than $fgbest$, $fgbest$ and $xgbest$ will be updated as follows:

$fgbest = fnew = 1.3354$

$xgbest = xnew = [0.9755, \quad -0.4022, \quad -0.4712]$

At the end of the twelfth step, the both bats are evaluated and the first bat $x1$ moved from 47.9140 to 20.7954, and the second bat $x2$ moved from 2.2755 to 1.3354. Until now, we spent $2 * (j \times k) + 10 = 32$ function of evaluations(FEs). And we have still around $9,700$ FEs to stop the algorithm.

In the thirteenth step, we return to the fourth step.

In the fourteenth step, after the function of evaluations (FEs) is exhausted we return the fgbest and xgbest as a result of the algorithm.

10.6 Conclusions

In this chapter we have aimed to show how a bat's behaviour in nature can be modeled and become an optimization algorithm. And we have shown step-by-step how the algorithm works. We have also provided source-codes in MAT-LAB and in C++ programming language which could help researchers to implement the algorithm on optimization problems and work on improving the algorithm. Finally, the step-by-step numerical example of DVBA has been presented in detail, so we can see how virtual bats improve the solution during the competition process.

References

1. A.O. Topal, O. Altun. "Dynamic virtual bats algorithm (DVBA) for global numerical optimization," in *Proc. of the IEEE International Conference on Congress on Intelligent Networking and Collaborative Systems (INCoS)*, 2014, pp. 320-327 .

2. A.O. Topal, O. Altun. "A novel meta-heuristic algorithm: Dynamic Virtual Bats Algorithm." *Information Sciences*, vol. 354 (2016): 222-235.

3. X.-S. Yang, "A new metaheuristic bat-inspired algorithm", in: *Nature Inspired Cooperative Strategies for Optimization (NICSO 2010)*, Springer, 2010, pp. 65-74.

4. C. Yu, L. Kelley, S. Zheng, Y. Tan, "Fireworks algorithm with differential mutation for solving the CEC 2014 competition problems", in *Proc. of the IEEE Congress on Evolutionary Computation (CEC)*, IEEE, 2014, pp. 3238-3245.

5. A.O. Topal, Y.E. Yildiz, M. Ozkul. "Improved dynamic virtual bats algorithm for global numerical optimization." *Proc. of the World Congress on Engineering and Computer Science*. Vol. 1, pp. 462-467, 2017.

11

Dispersive Flies Optimisation: A Tutorial

Mohammad Majid al-Rifaie

School of Computing and Mathematical Sciences
University of Greenwich, Old Royal Naval College
London, United Kingdom

CONTENTS

11.1 Introduction

The motivation for studying Dispersive Flies Optimisation (DFO) [1] is the algorithm's minimalist update equation which only uses the flies' position vectors for the purpose of updating the population. This is in contrast to several other population-based algorithms and their variants which besides using position vectors, use a subset of several other vectors: velocities and memories (personal best and global best) in particle swarm optimisation (PSO) [2], mutant and trial vectors in differential evolution (DE) [3], pheromone, heuristic vectors in Ant Colony Optimisation (ACO) [4], and so forth [5].

The inspiration for the algorithm comes from the swarming behaviour of flies over food sources, and their dispersing behaviour when facing a threat. The primary aim of the algorithm is numerical optimisation, which is effectively adjusting several parameters of a certain problem and getting a better solution over time. One of the strengths of DFO, and swarm intelligence algorithms in general, is that they can deal with noisy environments or dynamically changing environments, where the solutions are non-stationary. While

DFO has been applied to discrete problems, it is primarily proposed to deal with continuous search spaces. DFO has been applied to various problems, including but not limited to medical imaging [6], optimising machine learning algorithms [7, 8], training deep neural networks [9], computer vision and quantifying symmetrical complexities [10], analysis of autopoiesis in computational creativity [11] and identification of animation key points from 2D-medialness maps [12].

11.2   Dispersive flies optimisation

The swarming behaviour of the individuals in DFO consists of two tightly connected mechanisms, one is the formation of the swarms and the other is its breaking or weakening. The position vector of a fly in DFO is defined as:

$$\vec{x}_i^t = \left[x_{i0}^t, x_{i1}^t, ..., x_{i,D-1}^t \right], \qquad i \in \{0, 1, 2, ..., N\text{-}1\} \tag{11.1}$$

where i represents the ith individual, t is the current time step, D is the problem dimensionality, and N is the population size. For continuous problems, $x_{id} \in \mathbb{R}$ (or a subset of $\mathbb{R}$).

In the first iteration, when $t = 0$, d^{th} component of ith fly is initialised as:

$$x_{id}^0 = U(x_{\min,d}, x_{\max,d}) \tag{11.2}$$

This, effectively generates a random value between the lower ($x_{\min,d}$) and upper ($x_{\max,d}$) bounds of the respective dimension, d.

On each iteration, dimensions of the position vectors are independently updated, taking into account:

- current fly's position

- current fly's best neighbouring individual (consider ring topology, where each fly has a left and a right neighbour)

- and the best fly in the swarm.

Therefore, the update equation is

$$x_{id}^{t+1} = x_{i_nd}^t + u(x_{sd}^t - x_{id}^t) \tag{11.3}$$

where,

- x_{id}^t: position of the ith fly in d^{th} dimension at time step t

- $x_{i_nd}^t$: position of $\vec{x}_i^t$'s best *neighbouring* individual (in ring topology) in d^{th} dimension at time step t

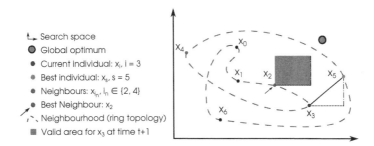

FIGURE 11.1
Sample update of x_i, where $i = 3$ in a 2D space.

- x_{sd}^t: position of the *swarm*'s best individual in the dth dimension at time step t and $s \in \{0, 1, 2, ..., N\text{-}1\}$

- $u \sim \text{U}(0, 1)$: generated afresh for each dimension update.

The update equation is illustrated in an example in Fig. 11.1 where $\vec{x}_3$ is to be updated. The algorithm is characterised by two main components: a dynamic rule for updating the population's position (assisted by a social neighbouring network that informs this update), and communication of the results of the best found individual to others.

As stated earlier, the position of members of the swarm in DFO can be restarted, with one impact of such restarts being the displacement of the individuals which may lead to discovering better positions. To consider this eventuality, an element of stochasticity is introduced to the update process. Based on this, individual dimensions of the population's position vectors are reset if a random number generated from a uniform distribution on the unit interval $\text{U}(0, 1)$ is less than the *restart threshold*, Δ. This guarantees a restart to the otherwise permanent stagnation over likely local minima.

Algorithm 10 summarises the DFO algorithm. In this algorithm, each member of the population is assumed to have two neighbours (i.e. ring topology).

Algorithm 10

1: **procedure** DFO $(N, D, \vec{x}_{\min}, \vec{x}_{\max}, f)^*$
2:     **for** $i = 0 \rightarrow N - 1$ **do**   ▷ **Initialisation**: Go through each fly
3:         **for** $d = 0 \rightarrow D - 1$ **do**   ▷ Initialisation: Go through each dimension
4:             $x_{id}^0 \leftarrow \text{U}(x_{\min,d}, x_{\max,d})$ ▷ Initialise d^{th} dimension of fly i
5:         **end for**
6:     **end for**
7:     **while** ! termination criteria **do**   ▷ Main DFO loop
8:         **for** $i = 0 \rightarrow N - 1$ **do**   ▷ **Evaluation**: Go through each fly
9:             $\vec{x}_i.\text{fitness} \leftarrow f(\vec{x}_i)$

```
10:        end for
11:        x⃗ₛ  = arg min [f(x⃗ᵢ)],   i ∈ {0, 1, 2, ..., N − 1}        ▷ Find best fly
12:        for i = 0 → N − 1 and i ≠ s do      ▷ Update each fly except the best
13:            x⃗ᵢₙ = arg min [f(x⃗₍ᵢ₋₁₎%N), f(x⃗₍ᵢ₊₁₎%N)]      ▷ Find best neighbour
14:            for d = 0 → D − 1 do              ▷ Update each dimension
15:                if U(0, 1) < Δ then                  ▷ Restart mechanism
16:                    xᵢd^(t+1) ← U(x_min,d, x_max,d)        ▷ Restart within bounds
17:                else
18:                    u ← U(0, 1)
19:                    xᵢd^(t+1) ← xᵢₙd^t + u(x_sd^t − xᵢd^t)      ▷ Update the dimension value
20:                    if xᵢd^(t+1) < x_min,d or xᵢd^(t+1) > x_max,d then    ▷ Out of bounds
21:                        xᵢd^(t+1) ← U(x_min,d, x_max,d)          ▷ Restart within bounds
22:                    end if
23:                end if
24:            end for
25:        end for
26:    end while
27:    return x⃗ₛ
28: end procedure
```

_____

* INPUT: swarm size, dimensions, lower/upper bounds, fitness function.

11.3   Source code

The source code in this section provides the standard implementation of DFO in three programming languages, Matlab (Listings 11.1 and 11.2), C++ (Listing 11.3) and Python (Listing 11.4). The fitness function used in the code is the Sphere function which is defined below in Eq. 11.4. The fitness function f takes a D-dimensional vector (i.e. one fly, or $\vec{x}_i$) and returns a single value (i.e. fitness value). DFO is tasked to find the optimal value for each parameter or dimension.

$$f(\vec{x}_i) = \sum_{d=1}^{D} x_{id}^2 \qquad \text{where } -5.12 \leqslant x_{id} \leqslant 5.12 \tag{11.4}$$

11.3.1   Matlab

Implementation of DFO in Matlab is provided below.

```matlab
1 function [sum]=f(X)
2    [N,D]=size(X);
3    sum=zeros(N,1);
4    for i=1:N
5      for d=1:D
```

```
6        sum(i,1)=sum(i,1)+X(i,d)^2;
7      end
8    end
9 end
```

Listing 11.1
Implementation of the fitness function (Sphere) in Matlab (see Eq. 11.4).

```
1 clear
2 % INITIALISE PARAMETERS (N: swarm size, D: dimensionality)
3 N=100; D = 30; delta = 0.001; maxIter=1000;
4 % LOWER AND UPPER BOUNDS
5 lowerB(1,1:D)=-5.12; upperB(1,1:D)=5.12;
6 s = 0;        % INITIAL INDEX OF BEST FLY
7 X = zeros(N,D); % FLIES MATRIX
8 fitness(1,1:N) = realmax; % FITNESS VALUES
9
10 for i=1:N
11   for d=1:D
12     X(i,d) = lowerB(d)+rand()*(upperB(d)-lowerB(d));
13   end
14 end
15
16 for itr=1:maxIter
17   % EVALUATE FITNESS OF THE FLIES AND FIND THE BEST
18   fitness = f(X);
19   [sFitness, s] = min(fitness);
20   disp(['Iteration: ', num2str(itr),'    Best fly index: ', num2str(s
       ),'      Fitness value: ', num2str(sFitness), ] )
21
22   % UPDATE EACH FLY INDIVIDUALLY
23   for i=1:N
24     % ELITIST STRATEGY (DON'T UPDATE BEST FLY)
25     if i == s continue; end
26
27     % FIND BEST NEIGHBOUR FOR EACH FLY
28     left=mod(i-2,N)+1; right=mod(i,N)+1; % INDICES: LEFT & RIGHT FLIES
29     if fitness(right)<fitness(left) bNeighbour = right;
30     else bNeighbour = left; end
31
32     for d=1:D
33       % DISTURBANCE MECHANISM
34       if rand()<delta X(i,d) = lowerB(d)+rand()*(upperB(d)-lowerB(d));
         continue; end
35
36       % UPDATE EQUATION
37       X(i,d) = X(bNeighbour,d) + rand()*( X(s,d)- X(i,d) );
38
39       % OUT OF BOUND CONTROL
40       if or( X(i,d) < lowerB(d), X(i,d) > upperB(d) )
41         X(i,d) = lowerB(d)+rand()*(upperB(d)-lowerB(d));
42       end
43     end
44   end
45 end
46
47 % EVALUATE FITNESS OF THE FLIES AND FIND THE BEST
48 fitness = f(X);
49 [sFitness, s] = min(fitness);
50
51 disp( ['Final best fitness=', num2str(sFitness)] )
```

Listing 11.2
Complete DFO code in Matlab.

11.3.2   C++

DFO implementation in C++ is presented here.

```cpp
1  #include <iostream>
2  #include <stdlib.h>
3  #include <cmath>
4  #include <ctime>
5  using namespace std;
6
7  // FITNESS FUNCTION (SPHERE FUNCTION)
8  float f(float x[], int D) { // x IS ONE FLY AND D IS DIMENSION
9    float sum=0;
10   for(int i=0; i<D; i++)
11     sum=sum+x[i]*x[i];
12   return sum;
13 }
14
15 // GENERATE RANDOM NUMBER IN RANGE [0, 1)
16 float r() {
17   return (float) rand() / (RAND_MAX);
18 }
19
20 int main() {
21   srand(time(NULL)); // TO GENERATE DIFFERENT RANDOM NUMBERS
22   // INITIALISE PARAMETERS
23   int N=100; int D = 30; float delta = 0.001; int maxIter=1000;
24   float lowerB[D]; float upperB[D]; // LOWER AND UPPER BOUNDS
25   int s = 0;        // INITIAL INDEX OF BEST FLY
26   float X[N][D];    // FLIES VECTORS
27   float fitness[N]; // FITNESS VALUES
28
29   // SET LOWER AND UPPER BOUND CONSTRAINTS FOR EACH DIMENSION
30   for (int d=0; d<D; d++) {
31     lowerB[d]=-5.12; upperB[d]=5.12;
32   }
33
34   // INITIALISE FLIES WITHIN BOUNDS. MATRIX SIZE: (N,D)
35   for(int i=0; i<N; i++)
36     for(int d=0; d<D; d++)
37       X[i][d] = lowerB[d] + r()*(upperB[d]-lowerB[d]);
38
39   // MAIN DFO LOOP
40   for (int itr=0; itr<maxIter; itr++) {
41     for(int i=0; i<N; i++) { // EVALUATE EACH FLY AND FIND THE BEST
42       fitness[i]=f(X[i],D);
43       if (fitness[i] <= fitness[s]) s = i;
44     }
45
46     if ( itr%100 == 0) // PRINT RESULT EVERY 100 ITERATIONS
47       cout << "Iteration: " << itr << "\t Best fly index: " << s
48       << "\t Fitness value: " << fitness[s] << endl;
49
50     // UPDATE EACH FLY INDIVIDUALLY
51     for(int i=0; i<N; i++) {
52       // ELITIST STRATEGY (i.e. DON'T UPDATE BEST FLY)
53       if (i==s) continue;
54
55       // FIND BEST NEIGHBOUR FOR EACH FLY
56       int left; int right; int bNeighbour;
57       left=(i-1)%N; right=(i+1)%N; // INDICES: LEFT & RIGHT FLIES
58       if (fitness[right]<fitness[left]) bNeighbour = right;
59       else bNeighbour = left;
60
61       // UPDATE EACH DIMENSION SEPARATELY
62       for (int d=0; d<D; d++) {
63         if (r() < delta) { // DISTURBANCE MECHANISM
```

```
64        X[i][d] = lowerB[d] + r()*(upperB[d]-lowerB[d]); continue;
65      }
66
67      // UPDATE EQUATION
68      X[i][d] = X[bNeighbour][d] + r()*( X[s][d]- X[i][d] );
69
70      // OUT OF BOUND CONTROL
71      if (X[i][d] < lowerB[d] or X[i][d] > upperB[d])
72        X[i][d] = (upperB[d]-lowerB[d])*r()+lowerB[d];
73      }
74    }
75  }
76  // EVALUATE EACH FLY'S FITNESS AND FIND BEST FLY
77  for(int i=0; i<N; i++) {
78    fitness[i]=f(X[i],D);
79    if (fitness[i] < fitness[s]) s = i;
80  }
81  cout << "Final best fitness: " << fitness[s] << endl;
82  return 0;
83 }
```

Listing 11.3
Complete DFO code in C++.

11.3.3 Python

This part provides the implementation of DFO in Python.

```
1  import numpy as np
2
3  # FITNESS FUNCTION (SPHERE FUNCTION)
4  def f(x): # x IS A VECTOR REPRESENTING ONE FLY
5    sum = 0.0
6    for i in range(len(x)):
7      sum = sum + np.power(x[i],2)
8    return sum
9
10 N = 100              # POPULATION SIZE
11 D = 30               # DIMENSIONALITY
12 delta = 0.001        # DISTURBANCE THRESHOLD
13 maxIterations = 1000 # ITERATIONS ALLOWED
14 lowerB = [-5.12]*D   # LOWER BOUND (IN ALL DIMENSIONS)
15 upperB = [ 5.12]*D   # UPPER BOUND (IN ALL DIMENSIONS)
16
17 # INITIALISATION PHASE
18 X = np.empty([N,D]) # EMPTY FLIES ARRAY OF SIZE: (N,D)
19 fitness = [None]*N  # EMPTY FITNESS ARRAY OF SIZE N
20
21 # INITIALISE FLIES WITHIN BOUNDS
22 for i in range(N):
23   for d in range(D):
24     X[i,d] = np.random.uniform(lowerB[d], upperB[d])
25
26 # MAIN DFO LOOP
27 for itr in range (maxIterations):
28   for i in range(N): # EVALUATION
29     fitness[i] = f(X[i,])
30   s = np.argmin(fitness) # FIND BEST FLY
31
32   if (itr%100 == 0): # PRINT BEST FLY EVERY 100 ITERATIONS
33     print ("Iteration:", itr, "\tBest fly index:", s,
34         "\tFitness value:", fitness[s])
35
```

```
36  # TAKE EACH FLY INDIVIDUALLY
37  for i in range(N):
38    if i == s: continue # ELITIST STRATEGY
39
40    # FIND BEST NEIGHBOUR
41    left = (i-1)%N
42    right = (i+1)%N
43    bNeighbour = right if fitness[right]<fitness[left] else left
44
45    for d in range(D): # UPDATE EACH DIMENSION SEPARATELY
46      if (np.random.rand() < delta):
47        X[i,d] = np.random.uniform(lowerB[d], upperB[d])
48        continue;
49
50      u = np.random.rand()
51      X[i,d] = X[bNeighbour,d] + u*(X[s,d] - X[i,d])
52
53      # OUT OF BOUND CONTROL
54      if X[i,d] < lowerB[d] or X[i,d] > upperB[d]:
55        X[i,d] = np.random.uniform(lowerB[d], upperB[d])
56
57  for i in range(N): fitness[i] = f(X[i,]) # EVALUATION
58  s = np.argmin(fitness) # FIND BEST FLY
59
60  print("\nFinal best fitness:\t", fitness[s])
61  print("\nBest fly position:\n", X[s,])
```

Listing 11.4
Complete DFO code in Python.

A sample output of the code is shown below, where the index and fitness value of the best fly in every 100 iterations are displayed. At the end, the best solution, containing the best parameters found, is also shown.

```
1  Iteration: 0    Best fly index: 60   Fitness value: 150.5914836416227
2  Iteration: 100  Best fly index: 68   Fitness value: 0.011859096868779489
3  Iteration: 200  Best fly index: 53   Fitness value: 2.2510252307575077e-05
4  Iteration: 300  Best fly index: 90   Fitness value: 6.779722749003586e-08
5  Iteration: 400  Best fly index: 24   Fitness value: 3.163652073209633e-10
6  Iteration: 500  Best fly index: 49   Fitness value: 4.503385943119482e-11
7  Iteration: 600  Best fly index: 28   Fitness value: 1.1633233211569493e-13
8  Iteration: 700  Best fly index: 20   Fitness value: 2.7396155007581633e-15
9  Iteration: 800  Best fly index: 15   Fitness value: 2.3009156731064727e-17
10 Iteration: 900  Best fly index: 10   Fitness value: 4.744041469725476e-19
11
12 Final best fitness:  1.214236579001144e-20
13
14 Best fly position:
15 [-3.75694837e-12 -5.51560681e-13  5.40769833e-12  6.60606978e-11
16  -4.01610956e-13 -2.74643196e-12 -9.51829640e-12  7.06787789e-12
17   2.19263924e-11  1.00330905e-11  3.99326555e-12 -4.08853263e-12
18   1.56206199e-11  9.80355426e-12  1.12747027e-11  6.40617484e-12
19  -1.61567359e-11 -3.34267744e-11  2.88367151e-11 -4.66779500e-12
20  -6.35363397e-12  3.79648533e-12 -2.97971162e-12 -1.58178299e-12
21  -5.42664323e-11  7.37563208e-12 -2.74749207e-11  8.82333052e-12
22  -1.54070116e-11  9.84007230e-12]
```

Listing 11.5
Sample output: optimising Sphere function using DFO.

11.4 Numerical example: optimisation with DFO

In order to understand the swarming behaviour of the flies in DFO, a step by step example is presented in this section. These steps illustrate how the swarm moves towards the optimal solution. In this example, the number of flies is set to 5 ($N = 5$), the number of dimensions or parameters is set to 2 ($D = 2$) and DFO is run for 5 iterations. The rest of the configuration is the same.

Initially, each fly's position vector (in this example containing only two dimensions) is initialised within the range (in this case, between -5.12 and 5.12). This process is shown in Table 11.1 (lines 0-4). Next, the fitness of each fly is calculated to quantify how "good" the solution provided by each fly is. Then, based on the fitness values (calculated using Eq. 11.4), each fly's best neighbour is selected; the standard DFO uses *ring topology*, which means each fly has a left and a right neighbour, which are based on indices (e.g. a fly with index 3, has two neighbours: fly 2 is the left neighbour and fly 4 is the right neighbour). Once these steps are taken, the update equation (Eq. 11.3) is used to update the value of each dimension in each fly. Prior to updating any of the dimensions, a random number between 0 and 1 is generated and if the randomly generated number is less than the disturbance threshold, Δ[1], a random value between the lower and upper bounds is generated for that dimension, otherwise the update equation is used[2]. Once the new value for each dimension is calculated, it is checked against the constraint, ensuring it fits within the bounds; if the value is outside the valid region, a random number is generated between the lower and upper bounds.

In Table 11.1, the index of each fly is shown on the leftmost column, then the value of the first dimension, $d = 0$, followed by the value of the second dimension, $d = 2$, and the fitness value which is calculated using the values of the first and second dimensions; for instance to calculate the fitness of the first fly, fly 0 or $\vec{x}_0$ in line 0, the following can be (using Eq. 11.4):

$$f(\vec{x}_0) = 2.280657^2 + -4.809789^2 = 28.335468$$

The last column in Table 11.1 represents the index of the best neighbouring fly. The index of the best fly in the population, s, is highlighted in the same column. Using the elitist approach, the best fly in the swarm is kept intact and therefore is not updated, neither by the disturbance mechanism, nor by the update equation. This makes sure that the population does not lose the best solution it has found so far.

For instance, in order to update the first dimension of the first fly, x_{00} (line 0, where in x_{00}, the first 0 refers to the fly index, and the second 0 refers to

---

[1] Note that the value of Δ is problem dependent.

[2] In this example, none of the random numbers generated was less than Δ.

TABLE 11.1
Running DFO for 5 iterations.

L	Fly Index	d=0	d=1	Fitness value	Best neighbour
0	0	2.2806566	-4.80978936	28.3354681843	4
1	1	-2.13229778	4.71940929	26.8195179113	2
2	2	1.25796613	4.70359891	23.7063214781	3
3	3	-2.43396026	-0.90766546	6.7480191207	4
4	4	0.06228462	1.33995725	1.7993648107	s = 4
5	Itr = 1	d=0	d=1	Fitness value	Best neighbour
6	0	-0.91245846	2.61537071	7.6727443985	4
7	1	2.13769454	4.19262811	22.1478683732	0
8	2	-2.84773347	-4.11611456	25.0519850402	3
9	3	0.6558135	3.10292087	10.058209244	4
10	4	0.06228462	1.33995725	1.7993648107	s = 4
11	Itr = 2	d=0	d=1	Fitness value	Best neighbour
12	0	1.01468105	0.53914852	1.3202587652	1
13	1	0.34869085	0.20424149	0.1632998901	s = 1
14	2	3.05708686	2.96239515 *	18.1215650971	1
15	3	-0.02827646	0.42791449	0.1839103648	4
16	4	0.06228462	1.33995725	1.7993648107	3
17	Itr = 3	d=0	d=1	Fitness value	Best neighbour
18	0	0.14988694	-0.0675497	0.0270290553	4
19	1	0.34869085	0.20424149	0.1632998901	2
20	2	-0.04768011	-0.11545245	0.0156026608	s = 2
21	3	0.33950867	1.17054436	1.4854402283	2
22	4	0.38255993	0.04292476	0.1481946319	0
23	Itr = 4	d=0	d=1	Fitness value	Best neighbour
24	0	0.20121156	0.00658197	0.0405294143	4
25	1	-0.13562958	-0.24398233	0.0779227572	2
26	2	-0.04768011	-0.11545245	0.0156026608	1
27	3	-0.38006057	-0.27793257	0.2216925515	4
28	4	-0.01290562	-0.08049597	0.0066461562	s = 4
29	Itr = 5	d=0	d=1	Fitness value	Best neighbour
30	0	-0.20867294	-0.13664574	0.0622164545	1
31	1	0.01618933	-0.02789224	0.0010400715	s = 1
32	2	0.0404749	-0.00654758	0.0016810879	1
33	3	0.153082	0.10787423	0.0350709497	2
34	4	-0.01290562	-0.08049597	0.0066461562	3

the dimension number), the update equation is used:

$$x_{00} = x_{40} + u(x_{40} - x_{00})$$
$$= 0.06228462 + 0.4394 \, (0.06228462 - 2.2806566)$$
$$= -0.91245846$$

where $u = 0.4394$ is a random number between 0 and 1. Note that in this instance, the best neighbouring fly and the best fly in the swarm are identical ($\vec{x}_4$). The same process can be repeated for x_{01}:

$$x_{01} = x_{41} + u(x_{41} - x_{01})$$
$$= 1.33995725 + u(1.33995725 - (-4.80978936))$$
$$= 2.61537071$$

The updated values can be seen in Table 11.1, line 6; looking at the fitness value, it is evident that the fly has a better fitness value after the update (i.e. from 28.34 to 7.67). Next, let's update both dimensions of the second fly $\vec{x}_1$ whose values can be seen in line 1:

$$x_{10} = x_{20} + u(x_{40} - x_{10})$$
$$= 1.25796613 + u(0.06228462 - (-2.13229778))$$
$$= 2.13769454$$
$$x_{11} = x_{21} + u(x_{41} - x_{11})$$
$$= 4.70359891 + u(1.33995725 - 4.71940929)$$
$$= 4.19262811$$

The same process can be repeated for the rest of the flies. Note that in iteration 2, the second dimension of the third fly, x_{21} in line 14, which is highlighted with an asterisk (*), had been out of bounds and therefore the algorithm generated a random value within the allowed bounds for that dimension.

The optimisation process from the initialisation of the flies and throughout each iteration is illustrated in Fig. 11.2, showing the population's convergence towards the known optimal solution, i.e. (0,0).

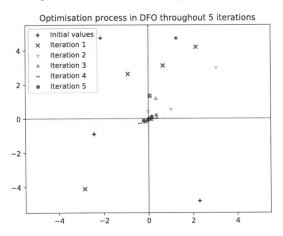

FIGURE 11.2
Illustrating the optimisation process in the example. The figure shows that after each iteration, the population is getting closer and closer ("converging") to the optimum point $(0, 0)$.

11.5 Conclusion

This paper provides an introduction to the main principles of the standard DFO algorithm. The high-level description of the algorithm is first provided, covering the main aspects of the algorithm and its update equation. Then an annotated pseudocode is presented, illustrating the detailed order of the steps. This is then complemented by providing a complete code of the DFO algorithm in Matlab, C++ and Python. The paper is finalised by presenting a step by step numerical example of how the algorithm guides the population to perform the optimisation. It is now possible to simply change the fitness function to that of another problem, specify the dimensionality of the problem (the number of the parameters to be optimised), the population size and disturbance threshold, and then run the code.

References

1. M.M. al-Rifaie. "Dispersive Flies Optimisation" in *Proc. of the 2014 Federated Conference on Computer Science and Information Systems*, vol. 2, pp. 529-538, 2014.

2. J. Kennedy. "The particle swarm: social adaptation of knowledge" in *Proc. of IEEE International Conference on Evolutionary Computation*, pp. 303-308, 1997.

3. R. Storn, K. Price. "Differential evolution–a simple and efficient heuristic for global optimization over continuous spaces". Journal of Global Optimization, vol. 11(4), pp. 341-359, 1997.

4. M. Dorigo, G.D. Caro, L.M. Gambardella. "Ant algorithms for discrete optimization". Artificial Life, vol. 5(2), pp. 137-172, 1999.

5. M.M. al-Rifaie. "Perceived simplicity and complexity in nature" in Proc. of AISB Annual Convention - Symposium VIII on Computational Architectures for Animal Cognition, pp. 299-305, Bath, United Kingdom, 2017.

6. M.M. al-Rifaie, A. Aber. "Dispersive Flies Optimisation and medical imaging" in Recent Advances in Computational Optimization, pp. 183-203, 2016.

7. H. Alhakbani. "Handling Class Imbalance Using Swarm Intelligence Techniques, Hybrid Data and Algorithmic Level Solutions". PhD Thesis, Goldsmiths, University of London, London, United Kingdom, 2018.

8. H.A. Alhakbani, M.M. al-Rifaie. "Optimising SVM to classify imbalanced data using dispersive flies" in *Proc. of the 2017 Federated Conference on Computer Science and Information Systems, FedCSIS 2017*, pp. 399-402, 2017.

9. H. Oroojeni, M.M. al-Rifaie, M.A. Nicolaou. "Deep neuroevolution: Training deep neural networks for false alarm detection in intensive care units" in *Proc. of European Association for Signal Processing (EUSIPCO) 2018*, pp. 1157-1161, 2018.

10. M.M. al-Rifaie, A. Ursyn, R. Zimmer, M.A.J. Javid. "On symmetry, aesthetics and quantifying symmetrical complexity" in *Proc. of International Conference on Evolutionary and Biologically Inspired Music and Art*, pp. 17-32, 2017.

11. M.M. al-Rifaie F.F. Leymarie, W. Latham, M. Bishop. "Swarmic autopoiesis and computational creativity". Connection Science, vol. 29(4), pp. 276-294, 2017.

12. P. Aparajeya, F.F Leymarie, M.M. al-Rifaie. "Swarm-Based Identification of Animation Key Points from 2D-medialness Maps" in Proc. of EvoMUSART - Computational Intelligence in Music, Sound, Art and Design, Lecture Notes in Computer Science, vol. 11453, Springer, pp. 69-83, 2019.

12

Elephant Herding Optimization

Nand K. Meena
School of Engineering and Applied Science
Aston University, Birmingham, United Kingdom

Jin Yang
School of Engineering and Applied Science
Aston University, Birmingham, United Kingdom

Adam Slowik
Department of Electronics and Computer Science
Koszalin University of Technology, Koszalin, Poland

CONTENTS

12.1 Introduction

Various researchers across the globe are impressed by the complex social and emotional family structure of elephants, as compared to other animals on the earth; elephant females form and lead the family. The elephants constitute profound clan bonding and like to stay in tight family groups headed by females called a herd. Generally, a herd comprises 8-100 elephants, depending on territory as well as on family size [1]. A herd is led by an old female, known as the matriarch. Mostly, a herd consist of females such as mother, her sisters and their calves. Sometimes, herd aggregation is also observed with 500–1000 elephants near a water and food source. It has been also researched that elephants makes lifelong emotional bonding with their family and friends, and even mourn the death of stillborn babies and their loved ones.

> The walking style of elephants is also unique and found to be very similar to humans. In walking, babies generally use trunks to hold the tails of their respective mothers and other females are surrounding them in order to protect them from hungry predators [2]. A new born calf is brought up and protected by the whole matriarchal herd. The male elephants (bulls) like to live a solitary life therefore, leave the family group between the age of 12-15 years to hang alone or with other males.

Sometimes, the herd is also separated even though they are closely related. The separation can be influenced by ecology, and social factors. Therefore, it may also possible that different herds found in a large territory can be from the same family. They always keep in touch with their blood groups by using different types of calls. The elephant has a great sense of hearing and can produce different sounds like roars, snorts, cries etc. to communicate with other groups but is specialised in subsonic rumbling.

O'Connell-Rodwell carried out experiments on captive elephants in the United States, Zimbabwe and India, over a period of 5 years. The outcome of these experimentations showed that elephants respond to the low-frequency sound waves that travel through and just above the ground [3]. A later research also confirmed that the elephants can respond to seismic waves even in the absence of low frequency oscillations. The mammals can detect the stress of a distant herd and incoming water-storms when thunder sounds a hundred miles away, as Sri Lankan and Thai elephants reportedly run away from their locations before the destructive tsunami in the year 2004 [4].

"The elephants can communicate through their feet, toenails, and trunks up to a distance of 20 miles. They have the ability to hear low frequency sound and seismic waves through their feet"

Dr. Caitlin O'Connell
Stanford University School of Medicine

The super intelligence and great memory of elephants have inspired a new nature inspired optimization technique. In 2015, Gai-Ge Wang et al. developed a swarm-intelligence based meta-heuristic optimization technique called elephant herding optimization (EHO). The method is inspired by the herding behaviour of elephants. As discussed, the elephant is considered to be a social animal and the herding consists several clans of elephants and their calves. Each clan moves under the influence of a leader, usually a matriarchal female. In the proposed algorithm, the leader elephant is representing the best solution of that clan whereas the male elephant is representing the worst solution. According to elephants' herding behaviour, the female elephants are used to living with their family groups while a male elephant separates when they grow up but lives in contact with their family groups using low frequency vibrations. For simplification, the number of elephants in each clan remains the same. For example, when the worst or male elephant leaves the clan, a new elephant can be produced to keep the number of elephants constant. The basic EHO is discussed in the following sections.

12.2 Elephant herding optimization

The complex herding behaviour of the elephants is modelled into mathematical equations. In order to search the global and local solutions, some set of rules is defined in basic EHO [5], as discussed below.

1. The number of elephants in each clan should be fixed. For example, if any elephant leaves the clan then a new elephant or a baby elephant can replace its position.

2. In each generation/iteration, a fixed number of male elephants will leave their family groups to live in isolation.

3. The elephants of each clan will live under the leadership of their respective matriarch.

An organization of an elephant herd consisting of multiple clans and elephants is presented in Fig. 12.1. In order to update the location of elephants in each generation, two position updating operators have been suggested in EHO namely, clan updating and separating operators. Based on these operators, the EHO method can be divided into two parts as discussed in the following sections.

12.2.1 Position update of elephants in a clan

In this stage, the positions of elephants remaining in the clan are updated. As discussed, the position of a elephant in a clan is affected by the matriarch of the respective clan. In EHO, the position of the jth elephant in clan c is updated as

$$p_{jc}^{t+1} = p_{jc}^{t} + \alpha \times (p_{best} - p_{jc}^{t}) \times r \tag{12.1}$$

where, p_{jc}^{t+1} and p_{jc}^{t} are the updated and previous positions, calculated in $t+1$ and t generations respectively. α is a scaling factor, varied between $[0, 1]$, used

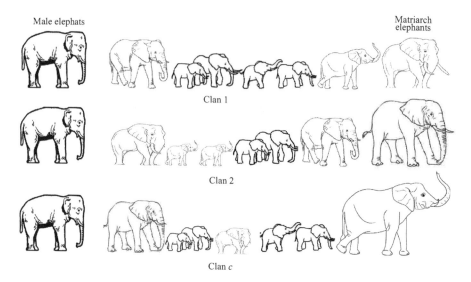

FIGURE 12.1
Elephant herd moving in clans.

to determine the influence of the matriarch on an individual elephant of clan c. p_{best} is the position of the leader elephant of the clan that holds the best fitness achieved so far. $r\epsilon[0,1]$ is the uniform distribution.

The position of matriarch elephants, i.e., $p_{jc}^{t} = p_{best}$ is updated as

$$p_{jc}^{t+1} = \beta \times p_{center,c} \qquad (12.2)$$

where, $p_{center,c}$ represents the central position of clan c, and $\beta\epsilon[0,1]$ is the scaling factor influenced by $p_{center,c}$. The matriarch makes use of β to update its own position by ensuring the security of that clan. The centre of a clan c can be determined as

$$p_{center,c} = \frac{\sum_{j=1}^{n_c} p_{jc}^{t}}{n_c} \qquad (12.3)$$

where, n_c is representing the number of elephants in clan c.

12.2.2   Separation of male elephants from the clan

As discussed in previous sections, the male elephants will leave their family when they grown-up, to live a solitary life or with male groups. The separation process needs to be mathematically modelled in the proposed optimization method. The elephant individuals with the worst fitness in each clan will leave their respective clans. In this separation, the new position of worst elephant j in cth clan can be determined as

$$p_{worst,jc}^{t+1} = p_{min,c} + rand \times (p_{max,c} - p_{min,c} + 1) \qquad (12.4)$$

where, $p_{min,c}$ and $p_{max,c}$ are the lower and upper bounds of elephant individuals in clan c. $rand\epsilon[0, 1]$ represents a uniform and stochastic distribution.

12.2.3 Pseudo-code of EHO algorithm

The pseudo-code of standard EHO is presented in Algorithm 11.

Algorithm 11 Pseudo-code of EHO.

1: determine the objective function $OF(.)$
2: set the number of clans N, where $c\epsilon[1, N]$, and number of elephants in each clan n_c
3: determine the values of scaling factors α, β, and maximum number of generations, G_{max}
4: set the lower and upper bounds for each variable/clan c, $[p_{min,c}, p_{max,c}]$
5: randomly generate the positions for all elephants in each clan, as follows
6: **for** each j-th elephant **do**
7:     **for** each c-th clan **do**
8:         $pp_{jc} = p_{min,c} + (p_{max,c} - p_{min,c}) \cdot rand$
9:     **end for**
10:     $Fitness_j = OF(pp_j)$
11: **end for**
12: determine the best and worst elephants with their locations *bestloc* and *worstloc*
13: set generation $t = 1$;
14: **while** $t \leq G_{max}$ **do**
15:     update the position of elephants in all clans, as follows

---

12.3 Source-code of EHO algorithm in Matlab

In Listing 12.1, the source-code for the EHO algorithm in presented. Here, the $OF(.)$ is representing the address of the objective function.

```
1  % EHO algorithm
2  clc ;
3  clear ;
4  %
5  % declaration of the parameters of EHO algorithm
6  %
7  nc=2;           % number of clans
8  N=10;           % number of elephants in each clan
9  alpha =0.5;     % scaling factor alpha
10 beta =0.1;      % scaling factor beta
11 LB=[-5  -5];    % lower bounds for all clans
12 UB=[5  5];      % upper bounds for all clans
```

16:     **for** each j-th elephant **do**
17:         **for** each c-th clan **do**
18:             **if** any$((j = bestloc)\&(j = worstloc)$ **then**
19:                 $pp\_new_{jc} = pp_{jc} + alpha \cdot (pbest_c - pp_{jc}) \cdot rand$
20:             **else if** $j == bestloc$ **then**
21:                 $pp_{center,c} = mean(pp_c)$
22:                 $pp\_new_{jc} = beta \cdot pp_{center,c}$
23:             **else if** $j == worstloc$ **then**
24:                 $pp\_new_{jc} = LB_c + (UB_c - LB_c + 1) \cdot rand$
25:             **end if**
26:         **end for**
27:         evaluate the fitness of new individual j as
28:         $Fitness_j = OF(pp\_new_j)$
29:     **end for**
30:     determine the new best and worst elephants
31:     **if** Is new best better than previous best **then**
32:         replace the best individual with new one
33:     **end if**
34:     set old population $pp = pp\_new$
35:     set iteration $t = t + 1$
36: **end while**
37: return the *pbest* as a result

```
13 Gmax = 50;     % maximum number of generations/iterations
14 %
15 % generate random population of elephants
16 %
17 for j=1:N
18 for c=1:nc
19 % random location for elephant j in clan c
20 pp(j,c)=LB(c)+rand*(UB(c)-LB(c));
21 end
22 % fitness of jth elephants pair in all clans
23 Fitness(j)=OF(pp(j,:));
24 end
25 % best fitness value and its location
26 fbest=min(Fitness)
27 bestloc=find(fbest==Fitness);
28 % worst fitness value and its location
29 fworst=max(Fitness);
30 worstloc=find(fworst==Fitness);
31 % position of fittest elephant or matriarch
32 pbest=pp(bestloc,:);
33 % position of weakest elephant or male
34 pworst=pp(worstloc,:) ;
35 %
36 % EHO generation starts .....
37 %
38 pp_new=pp;
39 for gen=1:Gmax
40 % clan updating and male separation operators
41 for j=1:N
42 for c=1:nc
43 if any((j\~=bestloc)&(j~=worstloc))
```

```
44 % update elephants positions except best and worst elephants
45 pp_new(j,c)=pp(j,c)+alpha*(pbest(c)-pp(j,c))*rand;
46 elseif j==bestloc
47 pp_center=mean(pp(:,c));
48 % update leader or matriarch as suggested
49 pp_new(j,c)=beta*pp_center;
50 elseif j==worstloc
51 % update worst elephant as suggested
52 pp_new(j,c)=LB(c)+(UB(c)-LB(c)+1)*rand;
53 end
54 end
55 % fitness calculation for new population
56 Fitness(j)=OF(pp_new(j,:));
57 end
58 % find the best fitness value
59 fbest_new=min(Fitness);
60 % determine the location of best elephant
61 bestloc_new=find(fbest_new==Fitness);
62 % find the worst fitness value
63 fworst=max(Fitness);
64 % location of worst elephant
65 worstloc=find(fworst==Fitness);
66 % fittest elephant or matriarch
67 pbest_new=pp_new(bestloc_new,:);
68 %weakest elephant or male
69 pworst=pp_new(worstloc,:) ;
70 % Preserve the fittest elephant and position
71 if fbest_new<fbest
72 pbest=pbest_new;
73 fbest=fbest_new;
74 bestloc=bestloc_new;
75 end
76 % replace the old population with new one
77 pp=pp_new;
78 % store the best elephant fitness of each generation
79 b(gen)=fbest;
80 % store the mean fitness values of elephants
81 m(gen)=mean(Fitness);
82 end
83 % display the position of best elephant or solution
84 disp(pbest);
85 % display the fitness of best elephant
86 disp( fbest );
87 % plot the best fitness values of all generations
88 subplot (1 ,2 ,1)
89 plot(b)
90 % plot the mean fitness values of all elephants
91 subplot (1 ,2 ,2)
92 plot (m)
```

Listing 12.1
Source-code of EHO in Matlab.

12.4   Source-code of EHO algorithm in C++

```
1 #include <iostream>
2 #include <algorithm>
3 #include <math.h>
4 #include <time.h>
```

```cpp
5  using namespace std;
6  // definition of the objective function OF(.)
7  float OF(float x[], int size_array)
8  {float t=0;
9  for(int i=0; i<size_array; i++){
10 t=t+(x[i]*x[i]-10*cos(2*M_PI*x[i]));}
11 return 10*size_array + t;}
12 // generate pseudo random values from the range [0, 1)
13 float r(){return (float)(rand()%1000)/1000;}
14 // main program function
15 int main()
16 {// initialization of the EHO algorithm parameters
17 int N=10, nc=2, Gmax=20; float alpha=0.1, beta=0.01;
18 float LB[nc], UB[nc], pp[N][nc], Fitness[N], pp_new[N][nc];
19 // initialization of pseudo random generator
20 srand (time(NULL));
21 // initialization of the constraints
22 for(int j=0; j<nc; j++){LB[j]=-5.12; UB[j]=5.12;}
23 // generate the random positions for elephants of all clans
24 for(int j=0; j<N; j++)
25 {
26 for(int c=0; c<nc; c++)
27 {
28 pp[j][c]=LB[c]+r()*(UB[c]-LB[c]);
29 }
30 // evaluate all positions of elephants (solutions)
31 Fitness[j]=OF(pp[j],nc);
32 }
33 // determine the matriarch elephant
34 // best fitness value
35 float fbest=*min_element(Fitness, Fitness+N);
36 // worst fitness value
37 float fworst=*max_element(Fitness, Fitness+N);
38 // position of the fittest elephant or matriarch
39 int bestloc=min_element(Fitness, Fitness+N)-Fitness;
40 // position of the weakest elephant or male who has to leave the clan
41 int worstloc=max_element(Fitness, Fitness+N)-Fitness;
42 // initialization of additional variables
43 float pbest[nc], pworst[nc], pp_center, sum, fbest_new, pbest_new[nc];
44 // remember the best "pbest" and the worst "pworst" solution
45 for(int i=0; i<nc; i++){pbest[i]=pp[bestloc][i];
46 pworst[i]=pp[worstloc][i];}
47 // EHO generation starts
48 for(int gen=0; gen<Gmax; gen++)
49 {
50 // clan updating and male separation operators
51 for(int j=0; j<N; j++)
52 {
53 for(int c=0; c<nc; c++)
54 {
55 if ((j!=bestloc) && (j!=worstloc))
56 {
57 // updating the position of the elephants except best and worst
       elephant
58 pp_new[j][c]=pp[j][c]+alpha*(pbest[c]-pp[j][c])*r();
59 }
60 else if (j==bestloc)
61 {
62 sum=0;
63 for(int i=0; i<N; i++){sum=sum+pp[i][c];}
64 pp_center=sum/N;
65 // updating the position of leader or matriarch
66 pp_new[j][c]=beta*pp_center;
67 }
68 else if (j==worstloc)
69 {
70 // updating the position of the worst elephant
```

```
71 pp_new[j][c]=LB[c]+(UB[c]-LB[c]+1)*r();
72 }
73 }
74 // fitness calculation for new positions
75 Fitness[j]=OF(pp_new[j],nc);
76 }
77 // determine the fittest elephant
78 // find the best fitness value
79 fbest_new=*min_element(Fitness, Fitness+N);
80 // find the worst fitness value
81 fworst=*max_element(Fitness, Fitness+N);
82 // determine the location of best elephant
83 int bestloc_new=min_element(Fitness, Fitness+N)-Fitness;
84 // location of worst elephant
85 worstloc=max_element(Fitness, Fitness+N)-Fitness;
86 for(int i=0; i<nc; i++){
87 // fittest elephant or matriarch
88 pbest_new[i]=pp_new[bestloc_new][i];
89 // weakest elephant or male
90 pworst[i]=pp_new[worstloc][i];}
91 // preserve the best elephant with its position and fitness
92 if (fbest_new<fbest)
93 {
94 for(int i=0; i<nc; i++){pbest[i]=pbest_new[i];}
95 fbest=fbest_new; bestloc=bestloc_new;
96 }
97 // replace the old population with new one
98 for(int j=0; j<N; j++){
99 for(int c=0; c<nc; c++){pp[j][c]=pp_new[j][c];}}
100 // display the best result at each iteration
101 cout<<"Iteration "<<gen<<" - The best: "<<fbest<<endl;
102 }
103 // display the final solution and final the best result
104 cout<<"Position of the best elephant (solution):"<<endl;
105 for(int c=0; c<nc; c++){cout<<pbest[c]<<endl;}
106 cout<<"Fitness of the best elephant (solution):"<<fbest<<endl;
107 getchar();
108 return 0;
109 }
```

Listing 12.2
Source-code of EHO algorithm in C++.

12.5 Step-by-step numerical example of EHO algorithm

Example 3 *Find the global minima of Rastrigin function $f(x)$*

$$f(x_j) = A \times N + \sum_{c=1}^{N} [x_{j,c}^2 - A\cos(2\pi x_{j,c})] \qquad (12.5)$$

where, $A = 10$ and $-5.12 \leq x_{j,c} \leq 5.12$

Solution: Here, x_j is a N-dimensional column vector. To realise the complexity of this function $f(x)$, a contour plot is presented in Fig. 12.2 which shows that $f(x)$ is a non-convex function. Finding of global minima, with conventional approaches, is fairly difficult due to its large search space and high number of multiple minimum values. Listing 12.3 presents the Matlab code to determine the function at any x.

```
1  function [Fit] = OF(x)
2  N = length(x);
3  A = 10;
4  for c = 1:N
5  ff(c) = x(c)^2-A*cos(2*pi*x(c));
6  end
7  Fit = A*N + sum(ff);
```

Listing 12.3
Definition of Restrigin function $OF(.)$ in Matlab.

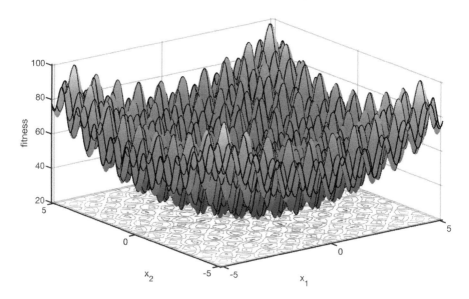

FIGURE 12.2
Contour plot of Restrigin function for $N = 2$.

Now, we demonstrate how the EHO algorithm can applied to determine the optimal minima of the Rastrigin function.

In the first step, let's assume that the value of dimension $N = 4$ which is analogous to number of clans. Each element of the vector, i.e., $x_{j,c} \; \forall j, c$, is analogous to an elephant in clan c.

In the second step, we assumed that the population size or number of elephants in a clan, $n_c = 5$ and values of $\alpha = 0.5$ and $\beta = 0.1$. For this function, the maximum and minimum bounds of elephants in all clans will be $x_{max} = [5.12, 5.12, 5.12, 5.12]$ and $x_{min} = [-5.12, -5.12, -5.12, -5.12]$ respectively.

In the third step, we generate random but feasible (predefined boundaries) population (positions) of $n_c = 5$ elephants, as presented below:
$x_1 = \{-2.542 \; -2.146 \; 1.199 \; -2.403\}$

$x_2 = \{3.322\ 4.943\ 2.358\ -1.599\}$
$x_3 = \{0.870\ -4.016\ 4.161\ 3.888\}$
$x_4 = \{3.254\ -2.450\ 0.966\ -4.890\}$
$x_5 = \{-0.765\ -1.918\ -3.466\ -3.290\}$

In the fourth step, we determine the value of function $f(x_j)\ \forall j$ by using OF(.). For $x_1 = \{-2.542\ -2.146\ 1.199\ -2.403\}$, it is determined as
$OF(x_1) = A.N + \sum_{c=1}^{N}[x_{1,c}^2 - A\cos(2\pi x_{1,c})] = 10 \times 4 + (-2.542)^2 - 10 \times \cos(2\pi \times (-2.542)) + (-2.146)^2 - 10 \times \cos(2\pi \times (-2.146)) + (1.199)^2 - 10 \times \cos(2\pi \times (1.199)) + (-2.403)^2 - 10 \times \cos(2\pi \times (-2.403)) = 66.903$
Similarly, we repeat the computation for all individuals or elephants, as presented below:
$Fit_1 = OF(x_1) = 66.903$
$Fit_2 = OF(x_2) = 92.995$
$Fit_3 = OF(x_3) = 59.591$
$Fit_4 = OF(x_4) = 73.720$
$Fit_5 = OF(x_5) = 69.718$

In the fifth step, the best (an individual with smallest fitness value, Fit_{best}) and worst (a individual with highest fitness value, Fit_{worst}) elephants are identified, as our goal is to minimize the function. It is observed that elephants x_3 and x_2 are the best and worst elephants respectively. Therefore, $x_{best} = \{0.870\ -4.016\ 4.161\ 3.888\}$ with $Fit_{best} = Fit_3 = 59.591$, and $x_{worst} = \{3.322\ 4.943\ 2.358\ -1.599\}$ with $Fit_{worst} = Fit_2 = 92.995$.

In the sixth step, the generation loop of the algorithm starts and the termination criteria is checked whether the number of generations has crossed its maximum allowed limit. If yes, we jump to step thirteen. If no, we move to the seventh step.

In the seventh step, the population of common elephants '1', '4', and '5' is updated by using (12.1), as illustrated below.
$x_{j,c}^{new} = x_{j,c} + \alpha.(x_{best,c} - x_{j,c}).r$
Let's assume that the value of random number $r = \{0.671\ 0.459\ 0.036\ 0.901\}$
then elephant '1' is updated as
$x_{1,1}^{new} = -2.542 + (0.5).(0.870 + 2.542).(0.671) = -1.397$
$x_{1,2}^{new} = -2.146 + (0.5).(-4.016 + 2.146).(0.459) = -2.575$
$x_{1,3}^{new} = 1.199 + (0.5).(4.161 - 1.199).(0.036) = 1.252$
$x_{1,4}^{new} = -2.403 + (0.5).(3.888 + 2.403).(0.901) = 0.431$
Therefore, $x_1^{new} = \{-1.397\ -2.575\ 1.252\ 0.431\}$. Similarly, x_4^{new} and x_5^{new} are calculated and presented below.
$x_4^{new} = \{2.242\ -3.181\ 2.050\ -1.564\}$
$x_5^{new} = \{-0.158\ -2.330\ -0.966\ -2.676\}$
The position of the best elephant is updated by using (12.2) and (12.3) where, $x_{center,c}$ is calculated as

$$x_{center,1} = \frac{\sum_j^{nc} -2.542+3.322+0.870+3.254-0.765}{5} = 0.828;$$

$$x_{center,2} = \frac{\sum_j^{nc} -2.146+4.943-4.016-2.450-1.918}{5} = -1.117;$$

$$x_{center,3} = \frac{\sum_j^{nc} 1.199+2.358+4.161+0.966-3.466}{5} = 1.044;$$

$$x_{center,4} = \frac{\sum_j^{nc} -2.403-1.599+3.888-4.890-3.290}{5} = -1.659;$$

Now, we update the position of the best elephant, i.e. '3', by using (12.2) as

$$x_{3,1}^{new} = (0.1).(0.828) = 0.083$$
$$x_{3,2}^{new} = (0.1).(-1.117) = -0.112$$
$$x_{3,3}^{new} = (0.1).(1.044) = 0.104$$
$$x_{3,4}^{new} = (0.1).(-1.659) = -0.166$$

Therefore, $x_3^{new} = \{0.083 \ -0.112 \ 0.104 \ -0.166\}$

Similarly, we update the position of the worst elephant, i.e. '2' by using (12.4), as determined below. Suppose, the random number in the equation is $rand = r_1 = \{0.706 \ 0.032 \ 0.277 \ 0.046\}$

$$x_{2,1}^{new} = -5.12 + (0.706).(5.12+5.12+1) = 2.815$$
$$x_{2,2}^{new} = -5.12 + (0.032).(5.12+5.12+1) = -4.760$$
$$x_{2,3}^{new} = -5.12 + (0.277).(5.12+5.12+1) = -2.007$$
$$x_{2,4}^{new} = -5.12 + (0.046).(5.12+5.12+1) = -4.603$$

Thus, $x_2^{new} = \{2.815 \ -4.760 \ -2.007 \ -4.603\}$.

In the eighth step, we compute the new fitness values of elephants for newly generated positions $x_j^{new} \ \forall j$ using $OF(.)$. The newly computed fitnesses are given below

$$Fit_1^{new} = OF(x_1^{new}) = 76.424$$
$$Fit_2^{new} = OF(x_2^{new}) = 89.186$$
$$Fit_3^{new} = OF(x_3^{new}) = 10.787$$
$$Fit_4^{new} = OF(x_4^{new}) = 56.782$$
$$Fit_5^{new} = OF(x_5^{new}) = 47.613$$

In step nine, the best x_{best}^{new}, and worst x_{worst}^{new}, elephants of the newly generated population are determined. From step eight, it is found that x_3^{new} and x_2^{new} are fittest (Fit_{best}^{new}) and weakest (Fit_{worst}^{new}) elephants respectively in this population. Therefore,

$$x_{best}^{new} = \{0.083 \ -0.112 \ 0.104 \ -0.166\}$$
$$x_{worst} = x_{worst}^{new} = \{2.815 \ -4.760 \ -2.007 \ -4.603\}$$

In the tenth step, Fit_{best}^{new} is compared with Fit_{best}; if $Fit_{best}^{new} < Fit_{best}$ then do following

$$Fit_{best} = Fit_{best}^{new}$$
$$x_{best} = x_{best}^{new}$$

If no, do nothing.

In step eleven, store $x_j^{new} \ \forall j$ into $x_j \ \forall j$ and $Fit_j^{new} \ \forall j$ into $Fit_j \ \forall j$ as

$$x_1 = x_1^{new} = \{-1.397 \ -2.575 \ 1.252 \ 0.431\}$$
$$x_2 = x_2^{new} = \{2.815 \ -4.760 \ -2.007 \ -4.603\}.$$

$$x_3 = x_3^{new} = \{0.083 \ -0.112 \ 0.104 \ -0.166\}$$
$$x_4 = x_4^{new} = \{2.242 \ -3.181 \ 2.050 \ -1.564\}$$
$$x_5 = x_5^{new} = \{-0.158 \ -2.330 \ -0.966 \ -2.676\}$$

and

$$Fit_1 = Fit_1^{new} = 76.424$$
$$Fit_2 = Fit_2^{new} = 89.186$$
$$Fit_3 = Fit_3^{new} = 10.787$$
$$Fit_4 = Fit_4^{new} = 56.782$$
$$Fit_5 = Fit_5^{new} = 47.613$$

In the twelfth step, we return to the sixth step.

In the thirteenth step, we print the x_{best} with Fit_{best} as an optimal solution and stop the algorithm.

12.6   Conclusions

The chapter presents a simple tutorial of the EHO algorithm to solve mathematical optimization problems. The herding behaviour of elephants is presented in some set of mathematical equations along with its algorithm. In order to understand the basic programming involved in EHO, source-codes of Matlab and C++ are also provided. The problem solving ability of EHO is demonstrated by step-by-step numerical examples in which we solved the Restrigin function. We feel that this chapter will help an individual who wants to make use of the EHO algorithm to solve mathematical optimization problems.

Acknowledgement

This work was supported by the Engineering and Physical Sciences Research Council (EPSRC) of United Kingdom (Reference Nos.: EP/R001456/1 and EP/S001778/1).

References

1. *Elephant Social Structure*, [Available Online] https://www.elephant-world.com/elephant-social-structure/

2. N. K. Meena, S. Parashar, A. Swarnkar, N. Gupta and K. R. Niazi. "Improved elephant herding optimization for multiobjective DER accommodation in distribution systems". *IEEE Transactions on Industrial Informatics*, vol. 14(3), 2018, pp. 1029-1039.

3. C.E. O'Connell-Rodwell. "Keeping an "ear" to the ground: seismic communication in elephants". *Physiology*, vol. 22(4), pp.287-294, , 2007.

4. C. Ernst. "Understanding elephants UH scholars study wild and working animals on two continents". *Malamalama*, vol. 32(2), May 2007.

5. G. Wang, S. Deb and L. d. S. Coelho. "Elephant herding optimization," *2015 3rd International Symposium on Computational and Business Intelligence (ISCBI)*, Bali, 2015, pp. 1-5. doi: 10.1109/IS-CBI.2015.8

13

Firefly Algorithm

Xin-She Yang

School of Science and Technology
Middlesex University, London, United Kingdom

Adam Slowik

Department of Electronics and Computer Science
Koszalin University of Technology, Koszalin, Poland

CONTENTS

13.1 Introduction

The original firefly algorithm (FA) was first developed by Xin-She Yang in late 2007 and early 2008 [1]; this mimics the main characteristics of flashing behaviour of tropical fireflies. Due to the rich characteristics of FA as a nonlinear system, it can solve multimodal optimization problems effectively [2, 3]. It can also deal with stochastic functions and non-convex problems [4, 5]. A detailed comparison by Senthilnath et al. [6] suggests that the FA can obtain the best results with the least amount of computing time for clustering. Other studies show that FA can solve various problems in different applications effectively, including software testing [7], modeling to generate design alternatives [8], and scheduling problems [9]. In addition, it is possible to enhance the performance of FA further by introducing other components such as chaotic maps [10].

The rest of the chapter provides all the fundamentals of the firefly algorithm with the main pseudo-code, and both Matlab and C++ demo codes.

13.2 Original firefly algorithm

For an optimization problem such as function optimization with an objective function $f(\boldsymbol{x})$, the decision variables $\boldsymbol{x}$ can be considered as a D-dimensional vector formed by D independent variables

$$\boldsymbol{x} = [x_1, x_2, x_3, ..., x_D], \tag{13.1}$$

where we have used a row vector. Obviously, it can be changed into a column vector by a transpose (T) action. The problem is to minimize an objective

$$\text{Minimize } f(\boldsymbol{x}), \quad \boldsymbol{x} \in \mathbb{R}^D. \tag{13.2}$$

This is an unconstrained optimization problem, which forms our starting point. Once we understand the main idea of the firefly algorithm and its implementation, we can easily extend it to solve constrained optimization problems by incorporating constraints properly using constraint-handling techniques to be discussed later.

13.2.1 Description of the standard firefly algorithm

The position of a firefly such as firefly i can be considered as a solution vector to an optimization problem. Thus, the movement of positions is equivalent to the search moves in the decision space. The search process is an iterative process with a pseudo-time counter t, starting with $t = 0$. For simplicity, we use the notation $\boldsymbol{x}_i^t$ to denote the position of firefly i at iteration t. Here t should not be confused with the exponent of an exponential function. In fact, in many textbooks, another notation $\boldsymbol{x}_i^{(t)}$ is commonly used. However, as both notations are popular, we will use $\boldsymbol{x}_i^t$ here as it causes no confusion in this context.

The main equation for the firefly algorithm is

$$\boldsymbol{x}_i^{t+1} = \boldsymbol{x}_i^t + \beta_0 e^{-\gamma r_{ij}^2} (\boldsymbol{x}_j^t - \boldsymbol{x}_i^t) + \alpha \boldsymbol{\epsilon}_i^t, \tag{13.3}$$

where β_0 is the attractiveness at zero distance and α is a scaling factor. In addition, γ is an absorption coefficient, and the distance r_{ij} is defined as

$$r_{ij} = ||\boldsymbol{x}_i - \boldsymbol{x}_j||_2 = \sqrt{\sum_{k=1}^{D} (x_{i,k} - x_{j,k})^2}, \tag{13.4}$$

which is the Cartesian distance of two fireflies. Here, $x_{i,k}$ means the kth component or variable of the solution vector $\boldsymbol{x}_i$.

The vector ϵ_i^t is a vector of random numbers that are normally distributed with a zero mean and unity variance. This means that this vector is drawn from a normal distribution $N(0,1)$ and updated at every iteration.

The second term $\beta_0 \exp[-\gamma r_{ij}^2]$ is the attractiveness term, which mimics the attractiveness seen by two fireflies i and j. This term comes from the combination of the light intensity variation due to the inverse-square law with distance and exponential absorption in the media.

13.2.2 Pseudo-code of FA

The main steps of the FA consists of two loops over all fireflies. The updates of positions or solution vectors are done iteratively, and evaluations of new solutions are carried out within the loops. The pseudo-code for the firefly algorithm is presented in Algorithm 12.

Algorithm 12 Pseudo-code of the firefly algorithm.

1: Define the objective function $f(\boldsymbol{x})$
2: Set all the parameters α_0, θ, β_0 and γ
3: Initialize the population of n fireflies
4: Set the iteration counter $t = 0$
5: **while** $(t < t_{\max})$ **do**
6:     **for** $i = 1 : n$ **do**
7:         **for** $j = 1 : i$ **do**
8:             Calculate the distance $r_{ij} = \|\boldsymbol{x}_i - \boldsymbol{x}_j\|$
9:             Draw a random vector $\boldsymbol{\epsilon}$ from a normal distribution
10:            Calculate the positions/solutions of fireflies i and j
11:            Evaluate $f(\boldsymbol{x}_i)$ and $f(\boldsymbol{x}_j)$
12:            **if** $f(\boldsymbol{x}_i) < f(\boldsymbol{x}_j)$ **then**
13:                Move firefly j towards i (for minimization) using Eq. (13.3)
14:            **end if**
15:         **end for**
16:     **end for**
17:     Rank the firefly population and find the current best solution $\boldsymbol{g}_*$
18: **end while**
19: Output the best solution
20: Visualize the results

13.2.3 Parameters in the firefly algorithm

There are three parameters in the FA, and they are: α, β_0 and γ.

The attraction strength or attractiveness β of two fireflies is governed by

$$\beta = \beta_0 \exp[-\gamma r_{ij}^2]. \tag{13.5}$$

Obviously, $\beta = \beta_0$ if $r_{ij} = 0$. If γ is too small (i.e., $\gamma \to 0$), we have $\beta \to \beta_0$, which means that the light intensity or attractiveness remains close to a constant β_0. The visibility distance is large, the nonlinear term in Eq. (13.3) becomes linear. This corresponds to the case where a flashing firefly can be seen by all the other fireflies in the search space. This also means that collective information is used by all the fireflies. On the other hand, if $\gamma \to +\infty$, then $\beta \to 0$. The visibility distance for fireflies is short, which corresponds to a situation in a dense fog where one firefly cannot see other fireflies. Thus, the system becomes highly nonlinear, and each firefly moves by performing local random walks. This means that the exploitation ability of the firefly system is at the minimum, but its exploration ability is at the maximum.

So a simple rule of setting γ is to allow $\gamma r_{ij}^2 = 1$. That is

$$\gamma = \frac{1}{L^2}, \tag{13.6}$$

where L is the average length scale of the problem. For example, if a variable varies from -5 to $+5$, its scale is $L = 10$, so we can use $\gamma = 1/10^2 = 0.01$.

The attractiveness β_0 can be taken as $\beta_0 = 1$ for most applications, while $\gamma > 0$ should relate to the scales of the modes of the problem. If there is no prior knowledge of the problem modality, $\gamma = 0.01$ to 1 can be used first.

The scaling factor α in the third term controls the step size of the movements of the fireflies. If α is large, the steps are large, which can explore a large space; however, the solution generated may be too far away from the current region, even potentially jumping out of the search region. If α is too small, the steps are also small, which limits the exploration ability, and the moves become primarily local. Thus, there is a fine balance: the steps should not be too large or too small. Overall, the third term in Eq. (13.3) controls the randomness in the algorithm. It can be expected that the overall random components should become smaller and smaller as the population may move towards the true optimal position. Thus, ideally, α should be gradually reduced. For example, we can use

$$\alpha = \alpha_0 \theta^t, \tag{13.7}$$

where $0 < \theta < 1$ is a constant, and α_0 is the initial value of α. In most cases, we can use $\theta = 0.9$ to 0.99 and $\alpha_0 = 1$.

13.3 Source code of firefly algorithm in Matlab

As the algorithm is simple with only one main governing equation, it is straightforward to implement in any programming language. Here, we present a detail description of a simple Matlab implementation to find the minimum of the function $f(\boldsymbol{x})$

$$\text{minimize} f(\boldsymbol{x}) = (x_1 - 1)^2 + (x_2 - 1)^2 + \dots + (x_D - 1)^2, \quad x_i \in \mathbb{R}, \tag{13.8}$$

which has a global minimum of $x_* = (1, 1, ..., 1)$. Though the algorithm can search the whole domain, it would be more realistic if we impose some simple regular bounds. For example, for the above function to be optimized, we can use the following lower bound (Lb) and upper bound (Ub):

$$Lb = [-5, -5, ..., -5], \quad Ub = [+5, +5, ..., +5]. \tag{13.9}$$

The Matlab code here consists of three parts: initialization, main loops, and the objective function. These lines of codes can simply be put together line by line in sequential order, which should run smoothly to give the optimal solution of the minimum $f(x)$.

The first part is mainly initialization of parameter values such as α_0, β_0 and γ, as well as the generation of the initial population of $n = 20$ fireflies. The cost function is the objective function to be given later in a Matlab function.

```
1  function fa_demo       % Start the FA demo
2  n=20;                  % Population size (number of fireflies)
3  alpha=1.0;             % Randomness 0--1 (highly random)
4  beta0=1.0;             % beta value
5  gamma=0.1;             % Absorption coefficient
6  theta=0.97;            % Randomness reduction factor for alpha
7  d=10;                  % Number of dimensions
8  tMax=500;              % Maximum number of iterations
9  Lb=-5*ones(1,d);       % Lower bounds/limits
10 Ub=5*ones(1,d);        % Upper bounds/limits
11 % Generating the initial locations of n fireflies
12    for i=1:n,
13       ns(i,:)=Lb+(Ub-Lb).*rand(1,d);
14       Lightn(i)=cost(ns(i,:));   % Evaluate objectives
15    end
```

Listing 13.1
FA demo initialization.

The second part is the main part, consisting of two loops and the update of the firefly position vectors by the algorithmic equation (13.3). Then, the solutions are ranked according to their fitness or objective values. The best solution is found. In addition, new solutions are checked to see if they are within the simple bounds.

```
1  for k=1:tMax,          %%%%% start iterations %%%%%
2    alpha=alpha*theta;   % Reduce alpha by a factor theta
3    scale=abs(Ub-Lb);    % Scaling of the system
4  % Two loops over all fireflies
5  for i=1:n,
6    for j=1:n,
7        % Evaluate the objective values of current solutions
8        Lightn(i)=cost(ns(i,:));   % Call the objective
9        % Update moves
10       if Lightn(i)>Lightn(j),    % Brighter/more attractive
11       r=sqrt(sum((ns(i,:)-ns(j,:)).^2));   % Attractiveness
12       beta=beta0*exp(-gamma*r.^2);   % Attractiveness
13       steps=alpha.*(rand(1,d)-0.5).*scale;
14       % The FA update equation
15       ns(i,:)=ns(i,:)+beta*(ns(j,:)-ns(i,:))+steps;
16       end
17     end % end for j
18  end % end for i
19  % Check if the updated solutions/locations are within limits
```

```
20  ns=findlimits(n,ns,Lb,Ub);
21  %% Ranking fireflies by their light intensity/objectives
22  [Lightn,Index]=sort(Lightn);
23  ns_tmp=ns;
24  for i=1:n,
25    ns(i,:)=ns_tmp(Index(i),:);
26  end
27  %% Find the current best and display outputs
28  fbest=Lightn(1), nbest=ns(1,:)
29  end % End of t loop (up to tMax)
```

Listing 13.2
The main part of the firefly algorithm.

The third part is the objective function and ways for implementing the lower and upper bounds. This will ensure that all solution vectors should be within the regular bounds, and any new solution can be evaluated by calling the objective or cost function.

```
1   % Make sure the fireflies are within the bounds/limits
2   function [ns]=findlimits(n,ns,Lb,Ub)
3   for i=1:n,
4       ns_tmp=ns(i,:);
5       % Apply the lower bound
6       I=ns_tmp<Lb;  ns_tmp(I)=Lb(I);
7       % Apply the upper bounds
8       J=ns_tmp>Ub;  ns_tmp(J)=Ub(J);
9       % Update this new move
10      ns(i,:)=ns_tmp;
11  end
12  %% Cost or Objective function
13  function z=cost(x)
14  z=sum((x-1).^2);  % Solutions should be (1,1,...,1)
```

Listing 13.3
Variable limits and objective function.

13.4   Source code in C++

The same FA steps can be implemented in C++. In order to be consistent, the same function and limits are used, and the C++ code is given as follows:

```
1   #include <iostream>
2   #include <time.h>
3   #include <tgmath.h>
4   #include <algorithm>
5   using namespace std;
6
7   // Definition of the objective function OF(.)
8   float cost(float x[], int size_array)
9   {
10  float t=0;
11  for(int i=0; i<size_array; i++){
12      t=t+(x[i]-1)*(x[i]-1);}
13  return t;
14  }
15  // Generate pseudo random values from the range [0, 1)
```

```cpp
16  float ra(){return (float)(rand()%1000)/1000;}
17
18  int main()
19  {
20  int n=20;//Population size (number of fireflies)
21  float alpha=1.0;   //Randomness 0--1 (highly random)
22  float beta0=1.0;   //beta value
23  float gamma=0.1;   //Absorption coefficient
24  float theta=0.97;  //Randomness reduction factor for alpha
25  int d=10;          //Number of dimensions
26  int tMax=500;      //Maximum number of iterations
27  float Lb[d], Ub[d]; //Lower bounds/limits, Upper bounds/limits
28  float ns[n][d], ns_tmp[n][d];
29  float Lightn[n], Lightn_tmp[n], fbest, nbest[d];
30  float scale[d], r, beta, sum, steps[d];
31  // Initialization of the pseudo-random generator
32  srand(time(NULL));
33  // Initialization of the bounds
34  for(int j=0; j<d; j++){Lb[j]=-5; Ub[j]=5;}
35  // Generating the initial locations of n fireflies
36  for(int i=0; i<n; i++)
37  {
38      for(int j=0; j<d; j++)
39      {
40          ns[i][j]=Lb[j]+(Ub[j]-Lb[j])*ra();
41      }
42      Lightn[i]=cost(ns[i],d); //Evaluate objectives
43  }
44  //Iterations for pseudo-time marching
45  for(int k=0; k<tMax; k++)      //Start iterations
46  {
47      alpha=alpha*theta;
48      //scaling of the system
49      for(int i=0; i<d; i++)
50      {
51          scale[i]=fabs(Ub[i]-Lb[i]);
52      }
53      //updating fireflies
54      for(int i=0; i<n; i++)
55      {
56          for(int j=0; j<n; j++)
57          {   //Evaluate objective value of current solutions
58                  Lightn[i]=cost(ns[i],d); //Call the objective
59              //Update moves
60              if(Lightn[i]>Lightn[j]) //Brighter and more attractive
61              {
62                  sum=0;
63                  for(int h=0; h<d; h++)
64                  {
65                      sum=sum+(ns[i,h]-ns[j,h])*(ns[i,h]-ns[j,h]);
66                  }
67                  r=sqrt(sum);
68                  beta=beta0*exp(-gamma*r*r); //Attractiveness
69                  for(int h=0; h<d; h++)
70                  {
71                      steps[h]=alpha*(ra()-0.5)*scale[h];
72                  }
73                  //The FA update equation
74                  for(int h=0; h<d; h++)
75                  {
76                      ns[i][h]=ns[i][h]+beta*(ns[j][h]-ns[i][h])+steps[h];
77                  }
78              } // end if
79          } // end for j
80      } // end for i
81  //Check if the updated solution/locations are within limits
82  for(int i=0; i<n; i++)
```

```
83 {
84      for(int j=0; j<d; j++)
85      {
86          if (ns[i][j]<Lb[j]){ns[i][j]=Lb[j];}
87          if (ns[i][j]>Ub[j]){ns[i][j]=Ub[j];}
88      }
89 }
90 //Ranking fireflies by their light intensity/objectives
91 for (int i=0; i<n; i++)
92 {
93      for(int j=0; j<d; j++)
94      {
95          ns_tmp[i][j]=ns[i][j];
96      }
97      Lightn_tmp[i]=Lightn[i];
98 }
99 float worst_fit=*max_element(Lightn_tmp,Lightn_tmp+d)+1;
100 for(int i=0; i<n; i++)
101 {
102     int best_fit=min_element(Lightn_tmp, Lightn_tmp+d)-Lightn_tmp;
103     for(int j=0; j<d; j++)
104     {
105         ns[i][j]=ns_tmp[best_fit][j];
106     }
107     Lightn[i]=Lightn_tmp[best_fit];
108     Lightn_tmp[best_fit]=worst_fit;
109 }
110 fbest=Lightn[0];
111 for(int i=0; i<d; i++)
112 {
113     nbest[i]=ns[0][i];
114 }
115 } // End for k
116 cout<<"#### fbest: "<<fbest<<endl;
117 cout<<"#### nbest: [ ";
118 for(int i=0; i<d; i++){cout<<nbest[i]<<" ";}
119 cout<<"]"<<endl;
120 getchar();
121 return 0;
122 }
```

Listing 13.4
Firefly algorithm in C++.

13.5 A worked example

Let us use the firefly algorithm to find the minimum of

$$f(\boldsymbol{x}) = (x_1 - 1)^2 + (x_2 - 1)^2 + (x_3 - 1)^2, \tag{13.10}$$

in the simple ranges of $-5 \le x_i \le 5$. For a purely demonstrative purpose, we only use $n = 5$ fireflies. Suppose the initial population is randomly initialized and their objective (fitness) values are as follows:

$$\begin{cases}
\boldsymbol{x}_1 = \begin{pmatrix} 2.00 & 2.00 & 3.00 \end{pmatrix}, & f_1 = f(\boldsymbol{x}_1) = 6.00, \\
\boldsymbol{x}_2 = \begin{pmatrix} 5.00 & 0.00 & 5.00 \end{pmatrix}, & f_2 = f(\boldsymbol{x}_2) = 33.00, \\
\boldsymbol{x}_3 = \begin{pmatrix} -3.00 & -2.00 & 0.00 \end{pmatrix}, & f_3 = f(\boldsymbol{x}_3) = 26.00, \\
\boldsymbol{x}_4 = \begin{pmatrix} -5.00 & 0.00 & 5.00 \end{pmatrix}, & f_4 = f(\boldsymbol{x}_4) = 53.00, \\
\boldsymbol{x}_5 = \begin{pmatrix} 3.00 & 4.00 & 5.00 \end{pmatrix}, & f_5 = f(\boldsymbol{x}_5) = 29.00,
\end{cases} \tag{13.11}$$

where we have only used two decimal places for simplicity.

The best solution with the lowest (or best) value of the objective function is x_1 with $f_1 = 6.00$ for this population at $t = 0$. We did this intentionally so that x_1 is the best solution. Otherwise, we should re-arrange the order of the population so that the first solution is the best solution.

For simplicity, we can use $\alpha_0 = 1$, $\beta_0 = 1$, $\theta = 0.97$ and $\gamma = 0.1$, though these parameters are not good and they are not used in the actual Matlab implementation. However, such settings allow us to show and calculate new solutions easily so that we can focus on the main principle and procedure of the firefly algorithm.

The first loop is to compare each pair of the fireflies, and update their positions at $t = 1$.

- By comparing x_2 and x_1, we know that x_1 is better because $f_1 < f_2$, so we should move x_2 towards x_1 by

$$x_2^{t+1} = x_2 + \beta_0 e^{-\gamma r_{21}^2}(x_1 - x_2) + \alpha \epsilon_2, \tag{13.12}$$

where all the values on the right-hand side are values at iteration t.

For simplicity, let us draw a random vector and we get $\epsilon_2 = [-0.5, 0.4, 0.1]$. Since the distance between $x_1 = [2, 2, 3]$ and $x_2 = [5, 0, 5]$ is

$$r_{21} = \sqrt{(2-3)^2 + (2-0)^2 + (3-5)^2} = \sqrt{17}, \tag{13.13}$$

we have

$$x_2^{t+1} = \begin{pmatrix} 5 \\ 0 \\ 5 \end{pmatrix}^T + 1 \times e^{-0.1 \times (\sqrt{17})^2} \left[\begin{pmatrix} 2 \\ 2 \\ 3 \end{pmatrix}^T - \begin{pmatrix} 5 \\ 0 \\ 5 \end{pmatrix}^T \right] + 0.97 \times \begin{pmatrix} -0.5 \\ 0.4 \\ 0.1 \end{pmatrix}^T$$

$$= \begin{pmatrix} 3.97 \\ 0.75 \\ 4.73 \end{pmatrix}^T, \tag{13.14}$$

where the transpose turns a row vector into a column vector. This new x_2 at $t+1$ gives the objective value

$$f(x_2^{t+1}) = 22.79, \tag{13.15}$$

which is better than the original f_2. Even though the new objective 22.79 is still higher than $f_1 = 6$, the move from x_2 to x_2^{t+1} is an improvement over the original f_2, and this move should be accepted. Therefore, the new solution x_2 should be immediately updated as

$$x_2^{t+1} = (\ 3.97 \quad 0.75 \quad 4.73 \ ), \quad f_2^{t+1} = 22.79. \tag{13.16}$$

- Now we do a similar update by comparing x_3 and x_1. The distance r_{31} is

$$r_{31} = \sqrt{(-3-2)^2 + (-2-2)^2 + (0-3)^2} = \sqrt{50}. \tag{13.17}$$

If we draw $\epsilon_3 = [-0.2, 0.1, -0.5]$, we have

$$x_3^{t+1} = \begin{pmatrix} -3 \\ -2 \\ 0 \end{pmatrix}^T + 1 \times e^{-0.1 \times (\sqrt{50})^2} \left[\begin{pmatrix} 2 \\ 2 \\ 3 \end{pmatrix}^T - \begin{pmatrix} -3 \\ -2 \\ 0 \end{pmatrix}^T \right] + 0.97 \times \begin{pmatrix} -0.2 \\ 0.1 \\ -0.5 \end{pmatrix}^T$$

$$= \begin{pmatrix} -3.16 \\ -1.88 \\ -0.46 \end{pmatrix}^T , \tag{13.18}$$

which gives $f(x_3^{t+1}) = 27.73$, which is higher than the original $f_3 = 26.00$, which means that we should not update this move.

- We use a loop over the rest of the population such as x_4 and x_5, and update them in a similar manner. If the new moves jump out of the range, we compare them with the nearest bounds and force the new solutions to be within the simple domain bounded by Lb and Ub.

Once the population has been updated once, the solutions are ranked according to their objective values so that the first solution x_1^{t+1} should be the best solution. Then, a new round is carried out by setting $t \leftarrow t+1$. Once a predefined maximum number of iterations such as $t_{max} = 1000$, the best solution to the problem under consideration can be obtained.

In general, the firefly algorithm can be efficient. For the objective function, the optimal solution $x_* = [1, 1, 1]$ can be obtained within about 100 iterations. For example, in one run after 100 iterations with a population $n = 20$, the best solution we got is

$$x_1^{100} = (\begin{matrix} 1.020 & 0.997 & 1.001 \end{matrix}), \quad f(x_1^{100}) = 4.1 \times 10^{-4}. \tag{13.19}$$

After 200 iterations with the same population, we can get $f_{min} = 2.9 \times 10^{-7}$. Similarly, after 500 iterations, we can typically get $f_{min} = 2.3 \times 10^{-14}$.

13.6   Handling constraints

It is worth pointing out that the above implementations or demo codes are mainly for unconstrained optimization problems. For solving nonlinear constrained optimization problems, constraints should be handled properly using constraint-handling techniques such as penalty methods and the method of Lagrangian multipliers. In the simplest cases, a penalty parameter can be introduced so as to incorporate all the constraints into the modified objective; thus the constrained problem becomes a corresponding unconstrained one.

In general, a mathematical optimization problem in a D-dimensional design space can be written as

$$\text{minimize} \quad f(x), \quad x = (x_1, x_2, ..., x_D) \in \mathbb{R}^D, \tag{13.20}$$

subject to

$$\phi_i(x) = 0, \quad (i = 1, 2, ..., M), \tag{13.21}$$

$$\psi_j(x) \le 0, \quad (j = 1, 2, ..., N), \tag{13.22}$$

where x is the vector of D design variables, and $\phi_i(x)$ and $\psi_j(x)$ are the equality constraints and inequality constraints, respectively. The penalty-based method transforms the objective $f(x)$ into a modified objective Θ in the following form:

$$\Theta(x) = f(x)[\text{objective}] + P(x)[\text{penalty}], \tag{13.23}$$

where the penalty term $P(x)$ can take different forms, depending on the actual ways or variants of constraint-handling methods. For example, a static penalty method uses

$$P(x) = \lambda \left[\sum_{i=1}^{M} \phi_i^2(x) + \sum_{j=1}^{N} \max\{0, \psi_j(x)\}^2 \right]. \tag{13.24}$$

Since $\lambda > 0$ is fixed, independent of the iteration t, we can extend the above Matlab code for unconstrained problems to solve this type of problem.

13.7 Conclusion

The firefly algorithm is a simple, flexible and yet efficient algorithm for solving multimodal optimization problems. As we have seen from the above explanation of the main steps, the implementation is relatively straightforward, which requires a minimum amount of memory and computation costs.

This algorithm has been extended to other forms with many variants. Interested readers can refer to more advanced literature [3, 11].

References

1. X.S. Yang, *Nature-Inspired Metaheuristic Algorithms*, Luniver Press, UK, 2008.

2. X.S. Yang, "Firefly algorithm for multimodal optimisation", in *Proceedings of 5th Symposium on Stochastic Algorithms, Foundation and Applications*, Lecture Notes in Computer Science, volume 5792, pp. 169-178, 2009.

3. I. Fister, I. Fister Jr., X.S. Yang, J. Brest, "A comprehensive review of firefly algorithms". *Swarm and Evolutionary Computation*, 13(1):34-46, 2013.

4. X.S. Yang, "Firefly algorithm, stochastic test functions and design optimisation", *Int. Journal of Bio-Inspired Computation*, 2(2): 78-84, 2010.

5. X.S. Yang, S.S. Hosseini, A.H. Gandomi, "Firefly algorithm for solving non-convex economic dispatch problems with valve loading effect", *Applied Soft Computing*, 12(3): 1180–1186, 2012.

6. J. Senthilnath, S.N. Omkar, V. Mani, "Clustering using firefly algorithm: performance study", *Swarm and Evolutionary Computation*, 1(3): 163-171, 2011.

7. P. R. Srivastava, B. Mallikarjun, X.S. Yang, "Optimal test sequence generation using firefly algorithm", *Swarm and Evolutionary Computation*, 8(1): 44-53, 2013.

8. R. Imanirad, X.S. Yang, J.S. Yeomans, "Modeling-to-generate-alternatives via the firefly algorithm", *J. Appl. Oper. Res.*, 5(1): 14-21, 2013.

9. A. Yousif, A.H. Abdullah, S.M. Nor, A. Abdelaziz, "Scheduling jobs on grid computing using firefly algorithm", *J. Theor. Appl.Inform. Technol.*, 33(2): 155-164, 2011.

10. A.H. Gandomi, X.S. Yang, S. Talatahari, A. H. Alavi, "Firefly algorithm with chaos", *Commun. Nonlinear Sci. Numer. Simulation*, 18(1): 89-98, 2013.

11. X.S. Yang, *Nature-Inspired Optimization Algorithms*, Elsevier, 2014.

14

Glowworm Swarm Optimization: A Tutorial

Krishnanand Kaipa

Department of Mechanical and Aerospace Engineering
Old Dominion University, Norfolk, Virginia, United States

Debasish Ghose

Department of Aerospace Engineering
Indian Institute of Science, Bangalore, India

CONTENTS

14.1 Introduction

This book chapter focuses on the Glowworm Swarm Optimization (GSO) algorithm [1-3], a swarm intelligence algorithm loosely based on the behavior of glowworms, which are also known as fireflies or lightning bugs. Although there are other swarm intelligence algorithms in the literature based on certain perceived behaviors of fireflies (for instance, the firefly algorithm [4] and its extensions), the GSO algorithm precedes these by several years. The GSO algorithm was originally intended to be implementable in robotic systems where a team of mobile robots, or glowworms, equipped with sensors that would enable them to mimic the sensing capabilities of glowworms, would explore a signal landscape to identify multiple signal sources. The mathematical formulation of this problem devolved into a numerical optimization problem requiring the computation of multiple optima of a multimodal function (as against computation of the global optimum of such a function that most other swarm intelligence algorithms aim for). Glowworm swarm optimization (GSO), inspired by the behavior of glowworms, is a swarm intelligence algorithm introduced by Kaipa and Ghose in 2005 [1]; it subsequently underwent several experiments, modifications and changes by the authors and other researchers to evolve into a stable algorithm that not only served the purpose of computing multiple optima of a multimodal function, but was also designed in a way that would make it suitable for implementation in a swarm of mobile robots. This evolution of GSO and its various versions is described in detail in the book by Kaipa and Ghose [3].

Most swarm intelligence algorithms seek out the global optimum of a multimodal function as it is considered to be the best solution to a mathematical formulation of an optimization problem. The GSO works on a somewhat different premise that arises from the real world problem which the optimization problem seeks to solve. Often, the dynamic nature of the constraints in the search space makes a previous optimum solution infeasible to implement, making an alternative locally optimum solution more feasible. One domain where identification of multiple optimum solutions is required is financial decision making where locally feasible examples are used as a basis for learning [5] and in cases where it is important to identify diverse rules as a basis for a classifier [6]. Perhaps the most easily understood example of the need to search for multiple local optima is the problem of identifying multiple sources of signals in a landscape.

GSO fosters decentralized decision-making and movement protocols that are implementable in swarm robotics applications. For instance, a robot swarm can use the search protocol of GSO to move in an unknown region to carry out disaster response tasks comprising searches for multiple unknown signal sources. Examples of such applications include identifying nuclear spill points and hazardous chemical spill locations, identifying leaks in pressurized systems, and locating forest fires and dowsing them. It has been recognized that the GSO's approach of explicitly having the capability of partitioning a swarm into sub-swarms that would lead to multiple source localization is a very effective device [7], and while other methods have addressed this problem in an indirect way, GSO is the first algorithm to do so directly. In this chapter, the basic working principle of GSO is introduced, which is followed by a description of the phases that constitute each cycle of the algorithm. Next, the pseudocode, MATLAB code, and C++ code of the algorithm are presented. Finally, the working of various steps of the algorithm is illustrated by using a numerical example.

14.1.1   Basic principle of GSO

GSO is developed based on the behavior pattern of glowworms by which they can change the intensity of bio-luminescence and appear to glow at different intensities. Real glowworms carry a luminous pigment called luciferin, which is the source of bio-luminiscence. In the GSO algorithm, the quantity of luciferin encodes the fitness of its location in the search space. This makes the agent glowworms glow at an intensity approximately proportional to the function value being optimized. It is assumed that agents glowing that are brighter attract those that glow with lower intensity. In the algorithm, each glowworm selects, using a probabilistic mechanism, a neighbor that has a luciferin value higher than its own and moves toward it.

A critical aspect of the GSO algorithm is that it incorporates an adaptive neighborhood range by which the effect of distant glowworms is discounted when a glowworm has a sufficient number of neighbors with brighter glow or when the range to a neighboring agent goes beyond the maximum range of perception. These movements, based only on local information and selective neighbor interaction, enable the swarm of glowworms to split into disjoint subgroups that converge to high function value points. It is this property of the algorithm that allows it to be used to identify multiple optima of a multi-modal function. It has been shown [3] that GSO can tackle the following class of multimodal functions: unequal peaks, equal peaks, peaks of concentric circles, peaks surrounded by regions with step-discontinuities (non-differentiable objective functions), peaks comprising plateaus of equal heights, peaks located at irregular intervals, change in landscape features with change in scale, and non-separability involving interdependence of objective function variables. In addition to these, other researchers have shown its applicability to a much larger class of problems involving multiple optima.

In real life, natural glowworms are attracted to brighter glowing mates and form a cluster of glowworms that collectively grow brighter than their more scattered neighbors, which are, in turn, attracted to this brighter glow. The general idea in GSO is similar in the following aspects:

1. Agents are assumed to be attracted to move toward other agents that have higher luciferin value (brighter luminescence).

2. The multiple peaks can be likened to the nuclei of clusters that serve as bright beacons.

14.1.2   The Glowworm Swarm Optimization (GSO) algorithm

GSO initially distributes a swarm of agents randomly in the search space. These agents will be called glowworms from now onwards. Further, they have other behavioral traits that are not found in their natural counterparts. Accordingly, the algorithm encapsulates the interplay between the following three mechanisms:

1. **Fitness broadcast**: Glowworms carry a luminescent pigment called *luciferin*, whose quantity encodes the fitness of their locations in the objective space. This allows them to glow at an intensity that is proportional to the function value being optimized. It is assumed

that the luciferin level of a glowworm as sensed by its neighbor does not reduce due to distance[1].

2. **Positive taxis**: Each glowworm is attracted by, and moves toward, a single neighbor whose glow is brighter than that of itself; when surrounded by multiple such neighbors, it uses a probabilistic mechanism (described in Section 14.1.3) to select one of them.

3. **Adaptive neighborhood**: Each glowworm uses an adaptive neighborhood to identify neighbors; it is defined by a local-decision domain that has a variable range r_d^i bounded by a hard-limited sensor range r_s ($0 < r_d^i \leq r_s$). A suitable heuristic is used to modulate r_d^i (described in Section 14.1.3). A glowworm i considers another glowworm j as its neighbor if j is within the neighborhood range of i and the luciferin level of j is higher than that of i.

Note that the glowworms depend only on information available in the local-decision domain to decide their movements. Each glowworm selects, using a probabilistic mechanism, a neighbor that has a luciferin value higher than its own and moves toward it. These movements, that are based only on local information and selective neighbor interactions, enable the swarm of glowworms to partition into disjoint subgroups that steer toward, and meet at, multiple optima of a given multimodal function.

The significant difference between GSO and most earlier approaches to multimodal function optimization problems is the adaptive local-decision domain, which is used effectively to locate multiple peaks.

The description given above has been abstracted from the description in [2, 3].

14.1.3 Algorithm description

In the following, the algorithm is explained through maximization problems. However, the algorithm can be easily modified and used to find multiple minima of multimodal functions. GSO starts by placing a population of n glowworms randomly in the search space so that they are well dispersed. Initially, all the glowworms contain an equal quantity of luciferin ℓ_0. Each cycle of the algorithm consists of a luciferin update phase, a movement phase, and a neighborhood range update phase. The GSO algorithm is given in Alg. 13. The following description has been abstracted from [2, 3].

Algorithm 13 Glowworm Swarm Optimization (GSO) Algorithm.

1: Set number of dimensions m
2: Set number of glowworms n
3: Set step size s

---

[1]In natural glowworms, the brightness of a glowworm's glow as perceived by its neighbor reduces with increase in the distance between the two glowworms.

4: **for** $(i = 1 : n)$ **do**

5:      $x_i(0) \leftarrow deploy\_glowworm\_randomly(i, m)$

6:      $\ell_i(0) \leftarrow \ell_0$

7:      $r_d^i(0) \leftarrow r_0$

8: **end for**

9: Set maximum iteration number $= t_{max}$

10: Set $t = 1$

11: **while** $(t \leq t_{max})$ **do**

12:      **for** $(i = 1 : n)$ **do**

13:          $\ell_i(t) \leftarrow (1 - \rho)\ell_i(t - 1) + \gamma J(x_i(t))$

14:      **end for**

15:      **for** $(i = 1 : n)$ **do**

16:          $N_i(t) \leftarrow \{j : d_{ij}(t) < r_d^i(t); \ell_i(t) < \ell_j(t)\}$

17:          **for** $(j \in N_i(t))$ **do**

18:             $p_{ij}(t) \leftarrow \frac{\ell_j(t) - \ell_i(t)}{\sum_{k \in N_i(t)} \ell_k(t) - \ell_i(t)}$

19:          **end for**

20:          $j \leftarrow select\_glowworm(\vec{p})$

21:          $x_i(t + 1) \leftarrow x_i(t) + s \left(\frac{x_j(t) - x_i(t)}{\|x_j(t) - x_i(t)\|} \right)$

22:          $r_d^i(t + 1) \leftarrow \min\{r_s, \max\{0, r_d^i(t) + \beta(n_t - |N_i(t)|)\}\}$

23:      **end for**

24:      $t \leftarrow t + 1$

25: **end while**

Luciferin update phase: The luciferin update depends on the function value at the glowworm position. During the luciferin-update phase, each glowworm adds, to its previous luciferin level, a luciferin quantity proportional to the fitness of its current location in the objective function space. Also, a fraction of the luciferin value is subtracted to simulate the decay in luciferin with time. The luciferin update rule is given by:

$$\ell_i(t + 1) \quad = \quad (1 - \rho)\ell_i(t) + \gamma J(x_i(t + 1)) \tag{14.1}$$

where $\ell_i(t)$ represents the luciferin level associated with glowworm i at time t, ρ is the luciferin decay constant $(0 < \rho < 1)$, γ is the luciferin enhancement constant and $J(x_i(t))$ represents the value of the objective function at agent i's location at time t.

Movement phase: During the movement phase, each glowworm decides, using a probabilistic mechanism, to move toward a neighbor that has a luciferin value higher than its own. That is, glowworms are attracted to neighbors that glow brighter. For each glowworm i, the probability of moving toward a neighbor j is given by:

$$p_{ij}(t) \quad = \quad \frac{\ell_j(t) - \ell_i(t)}{\sum_{k \in N_i(t)} \ell_k(t) - \ell_i(t)} \tag{14.2}$$

where $j \in N_i(t)$ and

$$N_i(t) = \{j : d_{ij}(t) < r_d^i(t) \text{ and } \ell_i(t) < \ell_j(t)\} \qquad (14.3)$$

is the set of neighbors of glowworm i at time t, $d_{ij}(t)$ represents the Euclidean distance between glowworms i and j at time t, and $r_d^i(t)$ represents the variable neighborhood range associated with glowworm i at time t. Let glowworm i select a glowworm $j \in N_i(t)$ with $p_{ij}(t)$ given by (14.2). Then, the discrete-time model of the glowworm movements can be stated as:

$$x_i(t+1) = x_i(t) + s \left(\frac{x_j(t) - x_i(t)}{\|x_j(t) - x_i(t)\|} \right) \qquad (14.4)$$

where $x_i(t) \in R^m$ is the location of glowworm i, at time t, in the $m-$dimensional real space R^m, $\| \cdot \|$ represents the Euclidean norm operator, and s (> 0) is the step size.

Neighborhood range update phase: Each agent i is associated with a neighborhood whose radial range r_d^i is dynamic in nature $(0 < r_d^i \leq r_s)$. The fact that a fixed neighborhood range is not used needs some justification. When the glowworms depend only on local information to decide their movements, it is expected that the number of peaks captured would be a function of the radial sensor range. In fact, if the sensor range of each agent covers the entire search space, all the agents move to the global optimum and the local optima are ignored. Since we assume that *a priori* information about the objective function (e.g., number of peaks and inter-peak distances) is not available, it is difficult to fix the neighborhood range at a value that works well for different function landscapes. For instance, a chosen neighborhood range r_d would work relatively better on objective functions where the minimum inter-peak distance is more than r_d rather than on those where it is less than r_d. Therefore, GSO uses an adaptive neighborhood range in order to detect the presence of multiple peaks in a multimodal function landscape.

Let r_0 be the initial neighborhood range of each glowworm (that is, $r_d^i(0) = r_0 \; \forall \; i$). To adaptively update the neighborhood range of each glowworm, the following rule is applied:

$$r_d^i(t+1) = \min\{r_s, \max\{0, r_d^i(t) + \beta(n_t - |N_i(t)|)\}\} \qquad (14.5)$$

where β is a constant parameter and n_t is a parameter used to control the number of neighbors.

The quantities $\rho, \gamma, s, \beta, n_t$, and ℓ_0 are algorithm parameters for which appropriate values have been determined based on extensive numerical experiments and are kept fixed in this book (Table 14.1). The quantity r_0 is made equal to r_s in all the experiments. Thus, n and r_s are the only parameters that influence the algorithm behavior (in terms of the total number of peaks captured) and need to be selected.

TABLE 14.1
Values of algorithmic parameters that are kept fixed for all the experiments.

ρ	γ	β	n_t	s	ℓ_0
0.4	0.6	0.08	5	0.03	5

14.2   Source-code of GSO algorithm in Matlab

The Matlab source-code for optimizing the objective function $J(.)$ using the GSO algorithm is given in GSO.m below. In this code, GSO maximizes a two-dimensional objective function $J(.)$:

$$J(x,y) = 3(1-x)^2 e^{-[x^2+(y+1)^2]} - 10\left(\frac{x}{5} - x^3 - y^5\right) e^{-(x^2+y^2)}$$
$$- \left(\frac{1}{3}\right) e^{-[(x+1)^2+y^2]} \tag{14.6}$$

where the input is the set of locations of all the swarm members stored in A, $x = A_i(1)$ and $y = A_i(2)$, for $i \in 1, 2, \ldots, n$. The result of the GSO code is an n-dimensional column vector with the objective function values for each glowworm i.

```
1  %——GSO.m (main front—end code)————
2  global n m A_init A Ell gamma ro step1 r_d r_s ...
3          beta r_min n_t Ave_d bound
4  %————Parameter initialization————
5  m = 2;                      % No. of dimensions
6  n      = 100;               % No. of agents
7  r_s    = 3;                 % Sensor range
8  r_d    = r_s*ones(n,1);     % Local decision range
9  r_min = 0;                  % Threshold decisin range
10 gamma = 0.6;                % Luciferin enhancement constant
11 ro     = 0.4;               % Luciferin decay constant
12 step1 = 0.03;               % Distance moved by a glowworm at each
           step
13 beta  = 0.08;               % decision range gain
14 n_t   = 5;                  % Desired no. of neighbors
15 %——Initialization of variables——
16 bound = 3;                  % Workspace range parameter
17 DeployAgents;               % Deploy the glowworms randomly
18 Ell = 5*ones(n,1);          % Initialization of Luciferin levels
19 j = 1;                      % Iteration index
20 iter = 250;                 % No. of iterations
21 Ave_d = zeros(iter,1);      % Average distance
22 %——Main loop——
23 while (j <= iter)
24    UpdateLuciferin;         % Update luciferin levels of glowworms
25    Act;                     % Select a direction and move
26    for k = 1 : n            % store the state histories
27    agent_x(k,j,:) = A(k,1); agent_y(k,j,:) = A(k,2);
28    end
29    j = j + 1;
```

```
30  end
31  figure(1); % Plot of glowworm  trajectories
32  plot(A_init(:,1),A_init(:,2),'x'); xlabel('X'); ylabel('Y');
33  hold on;
34  DefineAxis;
35  for k = 1 : n plot(agent_x(k,:,:),agent_y(k,:,:)); end
36  plot([-0.0093;1.2857;-0.46], [1.5814;-0.0048;-0.6292],'ok');
37  figure(2); % Plot of final locations of glowworms
38  plot(A(:,1),A(:,2),'.'); DefineAxis; grid on; hold on;
39  plot([-0.0093;1.2857;-0.46], [1.5814;-0.0048;-0.6292],'ok');
40  % Functions
41  function DeployAgents % Fun 1: DeployAgents.m
42  global n m A_init A bound
43  B = -bound*ones(n,m); A_init = B + 2*bound*rand(n,m); A = A_init;
44  function UpdateLuciferin % Fun 2: UpdateLuciferin.m
45  global n A J Ell gamma ro
46  for i = 1 : n
47          x = A(i,1); y = A(i,2);
48          J(i,:) = 3*(1-x)^2*exp(-(x^2) - (y+1)^2) ...
49                  - 10*(x/5 - x^3 - y^5)*exp(-x^2-y^2) ...
50                  - 1/3*exp(-(x+1)^2 - y^2);
51          Ell(i,:) = (1-ro)*Ell(i,:) + gamma*J(i,:);
52  end %
53  function Act % Fun 3: Act.m
54  global n r_s r_d N N_a beta n_t
55  N(:,:) = zeros(n,n); N_a(:,:)= zeros(n,1);
56  for i = 1 : n
57      FindNeighbors(i); FindProbabilities(i);
58      Leader(i) = SelectAgent(i);
59  end %
60  for i = 1 : n
61      Move(i,Leader(i));
62      r_d(i) = max(0, min(r_s,r_d(i) + beta*(n_t-N_a(i))));
63  end %
64  function FindNeighbors(i) % Fun 4: FindNeighbors.m
65  global n m A N r_d N_a Ell
66  n_sum = 0;
67  for j = 1 : n
68      if (j~=i) square_sum = 0;
69          for k = 1 : m
70              square_sum = square_sum + (A(i,k)-A(j,k))^2;
71          end
72          d = sqrt(square_sum);
73          if (d <= r_d(i)) & (Ell(i) < Ell(j))
74              N(i,j) = 1; n_sum = n_sum + 1;
75          end
76      end
77      N_a(i) = n_sum;
78  end %
79  function FindProbabilities(i) % Fun 5: FindProbabilities.m
80  global n N Ell pb
81  Ell_sum = 0;
82  for j=1:n Ell_sum = Ell_sum + N(i,j)*(Ell(j) - Ell(i)); end
83  if (Ell_sum == 0) pb(i,:) = zeros(1,n); else
84      for j=1:n pb(i,j) = (N(i,j)*(Ell(j)-Ell(i)))/Ell_sum; end
85  end %
86  function j = SelectAgent(i) % Fun 6: SelectAgent.m
```

```
87  global n pb
88  bound_lower = 0; bound_upper = 0; toss = rand; j = 0;
89  for k = 1 : n
90    bound_lower = bound_upper; bound_upper = bound_upper + pb(i,k);
91    if (toss > bound_lower) & (toss < bound_upper) j = k; break; end
92  end %————————————————————————————————
93  function Move(i,j) % Fun: Move.m
94  global A m step1 Ell bound
95  if (j~=0) & (Ell(i) < Ell(j))
96    temp(i,:) = A(i,:) + step1*Path(i,j); flag = 0;
97    for k = 1 : m
98      if (temp(i,k) < −bound)|(temp(i,k) > bound) flag=1;break;end
99    end
100   if (flag == 0) A(i,:) = temp(i,:); end
101 end %————————————————————————————————
102 function Del = Path(i,j) % Fun 8: Path.m
103 global A m
104 square_sum = 0;
105 for k = 1 : m
106   square_sum = square_sum + (A(i,k)−A(j,k))^2;
107 end
108 hyp = sqrt(square_sum);
109 for k = 1 : m Del(:,k) = (A(j,k) − A(i,k))/hyp;
110 end %————————————————————————————————
111 function DefineAxis % Fun: DefineAxis.m
112 global bound
113 axis([−bound bound −bound bound]); grid on;
```

Listing 14.1
GSO code in Matlab.

14.3 Source-code of GSO algorithm in C++

```
1  #include <iostream>
2  #include<fstream>
3  #include <algorithm>
4  #include <math.h>
5  using namespace std;
6  // Constant Parameters
7  #define n 1000      // n − Number of glowworms
8  #define r 125.0     // r − Sensor range
9  #define rho 0.4     // rho − Luciferin decay constant
10 #define gama 0.6    // gama − Luciferin enhancement constant
11 #define beta 0.08   // beta − neighborhood ehancement constant
12 #define s 0.03
13 #define d 2         // d − dimension of the search space
14 #define nd 5        // nd − Desired number of neighbors
15 #define PI 3.14159
16 #define W 3         // W − Workspace size
17 #define IterMax 500
18 /*Variables: Lc − Luciferin value, Rd − Neighborhood range,
19 P − Probability matrix, Ld − Leader set, N − Neighborhood matrix,
20 Na − Actual number of neighbors, X − Glowworm positions */
21 static int Ld[n], N[n][n], Na[n], randSeed = 1;
```

```cpp
22  static double X[n][d], Lc[n], Rd[n], P[n][n],
23                Sol[]={-PI/2, PI/2};
24  static ofstream outFile ("peaks_sol.dat",ios::app);
25  static double Distance(int i, int j) {double dis = 0;
26   for(int k = 0; k < d; k++) dis = dis + pow((X[i][k]-X[j][k]),2);
27   return sqrt(dis); }
28  static void DeployGlowworms(float lim) {
29   for(int i = 0; i < n; i++) {for(int j = 0; j < d; j++)
30  {X[i][j]=-lim+2*lim*rand()/(RAND_MAX+1.0);}}}
31  static void UpdateLuciferin() {
32   for (int i = 0; i < n; i++) {double x = X[i][0], y = X[i][1];
33    double J = 3*pow((1-x),2)* exp(-pow(x,2) - pow((y+1),2))
34     - 10*(x/5 - pow(x,3) - pow(y,5))*exp(-pow(x,2) - pow(y,2))
35     - 1/3 * exp(-pow((x+1),2) - pow(y,2));
36    Lc[i] = (1-rho)*Lc[i] + gama*J;}}
37  static void FindNeighbors() {for(int i = 0; i < n; i++) {
38  N[i][i] = 0; Na[i] = 0;for(int j = 0; j < n; j++){if (j!=i){
39  if ((Lc[i] < Lc[j]) && (Distance(i,j) < Rd[i])) N[i][j] = 1;
40   else N[i][j] = 0; Na[i] = Na[i] + N[i][j];}}}}
41  static void FindProbabilities() {for(int i = 0; i < n; i++)
42  {double sum = 0; for (int j = 0; j < n; j++)
43   sum = sum + N[i][j]*(Lc[j] - Lc[i]);
44   for(int j = 0; j < n; j++) {if (sum != 0)
45   P[i][j] = N[i][j]*(Lc[j] - Lc[i])/sum;else P[i][j] = 0;}}}
46  static void SelectLeader() {for (int i = 0; i < n; i++) {
47    double b_lower = 0; Ld[i] = i;
48    double toss = rand()/(RAND_MAX + 1.0);
49    for (int j = 0; j < n; j++) {if (N[i][j] == 1) {
50    double b_upper = b_lower + P[i][j];
51     if ((toss >= b_lower) && (toss < b_upper))
52    {Ld[i] = j;break;} else b_lower = b_upper;}}}}
53  static void Move() {for (int i = 0; i < n; i++) {if (Ld[i]!=i){
54  int flag = 0;double temp[d]; double dis = Distance(i,Ld[i]);
55   for (int j = 0; j < d; j++) {
56    temp[j] = X[i][j] + s*(X[Ld[i]][j] - X[i][j])/dis;
57    if (fabs(temp[j]) >W) {flag = 1;break;}} if (flag == 0)
58    for (int j = 0; j < d; j++) X[i][j] = temp[j];}}}
59  static void UpdateNeighborhood() {
60  for (int i = 0; i < n; i++)
61   Rd[i] = max(0.0, min(r, Rd[i] + beta*(nd - Na[i])));  }
62  int main (int argc, char * const argv[]) {
63    for (int currParam = 0; currParam < argc; currParam++) {
64    if (strcmp(argv[currParam], "-r") == 0)
65     randSeed = atoi(argv[currParam+1]);} srand(randSeed);
66   DeployGlowworms(W);
67    for (int i = 0; i < n; i++) {Lc[i] = 5; Rd[i] = r;}
68    for (int t = 0; t < IterMax; t++ ) {
69   UpdateLuciferin(); FindNeighbors(); FindProbabilities();
70   SelectLeader(); Move(); UpdateNeighborhood(); }
71    for(int i = 0; i < n; i++) {
72     outFile << i << ' ' << X[i][0] << ' ' << X[i][1] << "\n";}
73   outFile.close(); exit(0); return 0;
74  }
```

Listing 14.2
Source-code of GSO in C++.

14.4 Step-by-step numerical example of GSO algorithm

Let us assume that we want to maximize the two-dimensional multi-modal function given in (14.6), which has three peaks at $(-0.0093, 1.5814)$, $(1.2857, -0.0048)$, and $(-0.46, -0.6292)$. As mentioned earlier, GSO attempts to capture all the function peaks. The result of applying the GSO algorithm is shown in Fig. 14.1, in which 100 glowworms move from their initial locations to one of the peak locations. The steps of the algorithm are shown below:

1. The GSO algorithm parameters are determined: $n = 100$, $r_s = 3$, $\gamma = 0.6$, $\rho = 0.4$, $\beta = 0.08$, $n_t = 5$, $s = 0.03$, and $\ell_0 = 5$. The initial decision range $r_d(0)$ is set to 1 for all glowworms ($1/3$ rd of r_s) to limit the number of initial neighbors of all glowworms.

2. A swarm of 100 glowworms $\{G_1, G_2, \ldots, G_{100}\}$ with a corresponding set of initial locations $A = A_{init}$ is created by randomly deploying all the glowworms in a 2-D space, where $-3 \leq A(i, j) \leq 3$, for all $i \in \{1, 2, \ldots 50\}$ and $j \in \{1, 2\}$. We illustrate the working of the algorithmic steps of GSO by tracking how the state of one of the

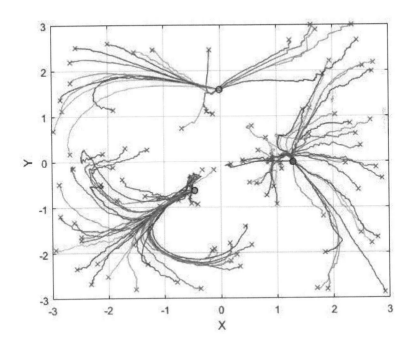

FIGURE 14.1
Paths traversed by all 100 glowworms from their initial locations to one of the peak locations.

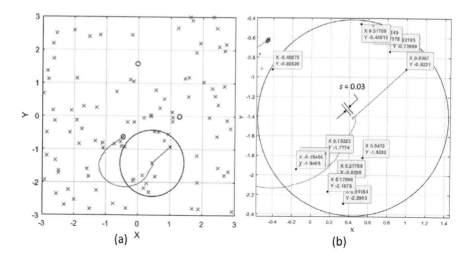

(a)   (b)

FIGURE 14.2

(a) Initial state (iteration $i = 0$) of the glowworm G_{45}: decision range $r_d(45,0)$ $= 1$, number of neighbors $Nbr(45,0) = 11$, and the selected leader glowworm G_{31}. The overall path of G_{45} from its initial location to one of the peaks and the initial direction of movement toward the selected leader are also shown. (b) Zoom-in view of the initial state of G_{45} to show the locations of all its neighbors locating within an adaptive range of $r_d(45,0) = 1$. Note that G_{45} makes a movement of step-size $s = 0.03$ toward its selected leader G_{31}.

glowworms G_{45} evolves as a function of iterations. Figure 14.2(a) shows the initial locations of all the 100 glowworms and the initial state (at iteration $i = 0$) of the glowworm G_{45}. For illustration purposes, the state of G_{45} at iterations $i = 0, 1, 2, 5, 50, 100, 150, 200$, and 249 are tabulated as shown in Table 14.2. The state at each iteration i includes the following variables:

(a) Current location, $A(45, i)$
(b) Objective function value, $J(45, i)$
(c) Luciferin value, $\ell(45, i)$
(d) Adaptive range radius, $r_d(45, i)$
(e) Number of neighbors, $Nbr(45, i)$
(f) Leader glowworm, $L(45, i)$
(g) Leader location, $A(L(45, i), i)$
(h) Updated location, $A(45, i + 1)$
(i) Updated adaptive range radius, $\ell(45, i + 1)$

3. The initial objective function value associated with each glowworm's position $J(i, 0)$, $i = 1, 2, \ldots, 100$ is computed. For example, the

TABLE 14.2
The state of glowworm G_{45} at different iterations.

i	$A(i)$	$J(i)$	$\ell(i)$	$r_d(i)$	$Nbr(i)$	$L(i)$	$A(i)$	$A(L(i))$	$r_d(i+1)$
0	(0.438, -1.416)	-5.682	-0.409	1	11	31	(0.997, -0.922)	(0.460, -1.396)	0.52
1	(0.459, -1.396)	-5.506	-3.549	0.52	1	77	(0.529, -1.849)	(0.464, -1.425)	0.84
5	(0.456, -1.483)	-5.891	-7.590	0.92	7	11	(0.140, -2.190)	(0.443, -1.510)	0.76
50	(-0.621, -2.110)	-1.760	-2.883	0.52	5	11	(-1.005, -2.028)	(-0.651, -2.103)	0.52
100	(-1.054, -1.065)	1.573	2.234	0.44	7	11	(-0.803, -0.830)	(-1.032, -1.045)	0.28
149	(-0.489, -0.624)	3.766	5.656	0.4	29	7	(-0.459, -0.611)	(-0.460, -0.630)	0
150	(-0.460, -0.630)	3.777	5.660	0	0	–	–	(-0.460, -0.630)	0.4
151	(-0.460, -0.630)	3.777	5.662	0.4	6	100	(-0.467, -0.636)	(-0.484, -0.648)	0.32
200	(-0.469, -0.630)	3.776	5.663	1.92	0	–	–	(-0.469, -0.630)	2.32
248	(-0.445, -0.626)	3.774	5.661	0.16	6	79	(-0.458, -0.625)	(-0.475, -0.625)	0.08
249	(-0.475, -0.625)	3.773	5.661	0.08	14	10	(-0.470, -0.647)	–	–

glowworm G_{45} is located at $A(45,0) = (0.4378, -1.4164)$. The objective function value at this location is $J(45,0) = -5.6816$.

4. The objective function values are used to then compute the luciferin values of each glowworm using (14.1). The luciferin value of G_{45} is $\ell(45,0) = -0.409$.

5. The neighbors of each glowworm are determined by using (14.1.3). For example, G_{45} has twelve glowworms within an adaptive range $r_d(45,0) = 1$: G_{86}, G_{91}, G_{34}, G_{31}, G_{77}, G_{15}, G_{54}, G_{11}, G_7, G_{23}, G_{18}, and G_{87}. Their locations are shown in Fig. 14.2(b). Note that a glowworm k is classified as a neighbor for glowworm j, if it is located within a circular neighborhood defined by the current adaptive range $r_d(j,i)$ and if $\ell(k,i) > \ell(j,i)$ as well. The luciferin values of all the twelve glowworms are shown in Table 14.3. Note from the table that except for one glowworm G_7 ($\ell(7,0) < \ell(45)$), all other eleven glowworms have a luciferin value more than that of G_{45}. Therefore, G_{45} has eleven neighbors at the initial iteration.

6. For each glowworm, the probability of selecting to move toward each of its neighbors is computed by using (14.2). A leader-neighbor toward which movement will be made is determined using these probabilities in a roulette-wheel fashion. For G_{45}, the probabilities computed for all the eleven neighbors found in the previous step are shown in Table 14.3 and G_{31} is selected as a leader. Since the

TABLE 14.3
Luciferin and probability values of glowworms within a adaptive range $r_d(45,0)$ of 1 unit of glowworm G_{45}'s location at the first iteration. For comparison purpose, the luciferin value G_{45} is provided here: $\ell(45,0) = $ -0.409.

G_i	G_{86}	G_{91}	G_{34}	G_{31}	G_{77}	G_{15}	G_{54}	G_{11}
$\ell(i,0)$	3.278	3.459	3.321	3.116	-0.099	1.357	0.188	0.767
$P(i,0)$	0.149	0.157	0.151	0.143	0.013	0.072	0.024	0.048

G_i	G_7	G_{23}	G_{18}	G_{87}				
$\ell(i,0)$	-0.735	0.036	0.237	4.512				
$P(i,0)$	–	0.018	0.026	0.199				

leader selection is probabilistic, the chosen leader was not necessarily the one with highest luciferin, and hence the one with highest probability (G_{87}, $\ell(87,0) = 4.512$, $P(87,0) = 0.199$), but G_{31} with a relatively less luciferin ($\ell(31,0) = 3.116$) and associated probability ($P(31,0) = 0.143$).

7. Each glowworm moves toward the selected neighbor by using (14.4). For example, G_{45} makes a movement of step-size $s = 0.03$ toward G_{31} and stops at a new location (0.4596, -1.3958) (Refer to Fig. 14.2(b)).

8. The adaptive range of each glowworm gets updated by using (14.5). For example, since the actual number of neighbors of G_{45} ($Nbr(45,0) = 11$) is more than the desired number of neighbors for any glowworm ($n_t = 5$), the update rule (14.5) causes the adaptive range of G_{45} to shrink from $r_d(45,0) = 1$ to $r_d(45,1) = 0.52$.

9. In the second iteration, all the steps 3 - 8 are executed to update the states of all the glowworms. For example, the objective function value at G_{45}'s new location is $J(45,1) = $ -5.506. The corresponding luciferin value $\ell(45,1) = $ -3.549. Figure 14.3 shows the updated locations of all the glowworms and the state of G_{45} at the second iteration. There are two glowworms G_7 and G_{77}) within G_{45}'s current adaptive range of 0.52 units. Their luciferin values are ranked as $\ell(7,1)(= -4.116) < \ell(45,1)(= -3.549) < \ell(77)(= -3.117)$. Therefore, G_{77} is the lone neighbor of G_{45} at the second iteration. G_{45} moves 0.03 units toward G_{77} and stops at a new location as shown in Table 14.2. Similarly, Fig. 14.4 shows the updated locations of all the glowworms and the state of G_{45} at iteration $i = 5$.

10. The algorithm continues in this way until the end of iterations ($i_{final} = 249$). Note from Table 14.2 that the final location of G_{45} is $(-0.4749, -0.6246)$, which corresponds to one of the peaks $(-0.46, -0.6292)$ within a tolerance of 0.02 units (the distance between the two locations is 0.0156 units).

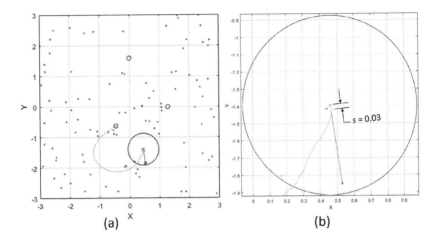

FIGURE 14.3
(a) State of the glowworm G_{45} at the second iteration ($i = 1$): decision range $r_d(45,0) = 0.52$, number of neighbors $Nbr(45,0) = 1$, and the newly selected leader glowworm G_{77}. The overall path of G_{45} from its initial location to one of the peaks and the direction of movement toward the selected leader are also shown. (b) Zoom-in view of the state of G_{45} to show its neighbors locating within an adaptive range of $r_d(45,0) = 0.52$. Note that G_{45} makes a movement of step-size $s = 0.03$ toward its selected leader G_{77}.

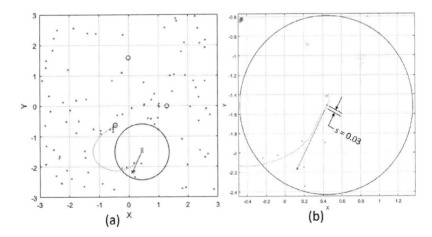

FIGURE 14.4
(a) State of the glowworm G_{45} at iteration $i = 5$: decision range $r_d(45,0) = 0.92$, number of neighbors $Nbr(45,0) = 7$, and the selected leader glowworm G_{11}. The overall path of G_{45} from its initial location to one of the peaks and the direction of movement toward the selected leader are also shown. (b) Zoom-in view of the state of G_{45} at iteration $i = 5$. Note that G_{45} makes a movement of step-size $s = 0.03$ toward its selected leader G_{11}.

A few observations are made to provide an insight into situations when a glowworm becomes leaderless. Note from Table 14.2 that the glowworm G_{45} has no leaders at iteration $i = 150$. This can be attributed to the fact that its adaptive range $r_d(45, 150)$ is zero, and hence it has zero neighbors. In order to find out why the range became zero, the previous state of G_{45} was analyzed. Note that the glowworm has a larger number of neighbors than the desired number of neighbors ($29 \gg n_t = 5$) at iteration $i = 149$ within an adaptive range $r_d(45, 149)$ of 0.4 units. This causes the range update rule (14.5) to shrink the adaptive range steeply. The reason for G_{45} having a high number of neighbors is due to the fact that it is already very close to the peak location $(-0.46, -0.6292)$ at iteration $i = 149$, where glowworms get crowded as increasingly more numbers of them move toward one of the three peaks. Similarly, G_{45} becomes leaderless at iteration $i = 200$, although it has a non-zero adaptive range of $r_d(45, 200) = 1.92$. This can be attributed to the fact that G_{45} has the highest within the adaptive range, and hence has no neighbors.

14.5   Conclusions

This chapter presented glowworm swarm optimization (GSO) that achieves simultaneous capture of multiple optima of multimodal functions and shows its special suitability for multiple signal source localization tasks. The underlying ideas behind the GSO technique were presented and the steps of the algorithm were illustrated using a numerical example involving simultaneous computation of multiple peaks of a multimodal function.

References

1. K.N. Kaipa and D. Ghose. Detection of multiple source locations using a glowworm metaphor with applications to collective robotics. In *Proceedings of IEEE Swarm Intelligence Symposium*, Pasadena, California, June 8–10, 2005, pp. 84–91.

2. K.N. Kaipa and D. Ghose. Glowworm swarm optimization for simultaneous capture of multiple local optima of multimodal functions. *Swarm Intelligence*, Vol. 3, No. 2, pp. 87–124, 2009.

3. K. N. Kaipa and D. Ghose. *Glowworm Swarm Optimization: Theory, Algorithms, and Applications, Studies in Computational Intelligence*, Vol. 698, Springer-Verlag, 2017.

4. X. S. Yang. Firefly algorithms for multimodal optimization, In: Watanabe O., Zeugmann T. (eds) *Stochastic Algorithms: Foundations and Applications. SAGA 2009. Lecture Notes in Computer Science*, Vol. 5792. Springer, Berlin, Heidelberg

5. R. Sikora and M.J. Shaw. A double-layered learning approach to acquiring rules for classification: Integrating genetic algorithms with similarity-based learning. *ORSA Journal on Computing*, Vol. 6, No. 2, pp. 174–187, 1994.

6. J. Horn, D.E Goldberg, and K. Deb. Implicit niching in a learning classifier system: Nature's way. *Evolutionary Computation*, Vol. 2, No. 1, pp. 37–66, 1994.

7. K. McGill and S. Taylor. Robot algorithms for localization of multiple emission sources. *ACM Computing Surveys*, 43, 3, Article 15, 25 pages, 2011.

15

Grasshopper Optimization Algorithm

Szymon Łukasik

Faculty of Physics and Applied Computer Science
AGH University of Science and Technology, Kraków, Poland

CONTENTS

15.1   Introduction

Grasshopper Optimization Algorithm (GOA) is an optimization technique introduced by Saremi, Mirjalili and Lewis in 2017 [9]. It belongs to the class of swarm optimization strategies [5]. The GOA procedure includes both social interaction between ordinary agents (grasshoppers) and the attraction of the best individual. Initial experiments performed by authors demonstrated promising exploration abilities of the algorithm – and they will be further examined in the course of our study.

GOA is reported to implement two components of grasshopper movement strategies. First it is the interaction of grasshoppers which demonstrates itself through slow movements (while in larvae stage) and dynamic motion (while in insect form). The second corresponds to the tendency to move towards the source of food. What is more, deceleration of grasshoppers approaching food and eventually consuming it is also taken into account [12].

The rest of this paper is organized as follows: in Section 15.2 we are providing more detailed description of the algorithm along with the pseudo-code of its implementation. Section 15.3 contains the implementation of the algorithm using the Matlab environment. It is followed by the detailed analysis of the algorithm dynamics – presented on the simple Sphere function benchmark in Section 15.5. Finally in Section 15.6 some final comments on the algorithm's properties are being provided.

15.2   Description of the Grasshopper Optimization Algorithm

This chapter, as others in this monograph, deals with continuous optimization, i.e. the task of finding a value of x – within the feasible search space $S \subset R^D$ – denoted as x^* such as $x^* = \mathrm{argmin}_{x \in S} f(x)$, assuming that the goal is to minimize cost function f. GOA represents a typical population based metaheuristics [10]. It means that the aforementioned problem is tackled with the population consisting of P agents of the same type. Each agent is represented as a solution vector x_p, $p = 1, ..., P$ and represents exactly one solution in the domain of tested function f [6].

The movement of individual p in iteration k can be presented using the following equation:

$$x_{pd} = c \left(\sum_{q=1, q \neq p}^{P} c \frac{UB_d - LB_d}{2} s(|x_{qd} - x_{pd}|) \frac{x_{qd} - x_{pd}}{dist(x_q, x_p)} \right) + x_d^*$$

(15.1)

with $d = 1, 2, ..., D$ representing search space dimensionality. Note that index k was omitted for the sake of readability. The equation contains two components: the first corresponds to the pairwise social interactions between grasshoppers, the second – to the movement attributed to the wind (in the algorithm – in the direction of the best individual). The impact of gravitation – though important for the real grasshopper swarms [4] – is not included in the basic algorithm scheme.

Function s, present in the first factor of 15.1, defines the strength of social forces, and was defined by creators of the algorithm as:

$$s(r) = f e^{\frac{-r}{l}} - e^{-r}$$

(15.2)

with default values of $l = 1.5$ and $f = 0.5$.

It divides the space between two considered grasshoppers into three separate zones, as demonstrated in Figure 15.1. Individuals being very close are found in the so called repulsion zone. On the other hand distant grasshoppers are located in the zone of attraction. The zone (or equilibrium state) between them – called the comfort zone – is characterized by the lack of social interactions. The first factor is additionally normalized by the upper and lower bounds of the feasible search space.

In addition to that parameter c – occurring twice in formula (15.1) – is decreased according to the the following equation:

$$c = c_{max} - k\frac{c_{max} - c_{min}}{K_{max}} \qquad (15.3)$$

with maximum and minimum values – c_{max}, c_{min} respectively – and K representing maximum number of iterations serving as the algorithm's termination criterion. The first occurrence of c in (15.1) reduces the movements of grasshoppers around the target – balancing between exploration and exploitation of the swarm around the target. It is analogous to the inertia weight present in the Particle Swarm Optimization Algorithm [3]. The whole component $c\frac{UB_d - LB_d}{2}$, as noted in [9], linearly decreases the space that the grasshoppers should explore and exploit.

The pseudo-code for the generic version of the GOA technique is presented in Algorithm 14.

Algorithm 14 Grasshopper Optimization Algorithm.

1: $k \leftarrow 1$, $f(x^*(0)) \leftarrow \infty$ ▷ initialization
2: **for** $p = 1$ to P **do**
3:      $x_p(k) \leftarrow$ Generate_Solution(LB, UB)
4: **end for**
5:   ▷ find best
6: **for** $p = 1$ to P **do**
7:      $f(x_p(k)) \leftarrow$ Evaluate_quality($x_p(k)$)
8:      **if** $f(x_p(k)) < f(x^*(k-1))$ **then**
9:          $x^*(k) \leftarrow x_p(k)$
10:      **else**
11:          $x^*(k) \leftarrow x^*(k-1)$
12:      **end if**
13: **end for**
14: **repeat**
15:      $c \leftarrow$ Update_c($c_{max}, c_{min}, k, K_{max}$)
16:      **for** $p = 1$ to P **do**
17:                            ▷ move according to formula (15.1)
18:          $x_p(k) \leftarrow$ Move_Grasshopper($c, UB, LB, x^*(k)$)
19:                                  ▷ correct if out of bounds
20:          $x_p(k) \leftarrow$ Correct_Solution($x_p(k), UB, LB$)
21:          $f(x_p(k)) \leftarrow$ Evaluate_quality($x_p(k)$)
22:          **if** $f(x_p(k)) < f(x^*k)$ **then**
23:              $x^*(k) \leftarrow x_p(k)$, $f(x^*k) \leftarrow f(x_p(k))$
24:          **end if**
25:      **end for**
26:      **for** $p = 1$ to P **do**
27:          $f(x_p(k+1)) \leftarrow f(x_pk)$, $x_p(k+1) \leftarrow x_p(k)$
28:      **end for**

29:     $f(x^*(k+1)) \leftarrow f(x^*k),\ x^*(k+1) \leftarrow x^*(k)$

30:     $k \leftarrow k+1$

31: **until** $k < K$ **return** $f(x^*(k)),\ x^*(k)$

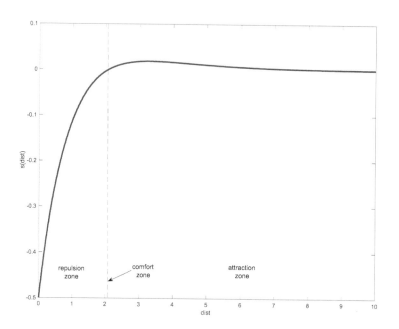

FIGURE 15.1

Function s for $l = 1.5$ and $f = 0.5$ and zones of grasshoppers' movement.

15.3   Source-code of GOA in Matlab

Listing 15.1 provides the source-code for the GOA technique in MATLAB. The code uses additional parameters to set up the algorithm's initialization. They include population size, maximum number of iterations, bounds for the search space and its dimensionality, and finally – the objective function. The location of the minimum found (denoted as X_best) along with the corresponding value of the objective function (fit_best) are being returned.

```
1
2 % P - population size
3 % k_max - max. iterations
4 % LB, UB - lower & upper bound of the search space
5 % N - search space dimensionality
```

```
 6  % fobj — objective function
 7
 8  if size(UB,1)==1
 9      UB=ones(N,1)*UB;
10      LB=ones(N,1)*LB;
11  end
12
13  % original implementation assumes even number of variables
14  % if odd number is used additional "dummy" variable is created
15  odd=false;
16  if (rem(N,2)~=0)
17      N = N+1;
18      UB = [UB;  100];
19      LB = [LB;  −100];
20      odd=true;
21  end
22
23  % initialize the population of grasshoppers
24  X=zeros(P,N);
25  for i=1:P
26      X(i,:)=[rand(N,1).*(UB−LB)+LB]';
27  endfit_X = zeros(1,P);
28
29  % default parameter values
30  cMax=1;
31  cMin=0.00004;
32
33  % calculate the fitness of initial population
34  for i=1:P
35      if odd == true
36          fit_X(1,i)=fobj(X(i,1:end−1));
37      else
38          fit_X(1,i)=fobj(X(i,:));
39      end
40  end
41
42  % sort and find the best
43  [sorted_fit_X, sorted_index]=sort(fit_X);
44      for i=1:P
45      sorted_X(i,:)=X(sorted_index(i),:);
46  end
47
48  X_best=sorted_X(1,:);
49  fit_best=sorted_fit_X(1);
50
51  % Main loop
52  iter=1;
53  while iter<k_max
54      c=cMax−iter*((cMax−cMin)/k_max);
55      for i=1:P
56          temp= X';
57          for k=1:2:N
58              S_i=zeros(2,1);
59              for j=1:P
60                  if i~=j
61                      % calculate the 2D distance between two grasshoppers
62                      Dist=pdist2(temp(k:k+1,j)', temp(k:k+1,i)');
63                      % get xj−xi/dij
64                      r_ij_vec=(temp(k:k+1,j)−temp(k:k+1,i))/(Dist+eps);
65                      % get |xjd − xid|
66                      xj_xi=2+rem(Dist,2);
67                      s_ij=((UB(k:k+1) − LB(k:k+1))*c/2)*S_func(xj_xi).*
68                      r_ij_vec;
69                      S_i=S_i+s_ij;
70                  end
71              end
72              S_i_total(k:k+1, :) = S_i;
```

```
73          end
74          X_new = c * S_i_total'+ (X_best); % new grasshopper position
75          X_temp(i,:)=X_new';
76       end
77       X=X_temp;
78       for i=1:P
79          % check for solutions out of bounds and fix
80          out_U=X(i,:)>UB';
81          out_L=X(i,:)<LB';
82          X(i,:)=(X(i,:).*(~(out_U+out_L)))+UB'.*out_U+LB'.*out_L;
83          % and get fitness
84          if odd == true
85             fit_X(1,i)=fobj(X(i,1:end-1));
86          else
87             fit_X(1,i)=fobj(X(i,:));
88          end
89          % Update the best solution (if needed)
90          if fit_X(1,i)<fit_best
91             X_best=X(i,:);
92             fit_best=fit_X(1,i);
93          end
94       end
95       iter = iter + 1;
96    end
97
98    if (odd==true)
99       X_best = X_best(1:N-1);
100   end
101
102   function o=S_func(r)
103      f=0.5;
104      l=1.5;
105      o=f*exp(-r/l)-exp(-r);
106   end
```

Listing 15.1

Source-code of the GOA technique in Matlab. This code represents only a minor modification of the original implementation by the algorithm co-author Seyedali Mirjalili.

The objective function which will be used in the following tests is known as the Sphere function. It is frequently used as an elementary unimodal benchmark function [2] and is described by the following formula:

$$f(x) = \sum_{i=1}^{N} x_i^2 \qquad \text{where } -5.12 \leqslant x_i \leqslant 5.12. \tag{15.4}$$

Its minimum $f(x^*) = 0$ is located at $x^* = [0, 0, ..., 0]^N$. It can be coded using MATLAB as described in the Listing 15.2.

```
1  function y = sphere(x)
2     y=sum(x.^2);
3  end
```

Listing 15.2

Definition of the Sphere objective function in Matlab.

15.4   Source-code of GOA in C++

Listing 15.3 provides the C++ implementation of GOA technique. For the linear algebra the Armadillo library was used [1]. It saves the user from the additional time overhead needed to implement vectorized operations. It allows us also to use automatic multi-threading.

```cpp
double Dist, c, cMax, cMin, xj_xi ;
int i, iter, j, k, odd ;
mat X, X_new, X_temp, fit_X, r_ij_vec, s_ij, sorted_X, temp ;
rowvec out_L, out_U ;
vec S_i, S_i_total, sorted_fit_X, sorted_index ;
i
f (UB.n_rows==1)
{
  UB = arma::ones<mat>(N, 1)*UB ;
  LB = arma::ones<mat>(N, 1)*LB ;
}

X = arma::zeros<mat>(P, N) ;
sorted_X = arma::zeros<mat>(P, N) ;
X_temp = arma::zeros<mat>(P, N) ;
S_i_total = arma::zeros<vec>(N) ;
c = double(arma::as_scalar(arma::zeros<rowvec>(1))) ;

for (i=1; i<=P; i++)
{
  X.row(i-1) = arma::trans(arma::randu<mat>(N, 1)%(UB-LB)+LB) ;
}

fit_X = arma::zeros<mat>(1, P) ;
cMax = double(1) ;
cMin = 0.00004 ;

for (i=1; i<=P; i++)
{
  fit_X(0, i-1) = fobj(X.row(i-1)) ;
}

[sorted_fit_X, sorted_index] = sort(fit_X) ;

for (i=1; i<=P; i++)
{
  sorted_X.row(i-1) = X.row((uword) sorted_index(i-1)-1) ;
}

X_best = sorted_X.row(0) ;
fit_best = sorted_fit_X(0) ;
iter = 1 ;
k = 1 ;
while (iter<k_max)
{
  c = cMax-iter*((cMax-cMin)/k_max) ;
  for (i=1; i<=P; i++)
  {
    temp = arma::trans(X) ;
    for (k=1; k<=N; k+=2)
    {
      S_i = arma::zeros<vec>(2) ;
      for (j=1; j<=P; j++)
      {
        if (i!=j)
```

```
57        {
58            Dist = pdist2(arma::trans(temp(arma::span(k-1, k), m2cpp::
59            span<uvec>(j-1, j-1))), arma::trans(temp(arma::span(k-1, k),
60            m2cpp::span<uvec>(i-1, i-1)))) ;
61            r_ij_vec = (temp(arma::span(k-1, k), m2cpp::span<uvec>(j-1,
62            j-1))-temp(arma::span(k-1, k), m2cpp::span<uvec>(i-1, i-1)))
63            /(Dist+datum::eps) ;
64            xj_xi = 2+rem(Dist, 2) ;
65            s_ij = ((UB(arma::span(k-1, k))-LB(arma::span(k-1, k)))*
66            c/2.0)*S_func(xj_xi)%r_ij_vec ;
67            S_i = S_i+s_ij ;
68        }
69    }
70    S_i_total(arma::span(k-1, k)-1, m2cpp::span<uvec>(0, S_i_total.
71    n_cols-1)-1) = S_i ;
72    }
73    X_new = c*S_i_total ;
74    X_temp.row(i-1) = arma::trans(X_new) ;
75 }
76
77 X = X_temp ;
78 for (i=1; i<=P; i++)
79 {
80    out_U = X.row(i-1)>arma::trans(UB) ;
81    out_L = X.row(i-1)<arma::trans(LB) ;
82    X.row(i-1) = (X.row(i-1)%(!(out_U+out_L)))+arma::trans(UB)%out_U+
       arma::trans(LB)%out_L ;
83    fit_X(0, i-1) = fobj(X.row(i-1)) ;
84 }
85
86 if (fit_X(0, i-1)<fit_best)
87 {
88    X_best = X.row(i-1) ;
89    fit_best = fit_X(0, i-1) ;
90 }
91 }
92 iter = iter+1 ;
93
94 double fobj(rowvec x)
95 {
96    double o ;
97    o = double(arma::as_scalar(arma::sum(arma::square(x)))) ;
98    return o ;
99 }
```

Listing 15.3
Source-code of the GOA technique in C++.

The following Section contains a numerical example of GOA run on a two-dimensional instance of the Sphere function minimization problem.

15.5 Step-by-step numerical example of GOA

As was already mentioned an illustrative example Grasshopper Optimization Algorithm will be used here to solve the minimization task realized for a simple two-dimensional variant of Sphere function (eq. 15.5):

$$f(x = [x_1, x_2]) = x_1^2 + x_2^2 \qquad (15.5)$$

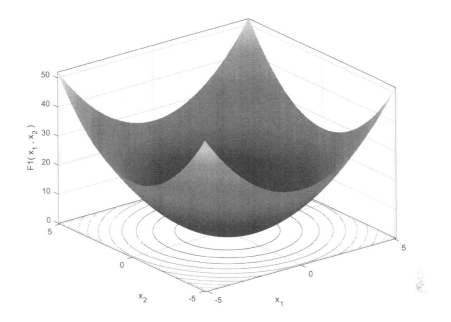

FIGURE 15.2
Two-dimensional Sphere function used in the experiments.

Function shape was demonstrated in Figure 15.2. The algorithm was executed with a population of $P = 10$ individuals. The maximum number of iterations k_{max} was set to 20. Default values of parameters $c_{max} = 1$ and $c_{min} = 0.00004$ were also employed, even though for some problems they were already demonstrated not to be the optimal ones [12].

The algorithm is starting with random dislocation of agents – obtained placement of population within the search space was demonstrated in Figure 15.3. It was obtained using a uniform random number generator with bounded output. Other methods could also be used [7].

In the subsequent iterations members of the swarm are being moved according to (15.1). The trajectory of the first grasshopper, initially located at [-4.24, -2.43] in the course of the algorithm's 20 iterations was demonstrated in Figure 15.4.

In the first iteration this swarm member is moved towards the rest of the population as was demonstrated in Table 15.1. It reports both the location of swarm members, their fitness, value of c, distance between the first grasshopper to the selected grasshopper p and contribution of each member p to the new location [0.38; 1.07] of the first grasshopper. In other words "New location" for grasshopper i gives a position of grasshopper 1 caused by the

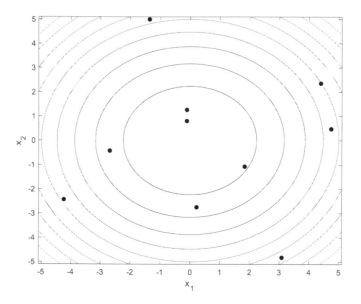

FIGURE 15.3
Initial configuration of the population.

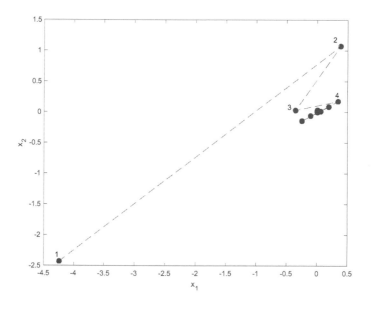

FIGURE 15.4
Trajectory of the first grasshopper.

TABLE 15.1

Location of swarm members in the first iteration and their impact on the new location of grasshopper no. 1 (with initial location [-4.24, -2.43]).

p	x_{p1}	x_{p2}	Fitness	c	$dist(x_1, x_p)$	$x_1 - x_p$	New location
1	-4.24	-2.43	23.93		0	[0,0]	—
2	3.08	-4.82	32.74		7.71	[-7.33, 2.39]	[0.08; 0.02]
3	4.39	2.36	24.84		9.87	[-8.64, -4.79]	[0.07; 0.04]
4	0.12	0.80	0.66		5.24	[-4.13, -3.23]	[0.07; 0.05]
5	-2.69	0.42	7.41	0.95	2.54	[-1.55, -2.01]	[0.04; 0.05]
6	4.74	0.48	22.72		9.45	[-8.98, -2.91]	[0.08; 0.03]
7	0.21	-2.75	7.60		4.47	[-4.46, 0.31]	[0.06; -0.01]
8	0.11	1.27	1.63		5.55	[-4.13, -3.70]	[0.07; 0.06]
9	1.83	-1.07	4.51		6.23	[-6.08, -1.36]	[0.03; 0.01]
10	-1.36	4.99	26.81		7.97	[-2.89, -7.42]	[0.03; 0.08]
					Sum:		[0.53; 0.28]

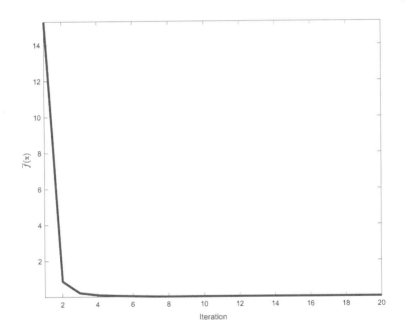

FIGURE 15.5

Mean population fitness during 20 iterations of GOA run for 2-dimensional Sphere function.

influence of grasshopper i). It is obtained by summing all entries in column "New location", scaling it by the value of c and adding the location of the best individual (grasshopper no. 4).

In the course of the algorithm the fitness of the population is rapidly improving. It is presented in Figure 15.5. The figure presents mean population fitness.

The final result of the optimization (with the fitness value of $8.38 \cdot 10^{-9}$) is very close to the global optimum – even though a very small number of individuals and algorithm's iterations were involved.

15.6 Conclusion

This chapter has studied the basic, off-the-shelf variant of the Grasshopper Optimization Algorithm. Besides the algorithm description we have provided a demonstration of its mechanics – using simple problem of minimizing Sphere function as our testbed. It can be seen that the Grasshopper Optimization Algorithm is a very effective optimizer, locating the minimum of the objective function within its first few iteration.

Finally, it is worth while to underline that the algorithm can be conveniently modified to include other factors, e.g. presence of multiple optimization criteria [8]. In addition to that it has already found many interesting applications and modifications to its standard scheme. They are covered in the forthcoming chapter [11].

References

1. Armadilo C++ library for linear algebra & scientific computing, url: http://arma.sourceforge.net/, accessed: June 1st, 2019.

2. N. Awad, M. Ali, J. Liang, B. Qu, and P. Suganthan. "Problem definitions and evaluation criteria for the cec 2017 special session and competition on single objective bound constrained real-parameter numerical optimization", Technical report, Nanyang Technological University Singapore, Nov 2016.

3. J. C. Bansal, P. K. Singh, M. Saraswat, A. Verma, S. S. Jadon, and A. Abraham. "Inertia weight strategies in particle swarm optimization", 2011 Third World Congress on Nature and Biologically Inspired Computing, Oct 2011, pp. 633-640.

4. R.F. Chapman and A. Joern. "Biology of Grasshoppers", A Wiley-Interscience publication, Wiley, 1990.

5. A. E. Hassanien and E. Emary. "Swarm Intelligence: Principles, Advances, and Applications", CRC Press, 2018.

6. S. Lukasik and P.A. Kowalski. "Study of flower pollination algorithm for continuous optimization" in P. Angelov et al. (eds Intelligent Systems'2014. Advances in Intelligent Systems and Computing, vol. 322, pp. 451-459, Springer, 2015.

7. H. Maaranen, K. Miettinen, and A. Penttinen. "On initial populations of a genetic algorithm for continuous optimization problems", Journal of Global Optimization, vol. 37(2006), no. 3, 405.

8. S. Z. Mirjalili, S. Mirjalili, S. Saremi, H. Faris, and I. Aljarah. "Grasshopper optimization algorithm for multi-objective optimization problems", Applied Intelligence vol. 48, no. 4, pp. 805-820, 2018.

9. S. Saremi, S. Mirjalili, and A. Lewis. "Grasshopper optimisation algorithm: Theory and application", Advances in Engineering Software, vol. 105, pp. 30-47, 2017.

10. X. S. Yang. "Nature-Inspired Optimization Algorithms", Elsevier, London, 2014.

11. S. Łukasik. "Grasshopper optimization algorithm - modifications and exemplary applications," Swarm Intelligence Algorithms: Modifications and Applications (A. Slowik, ed.), CRC Press, 2020.

12. S. Łukasik, P. A. Kowalski, M. Charytanowicz, and P. Kulczycki. "Data clustering with grasshopper optimization algorithm", 2017 Federated Conference on Computer Science and Information Systems (FedCSIS), pp. 71-74, 2017.

16

Grey Wolf Optimizer

Ahmed F. Ali
Department of Computer Science
Suez Canal University, Ismaillia, Egypt

Mohamed A. Tawhid
Department of Mathematics and Statistics
Faculty of Science, Thompson Rivers University, Kamloops, Canada

CONTENTS

16.1 Introduction

Grey wolf optimizer (GWO) is a population based swarm intelligence algorithm proposed by Mirjalili et al. in 2014 [4]. The GWO algorithm mimics the social dominant structure of the grey wolves pack. Due to the efficiency of the GWO algorithm, it has been applied in many works such as for CT liver segmentation [1], minimizing potential energy function [6], feature selection [2], [9], solving minimax and integer programming problems [7], solving a global optimization problem [8], solving a flow shop scheduling problem [3], solving an optimal reactive power dispatch problem [5], and casting production scheduling [10]. The rest of the paper is organized as follows. We describe the standard GWO algorithm in Section 16.2. We give and discuss the Matlab and C++ source codes of the GWO algorithm in Sections 16.3–16.4. In Section 16.5, we demonstrate a step by step numerical example of the GWO algorithm. Finally, we outline the conclusion in Section 16.6.

16.2 Original GWO algorithm

In the following subsections, we describe the main concepts of the GWO algorithm and how it works.

16.2.1 Main concepts and inspiration

Grey wolf optimizer (GWO) is a population based swarm intelligence algorithm, which mimics the dominant hierarchy of grey wolves. Grey wolves live in packs, each pack contains 5-12 members. In the pack, there are four levels of members in the dominant hierarchy as follows.

16.2.2 Social hierarchy

The leader of the group is called alpha α which can be male or female. The alpha is the highest member of the hierarchy in the pack and he/she is responsible for hunting, selecting a sleeping place and determining time to walk. The second type of wolves in the pack are the beta β, which help the alpha in their decisions. The third level in the dominant hierarchy is called delta δ which submits to alpha and beta members. The weakest members in the pack are called omega ω, which submit to the members in the prior top levels. The dominant hierarchy levels are shown in Figure 16.1.

16.2.3 Encircling prey

In this subsection, we give a mathematical model of the encircling prey process as follows.

$$\vec{D} = |\vec{C} \cdot \vec{X}_p(t) - \vec{A} \cdot \vec{X}(t)| \tag{16.1}$$

$$\vec{X}(t+1) = \vec{X}_p(t) - \vec{A} \cdot \vec{D} \tag{16.2}$$

where t is the current iteration, $\vec{A}$ and $\vec{C}$ are the coefficient vectors, $\vec{X}_p$ is the prey's position vector, and $\vec{X}$ indicates the grey wolf's position vector.

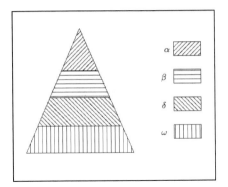

FIGURE 16.1
Social dominant hierarchy of grey wolf.

The vectors $\vec{A}$ and $\vec{C}$ are calculated as follows:

$$\vec{A} = 2\vec{a} \cdot \vec{r_1} \cdot \vec{a} \qquad (16.3)$$

$$\vec{C} = 2 \cdot \vec{r_2} \qquad (16.4)$$

where components of $\vec{a}$ are linearly decreased from 2 to 0 over the course of iterations and $\vec{r_1}$, $\vec{r_2}$ are random vectors in $[0, 1]$

16.2.4 Hunting process

The alpha guides the beta and delta in the hunting process. In the mathematical model, the alpha represents the overall best solutions, while the beta and delta represent the second and third best solutions in the population. All solutions (wolves) update their positions according to the position of the best first three solutions (alpha, beta and delta). The formula of the mathematical model of the hunting process is shown as follows.

$$\begin{aligned}
\vec{D_\alpha} &= |\vec{C_1} \cdot \vec{X_\alpha} - \vec{X}|, \\
\vec{D_\beta} &= |\vec{C_2} \cdot \vec{X_\beta} - \vec{X}|, \\
\vec{D_\delta} &= |\vec{C_3} \cdot \vec{X_\delta} - \vec{X}|,
\end{aligned} \qquad (16.5)$$

$$\begin{aligned}
\vec{X_1} &= \vec{X_\alpha} - \vec{A_1} \cdot (\vec{D_\alpha}), \\
\vec{X_2} &= \vec{X_\beta} - \vec{A_2} \cdot (\vec{D_\beta}), \\
\vec{X_3} &= \vec{X_\delta} - \vec{A_3} \cdot (\vec{D_\delta}),
\end{aligned} \qquad (16.6)$$

$$\vec{X}(t+1) = \frac{\vec{X_1} + \vec{X_2} + \vec{X_3}}{3}, \qquad (16.7)$$

where $\vec{X_1}$, $\vec{X_2}$ and $\vec{X_3}$ are the first three solutions in the population.

16.2.5   Attacking prey (exploitation)

The hunting process is finished by attacking the prey. This process can be mathematically modeled as follows. When $|A| < 1$, the wolves attack towards the prey, where the vector $\vec{A}$ is a random value in interval $[-2a, 2a]$ and a is decreased from 2 to 0 over the course of iterations.

16.2.6   Search for prey (exploration)

At the beginning of the search each wolf updates its position according to the position of α, β and δ in order to search for prey and converge to attack prey. This process is called diversification or exploration. The mathematical model of the exploration is defined as follows. When $|A| > 1$, the wolves (solutions) are forced to diverge from the prey to find a fitter prey. The vector $\vec{A}$ with random values greater than 1 or less than -1 to force the search agent to diverge from the prey.

16.2.7   Pseudo-code of GWO algorithm

Algorithm 15 Grey wolf optimizer algorithm.

1: Set the initial values of the population size n, parameter a and the maximum number of iterations Max_{itr}
2: Set $t := 0$
3: **for** $(i = 1 : i \leq n)$ **do**
4:     Generate an initial population $\vec{X_i}(t)$ randomly
5:     Evaluate the fitness function of each search agent (solution) $f(\vec{X_i})$
6: **end for**
7: Assign the values of the first, second and third best solutions $\vec{X_\alpha}$, $\vec{X_\beta}$ and $\vec{X_\delta}$, respectively
8: **repeat**
9:     **for** $(i = 1 : i \leq n)$ **do**
10:         Decrease the parameter a from 2 to 0
11:         Update the coefficients $\vec{A}$ and $\vec{C}$ as shown in Equations (16.3) and (16.4), respectively
12:         Update each search agent in the population as shown in Equations (16.5), (16.6), (16.7)
13:         Evaluate the fitness function of each search agent $f(\vec{X_i})$
14:     **end for**
15:     Update the vectors $\vec{X_\alpha}$, $\vec{X_\beta}$ and $\vec{X_\delta}$.
16:     Set $t = t + 1$
17: **until** $(t \geq Max_{itr})$.   ▷ Termination criteria are satisfied
18: Produce the best solution $\vec{X_\alpha}$

16.2.8 Description of the GWO algorithm

- **Parameters initialization.** The standard grey wolf optimizer algorithm starts by initializing the parameters of the population size n, the parameter a and the maximum number of iterations Max_{itr}.

- **Iteration initialization.** Initialize the iteration counter t.

- **Initial population generation.** The initial population n is randomly generated.

- **Solution evaluation.** In the population, each search agent (solution) $\vec{X}_i$ is evaluated by calculating its fitness function $f(\vec{X}_i)$.

- **Assign the overall best three solutions.** The overall best three solutions are assigned which are the alpha α, beta β and delta δ solution $\vec{X}_\alpha$, $\vec{X}_\beta$ and $\vec{X}_\delta$, respectively.

- **Main loop.** The following steps are repeated until the termination criterion is satisfied.

  Solution update. Each search agent (solution) in the population is updated according to the position of the α, β and δ solutions as shown in Equation (16.7).

  Parameter a update. Gradually decrease the parameter a from 2 to 0.

  Coefficients update. The coefficients $\vec{A}$ and $\vec{C}$ are updated as shown in Equations (16.3) and (16.4), respectively.

  Solution evaluation. Each search agent (solution) in the population is evaluated by calculating its fitness function $f(\vec{X}_i)$.

- **Update the overall best three solutions.** The first, second and third best solutions are updated $\vec{X}_\alpha$, $\vec{X}_\beta$ and $\vec{X}_\delta$, respectively.

- **Iteration counter increasing.** The iteration counter is increasing, where $t = t + 1$.

- **Termination criteria satisfied.** All the previous processes are repeated until termination criteria are satisfied.

- **Produce the overall best solution.** The overall best solution $\vec{X}_\alpha$ is produced.

16.3 Source-code of GWO algorithm in Matlab

In this section, we present the source code of the used fitness function which we need to minimize by using the GWO algorithm as shown in Listing 16.1.

The fitness function is shown in Equation 16.8. The function evaluates each solution in population X. Also, we present the source codes of the main GWO algorithm in Matlab [4] as shown in Listing 16.2.

$$f(X_i) = \sum_{j=1}^{D} X_{i,j}^2 \qquad \text{where } -10 \leqslant X_{i,j} \leqslant 10 \qquad (16.8)$$

```matlab
1   function [out]=fun(X)
2   [x,y]=size(X);
3   out=zeros(x,1); for i=1:x
4       for j=1:y
5       out(i,1)=out(i,1)+X(i,j)^2;
6       end
7   end
8
```

Listing 16.1
Definition of objective function $fun(.)$ in Matlab.

```matlab
1
2   % initialize alpha, beta, and delta_pos
3   Alpha_pos=zeros(1,dim);
4   Alpha_score=inf; %change this to -inf for maximization problems
5
6   Beta_pos=zeros(1,dim); Beta_score=inf; %change this to -inf for
        maximization problems
7   Delta_pos=zeros(1,dim); Delta_score=inf; %change this to -inf for
        maximization problems
8   %Initialize the positions of search agents
9   Positions=initialization(SearchAgents_no,dim,ub,lb);
10  Convergence_curve=zeros(1,Max_iter); l=0;
11  % Loop counter
12  % Main loop
13  while l<Max_iter
14  for i=1:size(Positions,1)
15  %Calculate objective function for each search agent fitness=fobj(
        Positions(i,:));
16  % Update Alpha, Beta, and Delta
17      if fitness<Alpha_score
18          Alpha_score=fitness; % Update alpha
19          Alpha_pos=Positions(i,:);
20      end
21      if fitness>Alpha_score && fitness<Beta_score Beta_score=fitness;
        % Update beta
22          Beta_pos=Positions(i,:);
23      end
24      if fitness>Alpha_score && fitness>Beta_score && fitness<
        Delta_score
25          Delta_score=fitness; % Update delta Delta_pos=Positions(i,:);
26      end
27  end
28  a=2-l*((2)/Max_iter); % a decreases linearly from 2 to 0
29  % Update the Position of search agents including omegas
30  for i=1:size(Positions,1)
31    for j=1:size(Positions,2)
32    r1=rand(); % r1 is a random number in [0,1]
33    r2=rand(); % r2 is a random number in [0,1]
34    A1=2*a*r1-a;
35    C1=2*r2;
36    D_alpha=abs(C1*Alpha_pos(j)-Positions(i,j));
37    X1=Alpha_pos(j)-A1*D_alpha;
38    r1=rand();
```

```
39    r2=rand();
40    A2=2*a*r1−a;
41    C2=2*r2;
42    D_beta=abs(C2*Beta_pos(j)−Positions(i,j));
43    X2=Beta_pos(j)−A2*D_beta;
44    r1=rand();
45    r2=rand();
46    A3=2*a*r1−a;
47    C3=2*r2;
48    D_delta=abs(C3*Delta_pos(j)−Positions(i,j));
49    X3=Delta_pos(j)−A3*D_delta;
50    Positions(i,j)=(X1+X2+X3)/3;
51    end % Return back the search agents that go beyond the boundaries
         of the search space
52    Flag4ub=Positions(i,:)>ub;
53    Flag4lb=Positions(i,:)<lb;
54    Positions(i,:)=(Positions(i,:).*(~(Flag4ub+Flag4lb)))
55                  +ub.*Flag4ub+lb.*Flag4lb;
56  end
57  l=l+1;
58  Convergence_curve(l)=Alpha_score;
59  end
```

Listing 16.2
The main code for the grey wolf optimization algorithm $GWO(.)$ in Matlab.

The rest of the matlab code is presented in
https://www.mathworks.com/matlabcentral/fileexchange/ 44974-grey-wolf-optimizer-gwo

16.4 Source-code of GWO algorithm in C++

In this section, we present the C++ code of the tested objective function and the GWO algorithm as shown in Listings 16.3–16.4.

```
1  class fun : public Problem
2  { public:
3  fun(unsigned int dimension) : Problem(dimension) { }
4  double eval(const std::vector<double>& solution)
5  {
6      double sum = 0.0;
7      for (int i = 0; i < solution.size(); ++i)
8          {
9              sum += solution[i] * solution[i];
10         }
11  return sum;
12  }
```

Listing 16.3
Definition of objective function $fun(.)$ and the main file in C++.

```
1  #include "grey_wolf_optimizer.hpp"
2  #include "benchmark_functions.hpp"
3  #include "optimization_utils.hpp"
4  #include <limits>
5  #include <iostream>
6  #include <exception>
7  #include <cmath>
```

```
 8  solution grey_wolf_optimizer(function f, calculation_type calc_type_a,
        calculation_type calc_type_c,
 9  int max_number_of_evaluations, int number_of_agents, double left_bound
        , double right_bound, int dimension)
10  {
11  std::vector<double> alpha_pos(dimension, 0.);
12  double alpha_score = std::numeric_limits<double>::infinity();
13  std::vector<double> beta_pos(dimension, 0.);
14  double beta_score = std::numeric_limits<double>::infinity();
15  std::vector<double> delta_pos(dimension, 0.);
16  double delta_score = std::numeric_limits<double>::infinity();
17  auto positions = get_initial_positions(left_bound, right_bound,
        dimension, number_of_agents);
18  solution s{};
19  int iteration{0};
20  const int max_number_of_iterations{max_number_of_evaluations /
        number_of_agents};
21  while (iteration++ < max_number_of_iterations)
22    {
23      clip_positions(positions, left_bound, right_bound);
24
25      for (auto &agent : positions)
26        {
27          double fitness = objective_function(f, agent);
28
29          if (fitness < alpha_score)
30            {
31              alpha_score = fitness;
32              alpha_pos = agent;
33            }
34
35          if (fitness > alpha_score and fitness < beta_score)
36            {
37              beta_score = fitness;
38              beta_pos = agent;
39            }
40
41          if (fitness > alpha_score and fitness > beta_score and fitness <
        delta_score)
42            {
43              delta_score = fitness;
44              delta_pos = agent;
45            }
46        }
47      double a = calculate_a(calc_type_a, iteration,
        max_number_of_iterations);
48
49      for (auto &agent : positions)
50        for (auto j = 0u; j < agent.size(); ++j)
51          {
52            // alpha
53            double r1 = get_random(0., 1.);
54            double r2 = get_random(0., 1.);
55
56            double A1 = 2. * a * r1 - a;
57            double C1 = calculate_c(calc_type_c, r2, iteration,
        max_number_of_iterations);
58            const double D_alpha = std::abs(C1 * alpha_pos[j] - agent[j
        ]);
59            const double X1 = alpha_pos[j] - A1 * D_alpha;
60
61            // beta
62            r1 = get_random(0., 1.);
63            r2 = get_random(0., 1.);
64            double A2 = 2. * a * r1 - a;
65            double C2 = calculate_c(calc_type_c, r2, iteration,
        max_number_of_iterations);
```

```
66    const double D_beta = std::abs(C2 * beta_pos[j] - agent[j]);
67    const double X2 = beta_pos[j] - A2 * D_beta;
68
69    // delta
70    r1 = get_random(0., 1.);
71    r2 = get_random(0., 1.);
72    double A3 = 2. * a * r1 - a;
73    double C3 = calculate_c(calc_type_c, r2, iteration,
  max_number_of_iterations);
74    const double D_delta = std::abs(C3 * delta_pos[j] - agent[j
  ]);
75    const double X3 = delta_pos[j] - A3 * D_delta;
76    agent[j] = (X1 + X2 + X3) / 3.;
77    }
78    s.convergence.push_back(alpha_score);
79    s.best = alpha_score;
80    }
81    return s;
82  }
```

Listing 16.4
SSA header file in C++.

The rest of the grey wolf optimization C++ code is represented in
https://github.com/czeslavo/gwo/blob/master/optimization/
   grey_wolf_optimizer.cpp

16.5 Step-by-step numerical example of GWO algorithm

In this section, we present the GWO algorithm when it applies to minimize
the objective function in Equation 16.8, at dimension $D = 5$. In step one,
the algorithm starts by setting the initial values of the population size $n = 6$,
parameter $a = 2$ and maximum number of iterations $Max_{itr} = 100$. In step
two, we initialize the iteration counter where $t = 0$. In step three, we start the
loop to create the initial population in the GWO algorithm. In step four, we
initialize the population. Each agent (solution) in the population is a vector
with five variables ($D = 5$). The population is represented as follows

$\vec{X}_1 = \{6.2945, -4.4300, 9.1433, 5.8441, 3.5747\}$
$\vec{X}_2 = \{8.1158, 0.9376, -0.2925, 9.1898, 5.1548\}$
$\vec{X}_3 = \{-7.4603, 9.1501, 6.0056, 3.1148, 4.8626\}$
$\vec{X}_4 = \{8.2675, 9.2978, -7.1623, -9.2858, -2.1555\}$
$\vec{X}_5 = \{2.6472, -6.8477, -1.5648, 6.9826, 3.1096\}$
$\vec{X}_6 = \{-8.0492, 9.4119, 8.3147, 8.6799, -6.5763\}$

In step five, we evaluate each agent (solution) in the population by calculating its objective function $OF(.)$ in Equation 16.8. For agent $X_1 =
\{6.2945, -4.4300, 9.1433, 5.8441, 3.5747\}$, the objective function of it is cal-
culated as the following $OF(\vec{X}_1) = (6.2945)^2 + (-4.4300)^2 + (9.1433)^2 +
(5.8441)^2 + (3.5747)^2 = 189.7788$. The objective function of each agent in
the population is presented as follows.

$OF(\vec{X}_1) = 189.7788$
$OF(\vec{X}_2) = 177.8568$
$OF(\vec{X}_3) = 208.7953$
$OF(\vec{X}_4) = 296.9700$
$OF(\vec{X}_5) = 114.7735$
$OF(\vec{X}_6) = 341.0943.$

In step seven, the overall best three solutions are assigned according to their objective functions. The overall best solution is $\vec{X}_\alpha$, where $\vec{X}_\alpha = \{2.6472, -6.8477, -1.5648, 6.9826, 3.1096\}$. The second overall best solution is $\vec{X}_\beta$, where $\vec{X}_\beta = \{8.1158, 0.9376, -0.2925, 9.1898, 5.1548\}$.

The third overall best solution in the population is the $\vec{X}_\delta$, where $\vec{X}_\delta = \{6.2945, -4.4300, 9.1433, 5.8441, 3.5747\}$.

In step eight, we start the main loop of the GWO algorithm. In steps nine and ten, for each agent in the population, the coefficient a is initialized, where $a = 2$. In step eleven, for the first three solutions in the population, the coefficients $\vec{A}$ and $\vec{C}$ are updated as shown in Equations 16.3–16.4. The values of vectors $\vec{A}$ and $\vec{C}$ for the first three agents (solutions) in the population are presented as follows.

$\vec{A}_1 = \{0.8242, 0.7793, 1.0621, 0.8375, -1.5240\}$
$\vec{C}_1 = \{0.0637, 0.6342, 1.5904, 1.5094, 0.9967\}$
$\vec{A}_2 = \{-0.8923, 1.8009, -1.2525, -0.8959, 1.8390\}$
$\vec{C}_2 = \{0.0923, 0.0689, 0.9795, 1.3594, 0.6808\}$
$\vec{A}_3 = \{-1.6115, -0.2450, -0.2177, 0.6204, 0.3411\}$
$\vec{C}_3 = \{1.6469, 0.7631, 1.2926, 0.3252, 0.4476\}$

In step twelve, we update the solutions in the population as shown in Equations 16.5–16.7 based on the updating of the first three solutions in the population as shown below

$\vec{D}_\alpha = |\{0.0637, 0.6342, 1.5904, 1.5094, 0.9967\}$
$\quad \cdot \{2.6472, -6.8477, -1.5648, 6.9826, 3.1096\}$
$\quad -\{6.2945, -4.4300, 9.1433, 5.8441, 3.5747\}|$
$\quad = \{6.1259, 0.0872, 11.6320, 4.6952, 0.4753\}.$
$\vec{D}_\beta = |\{0.0923, 0.0689, 0.9795, 1.3594, 0.6808\}$
$\quad \cdot \{8.1158, 0.9376, -0.2925, 9.1898, 5.1548\}$
$\quad -\{8.1158, 0.9376, -0.2925, 9.1898, 5.1548\}|$
$\quad = \{6.9834, 5.0604, 3.2607, 1.6099, 0.2644\}.$
$\vec{D}_\delta = |\{1.6469, 0.7631, 1.2926, 0.3252, 0.4476\}$
$\quad \cdot \{6.2945, -4.4300, 9.1433, 5.8441, 3.5747\}$
$\quad -\{-7.4603, 9.1501, 6.0056, 3.1148, 4.8626\}|$
$\quad = \{7.3214, 11.5253, 18.4015, 8.8641, 4.5395\}.$
$\vec{X}_1 = \{2.6472, -6.8477, -1.5648, 6.9826, 3.1096\}$
$\quad -\{0.8242, 0.7793, 1.0621, 0.8375, -1.5240\}$
$\quad = \{-2.4017, -6.9157, -13.9187, 3.0506, 3.8339\}.$

$\vec{X_2} = \{8.1158, 0.9376, -0.2925, 9.1898, 5.1548\}$
    $-\{-0.8923, 1.8009, -1.2525, -0.8959, 1.8390\}$
    $\cdot\{6.9834, 5.0604, 3.2607, 1.6099, 0.2644\}$
    $= \{-1.2289, 0.0369, 10.0896, 4.5571, 4.3765\}$
$\vec{X_3} = \{6.2945, -4.4300, 9.1433, 5.8441, 3.5747\}$
    $-\{-1.6115, -0.2450, -0.2177, 0.6204, 0.3411\}$
    $\cdot\{7.3214, 11.5253, 18.4015, 8.8641, 4.5395\}$
    $= \{20.0658, 12.1217, -3.1571, -14.7850, -3.7038\}.$

The new solutions in the population are generated as shown in Equation 16.7 and we obtain the following new $\vec{X_1} = \{5.4784, 1.7476, -2.3287, -2.3924, 1.5022\}$. In step thirteen, the objective function of each solution in the new population is calculated and we obtain the following values.

$OF(\vec{X_1}) = 61.2648$
$OF(\vec{X_2}) = 167.8913$
$OF(\vec{X_3}) = 54.6693$
$OF(\vec{X_4}) = 173.1462$
$OF(\vec{X_5}) = 24.3717$
$OF(\vec{X_6}) = 216.7138$

In step fifteen, the vectors $\vec{X_\alpha}, \vec{X_\beta}, \vec{X_\delta}$ are updated to be as the following.

$\vec{X_\alpha} = \{-3.4887, 0.0886, 0.0813, 2.1271, -2.7680\}$
$\vec{X_\beta} = \{4.4180, -3.7046, -4.2665, 1.5438, -0.9172\}$
$\vec{X_\delta} = \{-0.2841, -6.4442, -0.7633, -2.0841, 3.8379\}$

In step sixteen, the iteration counter is increased to be $t = 1$. In step seventeen, the termination criterion is tested and if it not satisfied, we return to step nine; otherwise the algorithm produces the overall best solution $\vec{X_\alpha}$ as shown in step eighteen.

16.6 Conclusion

In this chapter, we present the main steps of the GWO algorithm and how it works. We present the source code in Matlab and C++ language to help the user to implement it on various applications. In order to give a better understanding of the GWO algorithm, we demonstrate a step by step numerical example and we show how it can solve the global optimization problem.

References

1. A.F. Ali, A. Mostafa, G.I, Sayed, M.A. Elfattah and A.E. Hassanien. Nature inspired optimization algorithms for ct liver segmentation. In: Dey N., Bhateja V., Hassanien A. (eds) Medical Imaging in

Clinical Applications. Studies in Computational Intelligence, vol. 651, pp. 431-460, Springer, 2016.

2. E.Emary, H.M. Zawbaa and A.E. Hassanien. Binary grey wolf optimization approaches for feature selection. Neurocomputing, 172, 371-381, 2016.

3. G. M. Komaki and V. Kayvanfar. Grey Wolf Optimizer algorithm for the two-stage assembly flow shop scheduling problem with release time. Journal of Computational Science, 8, 109-120, 2015.

4. S. Mirjalili, S. M. Mirjalili, and A. Lewis, Grey Wolf Optimizer, Advances in Engineering Software, 69 (2014), 46–61.

5. M. H. Sulaiman, Z. Mustaffa, M. R. Mohamed, and O. Aliman. Using the gray wolf optimizer for solving optimal reactive power dispatch problem. Applied Soft Computing, 32, 286-292, 2015.

6. M.A. Tawhid and A.F. Ali. A hybrid grey wolf optimizer and genetic algorithm for minimizing potential energy function. Memetic Computing, 9(4), 347-359, 2017.

7. M.A. Tawhid and A.F. Ali. A simplex grey wolf optimizer for solving integer programming and minimax problems. Numerical Algebra,Control & Optimization, 7(3), 301-323, 2017.

8. M.A. Tawhid and A.F. Ali. Multidirectional grey wolf optimizer algorithm for solving global optimization problems. International Journal of Computational Intelligence and Applications, 1850022, 2018.

9. Q. Tu, X. Chen, and X. Liu. Multi-strategy ensemble grey wolf optimizer and its application to feature selection. Applied Soft Computing, 76, 16-30, 2019.

10. H. Qin, P. Fan, H. Tang, P. Huang, B. Fang and S. Pan. An effective hybrid discrete grey wolf optimizer for the casting production scheduling problem with multi-objective and multi-constraint. Computers & Industrial Engineering, 128, 458-476, 2019.

17

Hunting Search Algorithm

Ferhat Erdal
Department of Civil Engineering
Akdeniz University, Turkey

Osman Tunca
Department of Civil Engineering
Karamanoglu Mehmetbey University, Turkey

CONTENTS

17.1   Introduction

Firstly proposed by Oftadeh et al. in 2009 [1], the Hunting Search (HuS) technique is inspired by the behavior of animals hunting in a group. HuS has the distinguishing features of algorithm simplicity and search efficiency. During recent years, it has been successfully used in areas such as function optimization [1], simulated manufacturing condition improvement [2], hybrid PV-battery storage system optimization [3], a quadratic assignment problem [4], optimization of a traveling salesman problem [2] and steel and composite beam design [5, 6]. In this chapter, we first introduce the underlying inspiration and principles of the basic HuS method. The rest of this chapter is organized as follows: in Section 17.2.1 the pseudo code for the HuS algorithm is presented and discussed, in Section 17.3 the source-code for the HuS algorithm in Matlab is shown with more detail. Then, the pseudo-code of the HuS algorithm in

C++ programing language is demonstrated. In Sections 17.5 and 17.6, the numerical example of the HuS algorithm and some conclusions are provided, respectively.

17.2   Original HuS algorithm

In this section, the mentioned algorithm will be discussed with more detail. Compared to hunting process, in an optimization problem each 'hunter' is replaced with a 'solution' of the design problem. Note that group hunting and our stochastic search algorithm have a primary difference. In group hunting of lions, wolves, and dolphins, hunters can see the prey or when they hunt at night at least they can sense the smell of the prey and determine its position. In contrast, in optimization problems we have no indication of the optimum solution. In group hunting of animals, however, the solution (prey) is dynamic and the hunters (based on the current position of the prey) must correct their position. In optimization problems instead, the optimum solution is static and does not change its position during the search process. In fact, both real and artificial group hunting have their own difficulties. To resemble the dynamics of the hunting process in our algorithm, artificial hunters move towards the leader. The leader is the hunter which has the optimum solution among current solutions at hand. In fact, we assume that the leader has found the optimum point and other members move towards it. If any of them finds a point better than the current leader, it becomes leader in the next stage. Hunters not only gradually move toward the prey but also correct their position. Therefore, in this algorithm, after moving toward the previous leader, the hunters correct their position based on the position of other members. This is accomplished by introducing the 'hunting group consideration rate' (HGCR). In addition in the group hunting of real animals, if the prey escapes out of the ring, the hunters organize themselves to encircle the prey again. In the HuS algorithm, the ability will be given to hunters, so they can search out of the ring of siege. In the algorithm, if the positions of the hunters/solutions are too close to each other, the group is reorganized to find the optimum point in the next effort.

17.2.1   Pseudo-code and description of HuS algorithm

The pseudo-code for the global version of the HuS algorithm is presented in Algorithm 16, the flow chart of HuS algorithm is shown in Figure 17.1.

Algorithm 16 Pseudo code of original HuS.

1: Initialize optimization problem and parameters;
   HGS: number of hunters- hunting group size
   MML: maximum movement toward leader
   $HGCR$: hunting group consideration rate
   EN: number of epochs
   Ra: distance radius
   α and β reorganization parameters
2: Initialize hunting group - generate random population of HGS solutions;
3: Calculate the fitness values of initial members: $fitness(i) = f(x_i)$;
4: Set leader as the best fitness of all hunters
5: **while** (the termination conditions are not met) **do**
6:     **for** each hunter i **do**
7:         Change the position- move toward leader in view of MML
8:         Calculate $fitness(i)$
9:         **if** $fitness(i)$ is better than leader; leader $= fitness(i)$ **then**
10:        **end if**
11:    **end for**
12:    **for** each hunter i **do**;
13:        Correct position on the basis of $HGCR$ and local search
14:        Calculate $fitness(i)$
15:        **if** $fitness(i)$ is better than leader; leader = fitness(i) **then**
16:        **end if**
17:    **end for**
18:    **for** each hunter i **do**
19:        Update the position- reorganize the hunting group
20:        Calculate $fitness(i)$
21:        **if** $fitness(i)$ is better than leader; leader $= fitness(i)$ **then**
22:        **end if**
23:    **end for**
24: **end while**

Step 1. Initialize the design problem and algorithm parameters.
In this step, HuS algorithm parameters that are required to solve the optimization problem are also specified: number of solution vectors in hunting group (HGS), maximum movement toward the leader (MML), and hunting group consideration rate (HGCR), which varies between 0 and 1. The parameters MML and HGCR are parameters that are used to improvise the hunter solution vector.

Step 2. Initialize the hunting group.
Based on the solution vectors in the hunting group (HGS), the hunting group matrix is filled with feasible solution vectors. The values of objective function are computed and the leader is defined based on the values of objective functions of the hunters.

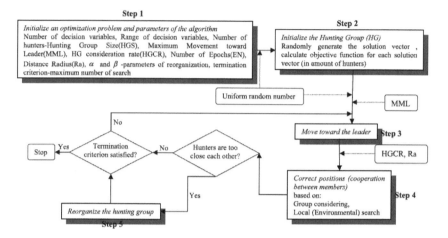

FIGURE 17.1
Displays the procedure of the Hunting Search algorithm, which consists of the steps discussed in the text.

Step 3. Moving toward the leader.
The new solution vectors $x' = (x'_1, x'_2, x'_3, ..., x'_N)$ are generated by moving toward the hunter that has the best position in the group as follows:

$$x'_i = x_i + rand \times MML \times (x_i^L - x_i) \qquad (17.1)$$

The MML is the maximum movement toward the leader, rand is a uniform random number which varies between 0 and 1, and x_i^L is the position value of the leader for the ith variable.
If the movement toward the leader is successful, the hunter stays in its new position. However, if each hunter's previous position is better than its new position it comes back to the previous position. This will give two main advantages. First, we do not compare the hunter with the worst hunter in the group, so we allow the weak members to search for other solutions; they may find better solutions. Secondly, for prevention of rapid convergence of the group the hunter compares its current position with its previous position; this means that good positions will not be eliminated in the cycle.
The value of parameter MML depends on the number of iterations in each epoch. According to different searches, the range between 0.05 (for epochs with large number of iterations) and 0.4 (for epochs with small number of iterations) gives the best results.
Step 4. Position correction-cooperation between members.
The cooperation among the hunters is required to be modeled in order to conduct the hunt more efficiently. After moving toward the leader, hunters choose another position to find better solutions. Hunters correct their position either following *"real value correction"* or *"digital value correction"*. In real value cor-

rection, the new hunter's position $x' = x'_1, x'_2, ..., x'_n$ is generated from HG, based on hunting group considerations or position corrections. For instance, the value of the first design variable for the jth hunter $x_i^{j'}$ for the new vector, can be selected as a real number from the specified $HG(x_i^1, x_i^2, ..., x_i^{HGS})$ or corrected using $HGCR$ parameter (chosen between 0 and 1). The variable is updated as follows:

$$x_i^{j'} \leftarrow \begin{cases} x_i^{j'} \in \{x_i^1, x_i^2, i = 1, ..., x_i^{HGS}\} & i = 1, ..., N \\ x_i^{j'} = x_i^j \pm Ra \text{ with probability } (1 - HGCR) & j = 1, ..., HGS \end{cases}$$

(17.2)

The parameter $HGCR$ is the probability of choosing one value from the hunting group stored in the HG. It is reported that selecting values between 0.1 and 0.4 produces better result. Ra is an arbitrary distance radius for the continuous design variable. It can be fixed or reduced during the optimization process. Several functions can be selected for reducing Ra. Eq. 17.3.

$$Ra(it) = Ra_{min}(x_i^{max} - x_i^{min}) \exp \left(\frac{\ln \left(\frac{Ra_{max}}{Ra_{min}} \right) \times it}{itm} \right)$$

(17.3)

where, it is the iteration number. x_i^{max} and x_i^{min} are maximum and minimum possible values for x_i. Ra_{max} and Ra_{min} are the maximum and minimum of relative search radius of the hunter, respectively, and itm is the maximum number of iterations in the optimization process.

In digital value correction, instead of using real values of each variable, hunters communicate with each other by the digits of each solution variable. For example, the solution variable with the value of 23.4356 has six meaningful digits. For this solution variable, the hunters choose a value for the first digits *(i.e.2)* based on hunting group consideration or position correction. After the quality of the new hunter position is determined by evaluating the objective function, the hunter moves to this new position; otherwise it keeps its previous position.

Step 5. Reorganizing the hunting group

In order to prevent being trapped in a local optimum, they must reorganize themselves to get another opportunity to find the optimum point. The algorithm does this in two independent conditions. If the difference between the values of objective function for the leader and the worst hunter in the group becomes smaller than a present constant ε_1 and the termination criterion is not satisfied, then the algorithm reorganizes the hunting group for each hunter. Alternatively, after a certain number of searches, the hunters reorganize themselves. The reorganising is carried out as follows: the leader keeps its position and the other hunters randomly choose their position in the design space

$$x'_i = x_i^L \pm rand \times (max(x_i) - min(x_i)) \times \alpha \exp(-\beta \times EN)$$

(17.4)

where x_i' is the position value of the leader for the ith variable, *rand* is a uniform random number between [0,1]. x_i^{max} and x_i^{min} are the maximum and minimum possible values of variable x_i, respectively. *EN* counts the number of times that the group has been trapped until this step. As the algorithm goes on, the solution gradually converges to the optimum point. Parameters α and β are positive real values.

Step 6. Repeat Steps 3–5 until the termination criterion is satisfied.
In this step, the computations are terminated when the termination criterion is satisfied. If not, Step 3 and Step 5 are then repeated. The termination criterion can be defined as the maximum number of searches. Alternatively, if after reorganizing the function for the leader and the worst hunter in the group remains smaller than a preset constant (ϵ), the search process ends.

17.3 Source code of HuS algorithm in Matlab

The Matlab source code for objective function which will optimize by HuS algorithm is shown below with detail.

```
1  % initialize the optimization problem and algorithm parameters
      [such as hunting group size (HGS), maximum movement toward the
      leader (MML), and hunting group consideration rate (HGCR)...
2  HGS=input(''), MML=input(''), HGCR=input(''),    Hmin=input(''),
      Hmax=input('');
3  % initialize the hunting group (HG) based on the number of
      hunters (HGS).
4  HG=zeros(N,HGS)+Hmin(1,:)+rand*(Hmax(1,:)-Hmin(1,:))
5  % starting main algorithm loop
6  while (iter<maxiteration);
7  % number of iteration is increasing by one
8  iter=iter+1;
9  % evaluating the hunters by using objective function and
      specifying the leader (HGL).
10 Eval=OF(HG);
11 A=[Eval,HG];
12 B=sort(A,1,'descend');
13 HGL=B(2,:);
14 % moving toward the leader.
15 HG=HG(i,j)+rand*MML*(HGL(1,j)-HG(i.j));
16 % position correction-cooperation between members.
17 for i=1:HGS;
18   if rand(0,1)<HGCR;
19   d=int(HGS*rand);
20   for j=0:N;
21   HG(i,j)=HG(d,j);
22   end
23 else
24   for j=0:N;
25   HG(i,j)=HG(i,j)+2*(rand-0.5))*Ramin*(max(HG(i))-min(HG(i)))*exp
      (log(Ramax/Ramin)*iter/maxiter);
```

```
26    end
27  end
28  end
29  % reorganizing the hunting group.
30  tt=iter/HGS*epochiter
31  for i=0:N;
32      HG(i,j)=best(j)+ 2*(rand-0.5))*(max(HG(i))-min(HG(i))*alfa*exp
            (-betha*tt);
33  end
34  % calculate the new fitness
35  Eval=OF(HG);
36  A=[Eval,HG];
37  B=sort(A,1,'descend');
38  NewHGL=B(2,:);
39  % if NewHGL is better than leader, NewHGL is leader
40  if NewHGL>HGL;
41  HGL=NewHGL
42  end %if
43  end %for
44  end %while
45  % the result of HuS algorithm
46  disp('HGL')
```

Listing 17.1
Source code of global version of the HuS in Matlab.

17.4   Source code of HuS algorithm in C++

The C++ source code for the objective function which will optimize by HS
algorithm is shown in below with detail.

```
1  //initialize the optimization problem and algorithm parameters [
          such as hunting group size (HGS), maximum movement toward the
          leader (MML), and hunting group consideration rate (HGCR)...
2  int main()
3  {
4      int HGS, MML, HGCR, Hmin, Hmax;
5      cout << "Enter HGS, MML, HGCR, Hmin, Hmax:";
6      cin >>HGS >>MML >>HGCR >>Hmin >>Hmax;
7  }
8  //initialize the hunting group (HG) based on the number of
          hunters (HGS).
9  {\bf float} HG[N][HGS];    {\bf float} Eval[N];
10 {\bf float} HGL[N];        {\bf float} EvalHGLbest[N];
11 HG[i][j]= Hmin(j)+r()*(Hmax(j)-Hmin(j));
12 HGL[i][j]=HG[i][j];
13 //evaluating the hunters by using objective function and
          specifying the leader (HGL).
14 Eval[i]=OF(HG[i],HGS);
15 EvalHGLbest=Eval(i);
16 If (Eval[i]<Eval[Best]) Best=i;
17 }
18 //starting main loop
19 while (iter<maxiteration)
```

```
20 {
21 //number of iterations is increasing by one
22 iter=iter+1;
23 // moving toward the leader.
24 HG [i][j]= HG [i][j]+r()*MML*(HGL[j]−HG[j]);
25 //position correction−cooperation between members.
26 for (i=1,i<HGS,i++)
27 {
28 if (r()<HGCR)
29 {
30 d=int(HGS*rand);
31 for (j=0,j<N,j++);
32 HG [i][j]=HG[d][j];
33 end
34 }
35 else
36 {
37 for (j=0,j<N,j++);
38 HG[i][j]= HG[i][j]+2*(r()−0.5))*Ramin*(max(HG[i])−min(HG[i]))*exp
      (log(Ramax/Ramin)*iter/maxiter);
39 }
40 }
41 //reorganizing the hunting group.
42 tt=iter/HGS*epochiter;
43 for (i=0, i<N, i++)
44 {
45   HG[i][j]=best[j]+ 2*(r()−0.5))*(HGmax[i]−HGmin[i])*alfa*exp(−
      betha*tt);
46 //calculate the fitness
47 Eval[i]=OF(HG[i],HGS);
48 NewEvalHGLbest=Eval(i);
49 If (Eval[i]<Eval[Best]) Best=i;
50 {
51 //if NewEvalHGL is better than leader, NewEvalHGL is leader
52 if NewEvalHGL>EvalHGL
53 {
54 EvalHGL=NewEvalHGL;
55 }
56 }
57 }
58 //the result of HuS algorithm
59 count<<EvalHGL;
```

Listing 17.2
Source code of global version of the HuS in C++.

17.5   Elaboration on HuS algorithm with constrained minimization problem

The hunting search algorithm explained in the previous subsections is used to determine the optimum solutions of the Kuhn-Tucker problem. This benchmark problem is a minimization with two design variables (x_1 and x_2) and one inequality constraint.

The objective function of the minimization problem is

$$f(x) = 5x_1^2 - 9x_1x_2 + 5x_2^2$$

which is subjected to

$$g(x) = 25 - 16x_1x_2 \leq 0.$$

The variables can be selected from following set of values

$$\{0.5, 1.0, 1.5, 2.0, 2.5, 3.0, \ldots, 10.0\}.$$

The constrained minimum solution is located in a narrow crescent-shaped region. The parameterization of the technique is conducted in Table 17.1 (Step 1) with its recommended settings in preceding studies, as well as extensive numerical experimentations conducted in the present study [1].

TABLE 17.1
HuS parameter values for constrained minimization problem.

Parameters	Values
Number of epochs (NE)	2
Iteration per epoch (IE)	30
Hunting Group Size (HGS)	10
Maximum movement toward to leader (MML)	0.3
Hunting group consideration rate (HGCR)	0.3
Ra_max, Ra_min	1e-2, 1e-7
Reorganization parameters-(α, β)	0.1, -1

The HG was initially structured with randomly generated solution vectors within the bounds prescribed for this example (i.e., 0 to 6.0) and the leader is defined (Step 2). After 10 searches the initial hunting group matrix is obtained as given in Table 17.2. The solution vectors are sorted according to the values of the objective function. Next, based on Eq. (17.1), the hunters move toward the leader, and if their new positions are better than the previous positions, they stay there. Otherwise, they come back to their previous positions (Step 3). The 11th, 12th and 13th iteration cannot find a better solution than the ones shown in Table 17.2. The 14th search gives a better hunting group matrix as shown in Table 17.3. A new hunting solution vector was improvised based on followed rules: group considerations with a 30.0% probability and position corrections with a 70.0% probability. As the objective function value of the new

TABLE 17.2

After 10 searches the initial hunting group matrix.

Row Number	x_1	x_2	$f(x)$
1	3.0	4.5	24.75
2	2.0	4.0	28.00
3	7.5	6.0	56.25
4	4.5	7.0	62.75
5	5.5	8.0	75.25
6	8.5	5.5	91.75
7	10.0	9.0	95.00
8	10.0	10.0	100.00
9	5.5	1.0	106.75
10	1.0	9.0	329.00

TABLE 17.3

After 14 searches the hunting group matrix.

Row Number	x_1	x_2	$f(x)$
1	3.0	4.5	24.75
2	2.0	4.0	28.00
3	7.5	6.0	56.25
4	4.5	7.0	62.75
5	5.5	8.0	75.25
6	8.5	5.5	91.75
7	10.0	9.0	95.00
8	6.0	9.0	99.00
9	10.0	10.0	100.00
10	5.5	1.0	106.75

harmony is 99.00, the new harmony is included in the hunting group matrix and the worst candidate is excluded from the matrix, as shown in Table 17.3 (Subsequent HM). After 20 iterations based on Eq. (17.4), the hunters reorganize themselves and improvise an optimal solution (Step 5). After 50 iterations (1000 function evaluations), the HuS algorithm still has the same optimal solution vector $x = (1.5; 1.5)$, which has a function value of 2.25, as shown in Table 17.4 and Figure 17.2. It is noticed that these design vectors remained the same even though the design cycles were continued, reaching 1000, which was the pre-selected maximum number of iterations. Consequently, the technique can be recommended for its application to optimize engineering design problems. It is interesting to see the function we have optimized and we give both a contour plot in Figure 17.2 and three-dimensional plot in Figure 17.3.

TABLE 17.4

After 20 iterations the hunting group matrix.

Row Number	x_1	x_2	$f(x)$
1	1.5	1.5	2.25
2	2.0	1.5	4.25
3	2.5	1.5	6.25
4	3.0	2.5	8.75
5	3.0	3.0	9.00
6	2.0	3.0	11.00
7	3.0	3.5	11.75
8	3.5	3.5	12.25
9	2.5	3.5	13.75
10	3.5	4.0	15.25

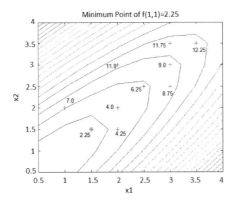

FIGURE 17.2

The contour plot of the Kuhn-Tucker function.

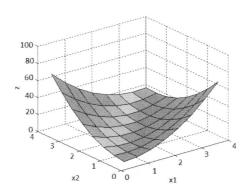

FIGURE 17.3

The three dimensional plot of the Kuhn-Tucker function.

17.6   Conclusion

In this study, besides the basic principles of the hunting search algorithm, operation principles of the algorithm have been explained. Then, source codes of the mentioned algorithm have been provided in Matlab and C++ mathematical programming languages. In the last part of the chapter, numerical example of the hunting search algorithm has been presented with more detail. We strongly claim that this chapter will make the implementation of one's own HuS technique in any programming easier for researchers to achieve.

References

1. R. Oftadeh, M.J. Mahjoob. "A new meta-heuristic optimization algorithm: Hunting Search". *Fifth International Conference on Soft Computing, Computing with Words and Perceptions in System Analysis, Decision and Control*, pp. 1-5, 2009.

2. R. Oftadeh, M.J. Mahjoob, M. Shariatpanahi. "A novel meta-heuristic optimization algorithm inspired by group hunting of animals: Hunting search". *Computers and Mathematics with Applications*, vol. 60(7), pp. 2087-2098, 2010.

3. A. Agharghor, M.E. Riffi. "First Adaptation of Hunting Search Algorithm for the Quadratic Assignment Problem". *Europe and MENA Cooperation Advances in Information and Communication Technologies*, pp. 263-267, Springer, 2017.

4. B. Naderi, M. Khalili, A.A. Khamseh. "Mathematical models and a hunting search algorithm for the no-wait flowshop scheduling with parallel machines". *International Journal of Production Research*, vol. 52(9), pp. 2667-2681, 2014.

5. O. Tunca, Erdal F., E. Dogan. "Optimum design of composite corrugated web beams using hunting search algorithm". *International Journal Of Engineering & Applied Sciences*, vol. 9(2), pp. 156-168, 2017.

6. F. Erdal, E. Doğan, O. Tunca, S. Taş. "Optimum Design of Corrugated Web Beams Using Stochastic Search Techniques". *Third International Conference on Advances in Civil, Structural and Environmental Engineering- ACSEE 2015*, pp. 121-125, 2015.

18

Krill Herd Algorithm

Ali R. Kashani
Department of Civil Engineering
University of Memphis, Memphis, Tennessee, United States

Charles V. Camp
Department of Civil Engineering
University of Memphis, Memphis, Tennessee, United States

Hamed Tohidi
Department of Civil Engineering
University of Memphis, Memphis, Tennessee, United States

Adam Slowik
Department of Electronics and Computer Science
Koszalin University of Technology, Koszalin, Poland

CONTENTS

18.1   Introduction

The krill herd (KH) algorithm is developed by mimicking the social behavior of krill in a herd [1]. This algorithm belongs to the swarm-intelligence algorithms category. Krill individuals tend to search for food in the form of large schools based on two major motivations: (1) increasing krill density, (2) reaching food. Therefore, the global optimum solution can be defined as a position with a shorter distance from the source of food. Obviously, such a point has a higher density of krill swarm. The KH algorithm considers krill individuals as potential solutions for a specific problem. Therefore, the krill individuals explore the solution space by changing their position during the time. Fundamentally, KH modifies the position of the krill based on two kinds of modifications: Lagrangian movements and evolutionary operations. Lagrangian movements are governed by three basic rules: (1) movements induced by other krill; (2) foraging activity; and (3) random diffusion. Evolutionary operators are crossover and mutation which modifies the solutions before moving to the new positions. As of now, the KH algorithm has received much attentions which has resulted in several variations and improvements in this algorithm such as chaotic KH [2], stud KH [3], fuzzy KH [4], oppositional KH [5], binary KH [6], improved KH [7-9]. The remainder of this chapter is organized accordingly. In Section 18.2 a step-by-step pseudo-code for the original KH algorithm is provided, in Sections 18.3 and 18.4 source-codes of the KH algorithm in both Matlab and C++ are presented, respectively, in Section 18.5 a detailed numerical example is solved by KH, and finally a conclusion is presented in Section 18.6.

18.2   Original KH algorithm

18.2.1   Pseudo-code of the original version of KH algorithm

The pseudo-code for the KH algorithm is depicted in Algorithm 17.

Algorithm 17 Pseudo-code of the original KH.

1: determine the $D - th$ dimensional objective function $OF(.)$
2: determine the range of variability for each $j - th$ dimension $\left[K_{i,j}^{min}, K_{i,j}^{max}\right]$
3: determine the KH algorithm parameter values such as NK – number of krill, MI – maximum iteration, V_f – foraging speed, D_{max} – maximum diffusion, N_{max} – maximum induced speed
4: randomly create swarm P which consists of NK krill individuals (each krill individual is a D-dimensional vector)
5: finding the best krill and its relevant vector
6: $Iter = 0$
7: **while** termination condition not met (here is reaching MI) **do**
8:     evaluate $X^{food} = \frac{\sum_{i=1}^{NK} \frac{1}{K_i} \cdot X_i}{\sum_{i=1}^{NK} \frac{1}{K_i}}$
9:     **for** each i-th krill in swarm P **do**
10:         $\alpha_i^{target} = 2 \cdot \left(rand + \frac{Iter}{MI}\right) \cdot \widehat{K}_{i,best} \cdot \widehat{X}_{i,best}$
11:         $R_{z,i} = \sum_{j=1}^{N} \|X_i - X_j\|$ and $d_{z,i} = \frac{1}{5 \cdot N} \cdot R_{z,i}$
12:         **if** $R_{z,i} < d_{z,i}$ and $K(i) \neq K(n)$ **then**
13:             $\alpha_i^{local} = \sum_{j=1}^{NN} \frac{K_i - K_j}{K^{worst} - K^{best}} \times \frac{X_j - X_i}{\|X_j - X_i\| + \epsilon}$

14:        **end if**

15:        $\omega = 0.1 + 0.8 \times \left(1 - \frac{1}{MI}\right)$

16:        $N_i^{new} = N^{max} \cdot \left(\alpha_i^{target} + \alpha_i^{local}\right) + \omega \cdot N_i^{old}$

17:        $\beta_i^{food} = 2 \cdot \left(1 - \frac{Iter}{MI}\right) \cdot \widehat{K}_{i,food} \cdot \widehat{X}_{i,food}$

18:        $\beta_i^{best} = \widehat{K}_{i,best} \cdot \widehat{X}_{i,best}$

19:        $F_i^{new} = V_f \cdot \left(\beta_i^{food} + \beta_i^{best}\right) + \omega \cdot F_i^{old}$

20:        $D_i = D^{max} \cdot \left(1 - \frac{Iter}{MI}\right) \cdot \delta$

21:        $\frac{dX_i}{dt} = N_i^{new} + F_i^{new} + D_i$

22:        $X_i = crossover\left(X_i, X_c\right)$

23:        $X_i = mutation\left(X_i, X_{best}, M_u\right)$

24:        $X_i^{new} = X_i + \frac{dX_i}{dt}$

25:        **end for**

26:        update best-found solution

27:        $Iter = Iter + 1$

28: **end while**

29: post-processing the results

18.2.2   Description of the original version of KH algorithm

Referring to Algorithm 17, KH algorithm handles a D-dimensional objective function $OF(.)$. Therefore, after defining the objective function, the valid boundary domains for the j-th variable of the i-th krill is defined as $\left[K_{i,j}^{min}, K_{i,j}^{max}\right]$ in the second step. The third step is devoted to the parameter setting of the KH algorithm such as a number of krill (NK), the maximum number of iteration (MI), the foraging speed (V_f), the maximum diffusion speed (D^{max}), the maximum induced speed (N_{max}). The original paper's recommendation for D^{max} is $[0.002, 0.010]$ (ms^{-1}), and for V_f and N_{max} are 0.02 and 0.01 (ms^{-1}), respectively. KH starts with a randomly produced initial population of NK number of krill. Each krill is represented by a D-dimensional vector as follows:

$$K_i = \{k_{i,1}, k_{i,2}, ..., k_{i,D}\} \tag{18.1}$$

where $k_{i,1}$ to $k_{i,D}$ are decision variables varying between $K_{i,j}^{min}$ and $K_{i,j}^{max}$.

In Step 5, the best-found solution is detected. The main loop of the KH algorithm is started in the next step. Within this loop, the location of the virtual food is calculated. In the original paper [1], a method based on the distribution of the krill individuals' fitness proposed to find the center of the food. As mentioned earlier, KH uses the following time-dependent Lagrangian model to simulate krill movements in a D-dimensional search space:

$$\frac{dX_i}{dt} = F_i + N_i + D_i \tag{18.2}$$

where F_i is the foraging action, N_i is movement induced by the other krill individuals, and D_i is random diffusion.

Step 9 in Algorithm 17 is utilized to address Equation (18.2). In step 11, sensing distance for every krill individual is calculated. In Equation (18.2), N_i depends on two different factors α_i^{target} (for the best neighbor effect) and α_i^{local} (for local neighbor effect). Those factors are computed in step 10 and 13, respectively. In step 15, inertia weights for induced motion by other krill individuals (ω_n) and inertia weight of the foraging motion (ω_f) are evaluated. The movement caused by other krill individuals (N_i^{new}) is computed in step 16. In steps 17 and 18, β_i^{food} and β_i^{best}, two necessary factors for evaluating the foraging motion (F_i) are calculated, respectively. After estimating F_i in step 19, physical diffusion of the krill individuals is computed in step 20. In steps 22 and 23, two genetic operators of crossover and mutation are applied to the current population. Finally, the new positions of the krill herd are calculated by adding the term of Δt, resulting from step 21 to their current positions. In step 26 the best-found solution is updated. Finally, the best-found solution is proposed when the termination criteria are satisfied.

18.3   Source-code of the KH algorithm in Matlab

In Listing 18.1 the source-code for the objective function handled by the KH algorithm is shown. In the objective function $OF(.)$, the input parameters are all the krill in a herd. The result of $OF(.)$ function is an N-dimensional column vector with the objective function values for each krill K_i from the whole population of K. The here tackled minimization objective function is given by Equation (18.3). The source-code for the KH algorithm in Matlab is presented in Listing 18.2.

$$OF(SA_i) = \sum_{j=1}^{D} SA_{i,j}^2; where - 5.12 < SA_{i,j} < 5.12 \qquad (18.3)$$

```
1   function [output]=OF(SA)
2   [x,y]=size(SA);
3   output=zeros(x,1);
4   for i=1:x
5       for j=1:y
6       output(i,1)=output(i,1)+SA(i,j)^2;
7       end
8   end
```

Listing 18.1
Definition of objective function $OF(.)$ in Matlab.

```
1   function [Best,Bestsolution,Kgb]= KH (NK,MI,UB,LB)
2   %% Initial Parameter Setting
3   C_flag = 1;                              % Crossover flag [Yes=1]
4   NP = length(LB); % Number of Parameter(s)
5   Dt = mean(abs(UB-LB))/2; % Scale Factor
6   F = zeros(NP,NK);D=zeros(1,NK);N=zeros(NP,NK); %R = zeros(NP,NK);
```

```
 7  Vf = 0.02; Dmax = 0.005; Nmax = 0.01; Sr = 0;
 8  %% Optimization & Simulation %Initial Krills positions
 9  X=rand(NK,NP).*(UB−LB)+LB;
10  for z1 = 1:NP
11      X(z1,:) = LB(z1) + (UB(z1) − LB(z1)).*rand(1,NK);
12  end
13  K=zeros(1,NK);
14  for z2 = 1:NK
15      K(z2)=OF(X(:,z2));
16  end
17  Kib=K;
18  Xib=X;
19  [Kgb,A]=min(K);
20  Xgb=zeros(NP,MI);
21  Xgb(:,1)=X(:,A);
22  Xf=zeros(NP,MI);
23  Kf=zeros(1,MI);
24  for j = 1:MI
25      %% Virtual Food
26      Sf=zeros(1,NP);
27      for ll = 1:NP
28          Sf(ll) = (sum(X(ll,:)./K));
29      end
30      Xf(:,j) = Sf./(sum(1./K)); %Food Location
31      Xf(:,j) =findlimits(Xf(:,j)',LB,UB,Xgb(:,j)');% Bounds Checking
32      Kf(j) = OF(Xf(:,j));
33      if 2<=j
34          if Kf(j−1)<Kf(j)
35              Xf(:,j) = Xf(:,j−1);
36              Kf(j) = Kf(j−1);
37          end
38      end
39      Kw_Kgb = max(K)−Kgb(j);
40      w=(0.1+0.8*(1−j/MI));
41      RR=zeros(NP,NK);
42      for i = 1:NK
43          % Calculation of distances
44          Rf = Xf(:,j)−X(:,i);
45          Rgb = Xgb(:,j)−X(:,i);
46          for ii = 1:NK
47              RR(:,ii) = X(:,ii)−X(:,i);
48          end
49          R = sqrt(sum(RR.*RR));
50          % Movement Induced % Calculation of BEST KRILL effect
51          if Kgb(j) < K(i)
52              alpha_b = −2*(1+rand*(j/MI))*(Kgb(j) − K(i)) /Kw_Kgb/ sqrt
    (sum(Rgb.*Rgb)) * Rgb;
53          else
54              alpha_b=0;
55          end
56          % Calculation of NEIGHBORS KRILL effect
57          nn=0;
58          ds = mean(R)/5;
59          alpha_n = 0;
60          for n=1:NK
61              condition=R<ds;
62              condition=sum(condition);
63              if (condition==NK && n~=i)
64                  nn=nn+1;
65                  if and(nn<=4,K(i)~=K(n))
66                      alpha_n=zeros(NP,1);
67                      alpha_n = alpha_n−(K(n) − K(i)) /Kw_Kgb/ R(n) * RR
    (:,n);
68                  end
69              else
70                  alpha_n = 0;
71              end
```

```
 72        end
 73        % Movement Induced
 74        N(:,i) = w*N(:,i)+Nmax*(alpha_b+alpha_n);
 75        % Foraging Motion % Calculation of FOOD attraction
 76        if Kf(j) < K(i)
 77            Beta_f=-2*(1-j/MI)*(Kf(j) - K(i)) /Kw_Kgb/ sqrt(sum(Rf.*Rf
    )) * Rf;
 78        else
 79            Beta_f=0;
 80        end
 81        % Calculation of BEST position attraction
 82        Rib = Xib(:,i)-X(:,i);
 83        if Kib(i) < K(i)
 84            Beta_b=-(Kib(i) - K(i)) /Kw_Kgb/ sqrt(sum(Rib.*Rib)) *Rib;
 85        else
 86            Beta_b=0;
 87        end
 88        % Foraging Motion
 89        F(:,i) = w*F(:,i)+Vf*(Beta_b+Beta_f);
 90        % Physical Diffusion %
 91        D = Dmax*(1-j/MI)*floor(rand+(K(i)-Kgb(j))/Kw_Kgb)*(2*rand(NP
    ,1)-ones(NP,1));
 92        % Motion Process %
 93        DX = Dt*(N(:,i)+F(:,i));
 94        % Crossover %
 95        if C_flag ==1
 96            C_rate = 0.8 + 0.2*(K(i)-Kgb(j))/Kw_Kgb;
 97            Cr = rand(NP,1) < C_rate ;
 98            % Random selection of Krill No. for Crossover
 99            NK4Cr = round(NK*rand+.5);
100            % Crossover scheme
101            add=X(:,NK4Cr).*(1-Cr)+X(:,i).*Cr;
102            for dim=1:NP
103                X(dim,i)=add(dim);
104            end
105        end
106        % Update the position
107        add=X(:,i)+DX;
108        for dim=1:NP
109            X(dim,i)=add(dim);
110        end
111        X(:,i)=findlimits(X(:,i)',LB,UB,Xgb(:,j)'); % Bounds Checking
112        K(i)=OF(X(:,i));
113        if K(i)<Kib(i)
114            Kib(i)=K(i);
115            Xib(:,i)=X(:,i);
116        end
117    end
118    % Update the current best
119    [Kgb(j+1), A] = min(K);
120    if Kgb(j+1)<Kgb(j)
121        Xgb(:,j+1) = X(:,A);
122    else
123        Kgb(j+1) = Kgb(j);
124        Xgb(:,j+1) = Xgb(:,j);
125    end
126 end
127 %% Post-Processing
128 Best = min(Kgb);
129 Bestsolution=Xgb(:,end);
130
131 function [ns]=findlimits(ns,Lb,Ub,best)
132 % Evolutionary Boundary Constraint Handling Scheme
133 n=size(ns,1);
134 for i=1:n
135    ns_tmp=ns(i,:);
136    I=ns_tmp<Lb;
```

```
137      J=ns_tmp>Ub;
138      A=rand;
139      ns_tmp(I)=A*Lb(I)+(1−A)*best(I);
140      B=rand;
141      ns_tmp(J)=B*Ub(J)+(1−B)*best(J);
142   ns(i,:)=ns_tmp;
143 end
```

Listing 18.2
Source-code of KH algorithm in Matlab.

18.4   Source-code of the KH algorithm in C++

The C++ source code for the objective function and the KH algorithm are presented in Listing 18.3 and Listing 18.4, respectively.

```
 1 #include <iostream>
 2 using namespace std;
 3 /* Function Definitions */
 4 double OF(double SA[], int size_array)
 5 {
 6    double output;
 7    int j;
 8    output = 0.0;
 9    for (j = 0; j < size_array; j++) {
10       output += SA[j] * SA[j];
11    }
12    return output;
13 }
```

Listing 18.3
Definition of objective function $OF(.)$ in C++.

```
 1 /* Function Definitions */
 2 /* Include files */
 3 #include <iostream>
 4 #include <stdlib.h>
 5 #include <math.h>
 6 using namespace std;
 7 KH(int NK, int MI, int dim, double lb[], double ub[])
 8 {
 9    /* Initial Krills positions */
10    double X[NK][dim]; double K[1][NK];
11 for (int i = 0; i < NK; i++) {
12    srand(time(0));
13    for (int j = 0; j < dim; j++) {
14       double r = (rand() % 10000) / 10000;
15       X[i][j] = (ub[j] − lb[j])*r + lb[j];
16    }
17    K[i] = OF(X[i], dim);
18 }
19 double Kib = K; double Xib = X; double Kgb; double Xgb[1][dim]; double
          RR[dim][NK];
20 Kgb = K[1];
21 for (int i = 0; i < NK; i++) {
22    if (K[i + 1] < Kgb) {
23       Kgb = K[i + 1]; A = i + 1;
24       for (int j = 0; j < dim; j++) {
```

```
25        Xgb[1][j] = X[i + 1][j];
26      }
27    }
28  }
29  double Xf[dim][MI]; double Kf[1][MI]; double Sf[1][dim]; double Kw_Kgb
        ; double Rf; double RR[dim][NK]; double R[1][dim]; double
        alpha_b_numerator; double R_ave;
30  double condition[1][NK]; double sum_condition; double alpha_n; double
        Food_multiplier_trans[dim][1]; double Food_multiplier; double Rib[
        dim][1]; double Sum_attraction_multipliers;
31  double D_multiplier[dim][1];
32  for (int t = 0; t < MI; t++) {
33    /* Virtual Food */
34    for (int int i = 0; i < dim; i++) {
35      double Null = 0;
36      for (int j = 0; j < NK; j++) {
37        Sf[i] += X[i][j] / K[j];
38        Null += 1 / K[j];
39        Xf[i][j] = Sf[i] / Null; /* Food Location */
40      }
41
42    }
43    for (int i = 0; i < NK; i++) {
44      for (int j = 0; j < dim; j++) {
45        /* Bounds Checking */
46        if (Xf[i][j] > ub[j]) Xf[i][j] = ub[j];
47        if (Xf[i][j] < lb[j]) Xf[i][j] = lb[j];
48      }
49      Kf[i] = OF(Xf[i], dim);
50    }
51    if (t >= 2) {
52      if (Kf[t-1] < Kf[t]) {
53        for (int i = 0; i < dim; i++) {
54          Xf[i][t] = Xf[i][t-1];
55        }
56      }
57    }
58    Kw_Kgb = max(K) - Kgb[t];
59    double w = (0.1 + 0.8*(1 - t / MI));
60    for (int i = 0; i < NK; i++) {
61      /* Calculation of distance */
62      for (int j = 0; j < dim; j++) {
63        Rf[j][1] = Xf[j][t] - X[j][i];
64        Rgb[1][j] = Xgb[1][j] - X[j][i];
65        alpha_b_numerator += (Rgb[1][j] * Rgb[1][j]);
66        Food_multiplier_trans[j][1] = Rf[j][1] * Rf[j][1];
67        Food_multiplier += Food_multiplier_trans[j][1];
68      }
69      for (int j = 0; j < NK; j++) {
70        for (int k = 0; k < dim; k++) {
71          RR[k][j] = X[k][j] - X[k][i];
72          R[1][j] += (RR[k][j] * RR[k][j]);
73        }
74        R[1][j] = sqrt(R[1][j]);
75        R_ave += R[1][j];
76      }
77      /* Movement Induced */
78      double alpha;
79      if (Kgb[t] < K[i]) {
80        srand(time(0));
81        r = (rand() % 10000) / 10000;
82        alpha_b = -2 * (1 + r*(t / MI))*(Kgb[t] - K[i]) / Kw_Kgb / sqrt(
        alpha_b_numerator) * Rgb;
83      }
84      else {
85        alpha_b = 0;
86      }
```

```cpp
87    /* Calculation of neighbors krill effect */
88    int nn = 0;
89    double ds = R_ave / (5 * NK);
90    for (int n = 0; n < NK; n++) {
91      for (int j = 0; j < NK; j++) {
92        condition[1][j] = R[1][j] < ds[1][j];
93        sum_condition += condition[1][j];
94      }
95      if (sum_condition == NK && n != i) {
96        nn += 1;
97        if (nn <= 4 && K(i) != K(n)) {
98          for (int j = 0; j < dim; j++) {
99            RR_multiplier[j][1] = RR[j][n];
100         }
101         alpha_n = alpha_n - (K[n] - K[i]) / Kw_Kgb / R[n] *
      RR_multiplier;
102       }
103     }
104   }
105   /* Movement Induced */
106   for (j = 0; j < dim; j++) {
107     N[dim][i] = w*N[dim][i] + Nmax*(alpha_b + alpha_n);
108   }
109   /* Calculation of food attraction */
110   Food_multiplier = sqrt(Food_multiplier);
111   if (Kf[j] < K[i]) {
112     double Beta_f = -2 * (1 - j / MI)*(Kf[j] - K[i]) / Kw_Kgb /
      Food_multiplier * Rf;
113   }
114   else {
115     double Beta_f = 0;
116   }
117   /* Calculation of best psition attraction */
118   for (int j = 0; j < dim; j++) {
119     Rib[j][1] = Xib[j][i] - X[j][i];
120     best_attraction_multipliers[j][1] = Rib[j][1] * Rib[j][1];
121     Sum_attraction_multipliers += best_attraction_multipliers[j][1];
122   }
123   if (Kib[i] < K[i]) {
124     double Beta_b = -(Kib[i] - K[i]) / Kw_Kgb / sqrt(
      Sum_attraction_multipliers) *Rib;
125   }
126   else {
127     double Beta_b = 0;
128   }
129   /* Foraging Motion */
130   for (int j = 0; j < dim; j++) {
131     F[j][i] = w*F[j][i] + Vf*(Beta_b + Beta_f);
132   }
133   /* Physical Diffusion */
134   for (int j = 0; j < dim; j++) {
135     srand(time(0)); double r = (rand() % 10000) / 10000;
136     D_multiplier[j][1] = 2 * r - 1;
137   }
138   srand(time(0));
139   double r = (rand() % 10000) / 10000;
140   double D = Dmax*(1 - t / MI)*floor(r + (K[i] - Kgb[t]) / Kw_Kgb)*
      D_multiplier;
141   double C_rate = 0.8 + 0.2*(K[i] - Kgb[t]) / Kw_Kgb;
142   /* Motion Process */
143   for (int j = 0; j < dim; j++) {
144     double Dx[j][i] = Dt*(N[j][i] + F[j][i] + D[j]);
145   }
146   if (C_flag == 1) {
147     for (int j = 0; j < dim; j++) {
148       srand(time(0));
149       R_vec[j][1] = rand() % 10000 / 10000;
```

```
150        double Cr = R_vec[j][1] < C_rate;
151        int NK4Cr = nearbyint(KN*((rand() % 10000) / 10000) + 0.5);
152        X[j][i] = X[j][NK4Cr] * (1 - Cr) + X[j][i] * Cr;
153      }

154    }
155    for (int j = 0; j < dim; j++) {
156      Delta[j][1] = X[j][i];
157      X[j][i] = Delta[j][1] + DX[j][i];
158      /* Bounds Checking */
159      if (X[i][j] > ub[j]) X[i][j] = ub[j];
160      if (X[i][j] < lb[j]) X[i][j] = lb[j];
161    }
162    K[i] = OF(X[i], dim);
163    if (K[i] < Kib[i]) {
164      Kib[i] = K[i];
165      for (int j = 0; j < dim; j++) {
166        Xib[j][i] = X[j][i];
167      }
168    }
169 }
```

Listing 18.4
Source-code of KH algorithm in C++.

18.5   Step-by-step numerical example of KH algorithm

In this section, a detailed computational process of the objective function defined by Equation (18.3) has been provided. In this study, the number of design variables and the number of krill in a herd are 5 and 6, respectively. The results are reported based on running the algorithm only for one iteration. Notably, the presented results may be slightly different from the real calculation. It has occurred because we used four decimal approximation here although the original algorithm uses the long format for variables. In the first step, foraging speed (V_f), maximum diffusion (D_{max}), and maximum induced speed (N_{max}) are initialized as 0.02, 0.005, and 0.01, respectively. In the second step, a herd of six krill is produced randomly within the acceptable boundaries domain.

$K_1 = \{3.2228, -2.2682, 4.6814, 2.9922, 1.8302\}$
$K_2 = \{4.1553, 0.4801, -0.1497, 4.7052, 2.6392\}$
$K_3 = \{-3.8196, 4.6848, 3.0749, 1.5948, 2.4897\}$
$K_4 = \{4.2330, 4.7604, -3.6671, -4.7543, -1.1036\}$
$K_5 = \{1.3553, -3.5060, -0.8012, 3.5751, 1.5921\}$
$K_6 = \{-4.1212, 4.8189, 4.2571, 4.4441, -3.3670\}$

In the third step, the relevant objective values for every krill is computed.

$OF(K_1) = 10.3862$
$OF(K_2) = 17.2666$

$OF(K_3) = 14.5898$
$OF(K_4) = 17.9180$
$OF(K_5) = 1.8370$
$OF(K_6) = 16.9842$

For calculating the best krill effect (α_b), in the fourth step, the objective values of the krill and their relevant solution vectors will be saved in Kib and Xib, respectively. Next, the best-found solution and its relevant vector are stored in Kgb and Xgb, respectively. In the next step, the main loop of the algorithm will be run iteratively until satisfying the termination criteria. To update the position of the krill school, movement induced by other krill (N_i), foraging activities (F_i), and random diffusion (D_i) are needed to be estimated. This procedure is simulated in the KH algorithm using a *for* loop for every i-th krill individual in step 14 (line 43 in Listing 18.2). However, before going through this step, we need a virtual food location and the distance from this food resource to compute N_i. In this way, S_f for every d-th dimension of the problem is defined as the accumulation of the d-th design variables for all the krill divided by their relevant objective values as follows (step 7):

$$S_f = \{1.0205, -1.2286, 0.2627, 2.6124, 1.1066\}$$

In order to calculate food location in step 8, each component of S_f is divided by the summation of the inverse of each objective value.

$$X_f = \{1.1573, -1.3933, 0.2979, 2.9626, 1.2549\}$$

The necessary step after estimating the food location is to check boundary constraints and bring the violated particle back to the valid domain. As there is no boundary constraint violation, the objective value for this food location is computed as follows (step 10):

$$K_f = OF(X_f) = 1.3395$$

It should be noted that since the second iteration, if there is no improvement in the food location KH will not update its position (step 11). In the next step, we evaluate the difference between the best and worst individuals, $K_w\_K_{gb}$, which is required for calculating α and β:

$$K_w\_K_{gb} = 17.9180 - 1.8370 = 16.0810$$

In step 13, the inertia weight is defined as follows:

$$\omega = \left(0.1 + 0.8 \times \left(1 - \frac{Iter}{MI}\right)\right) = (0.1 + 0.8 \times (1 - 1/1)) = 0.1$$

As mentioned previously step 14 evaluates N_i, F_i and D_i. Two necessary factors for computing N_i are α_b (effect of the best krill) and α_n (effect of neighbor). α_b needs R_{gb} that is specified by the difference between the best particle and i-th solution (i.e., $R_{gb} = X_{gb} - X_i$) in step 15 as follows:

$R_{gb1} = \{-1.8674, -1.2379, -5.4825, 0.5829, -0.2381\}$
$R_{gb2} = \{-2.7999, -3.9861, -0.6514, -1.1301, -1.0472\}$
$R_{gb3} = \{5.1750, -8.1909, -3.8760, 1.9803, -0.8976\}$
$R_{gb4} = \{-2.8776, -8.2665, 2.8659, 8.3294, 2.6957\}$
$R_{gb5} = \{0, 0, 0, 0, 0\}$
$R_{gb6} = \{5.4765, -8.3249, -5.0583, -0.8690, 4.9591\}$

Using R_{gb} and $K_w\_K_{gb}$ we can calculate α_b based on the following formula:

$$\alpha_b = -2 \times \left(1 + r \times \left(\frac{Iter}{MI}\right)\right) \times \frac{K_{gb} - K_i}{K_w\_K_{gb} \times \sqrt{R_{gb} \otimes R_{gb}}} \times R_{gb} \qquad (18.4)$$

where *Iter* represents the current iteration, and r is a uniform random number. Therefore, α_b values based on step 16 in this study are depicted as follows:

$\alpha_{b1} = \{-0.4256, -0.2822, -1.2498, 0.1329, -0.0543\}$
$\alpha_{b2} = \{-1.2767, -1.8176, -0.2970, -0.5153, -0.4775\}$
$\alpha_{b3} = \{1.3531, -2.1416, -1.0134, 0.5178, -0.2347\}$
$\alpha_{b4} = \{-0.8666, -2.4894, 0.8630, 2.5083, 0.8118\}$
$\alpha_{b5} = 0$
$\alpha_{b6} = \{1.6347, -2.4849, -1.5098, -0.2594, 1.4802\}$

Next, R is defined as the second square root of the summation of elements of the piece-wise multiplication of the RR vector. Therefore, in step 17, the RR vector is determined as the difference between the i-th and all the other krill.

$$RR_1 = \begin{bmatrix} 0 & 0.9325 & -7.0424 & 1.0102 & 1.0102 & -1.8674 \\ 0 & 2.7482 & 6.9530 & 7.0286 & 7.0286 & -1.2379 \\ 0 & -4.8311 & -1.6065 & -8.3485 & -8.3485 & -5.4825 \\ 0 & 1.7130 & -1.3974 & -7.7465 & -7.7465 & 0.5829 \\ 0 & 0.8090 & 0.6594 & 0.6594 & -2.9338 & -0.2382 \end{bmatrix}$$

$$RR_2 = \begin{bmatrix} -2.6710 & 0 & -7.9749 & 0.0776 & -2.8000 & -8.2765 \\ -2.7627 & 0 & 4.2048 & 4.2804 & -3.9861 & 4.3388 \\ 4.7671 & 0 & 3.2246 & -3.5173 & -0.6514 & 4.4069 \\ -2.5480 & 0 & -3.1104 & -9.4595 & -1.1301 & -0.2611 \\ 1.8784 & 0 & -0.1496 & -3.7428 & -1.0472 & -6.0063 \end{bmatrix}$$

$$RR_3 = \begin{bmatrix} 5.3040 & 5.1096 & 0 & 8.0526 & 5.1750 & -0.3015 \\ -6.9675 & -4.2979 & 0 & 0.0756 & -8.1909 & 0.1340 \\ 1.5425 & -3.2398 & 0 & -6.7419 & -3.8760 & 1.1822 \\ 0.5624 & 3.0840 & 0 & -6.3491 & 1.9803 & 2.8493 \\ 2.0280 & 0.1251 & 0 & -3.5933 & -0.8976 & -5.8567 \end{bmatrix}$$

$$RR_4 = \begin{bmatrix} -2.7486 & -2.9430 & -0.2208 & 0 & -2.8776 & -8.3541 \\ -7.0431 & -4.3734 & -0.1852 & 0 & -8.2665 & 0.0584 \\ 8.2845 & 3.5021 & 0.8760 & 0 & 2.8659 & 7.9242 \\ 6.9115 & 9.4331 & 6.3756 & 0 & 8.3294 & 9.1984 \\ 5.6213 & 3.7184 & 3.5813 & 0 & 2.6957 & -2.2634 \end{bmatrix}$$

$$RR_5 = \begin{bmatrix} 0.1290 & -0.0654 & 2.6568 & 2.8332 & 0 & -5.4765 \\ 1.2234 & 3.8930 & 8.0812 & -0.0982 & 0 & 8.3249 \\ 5.4186 & 0.6362 & -1.9899 & 4.0063 & 0 & 5.0583 \\ -1.4178 & 1.1037 & -1.9538 & 0.2239 & 0 & 0.8690 \\ 2.9256 & 1.0227 & 0.8855 & -1.0902 & 0 & -4.9591 \end{bmatrix}$$

$$RR_6 = \begin{bmatrix} 5.6055 & 5.4112 & 8.1333 & 8.3098 & 5.4765 & 0 \\ -7.1015 & -4.4319 & -0.2437 & -8.4232 & -0.2437 & 0 \\ 0.3603 & -4.4221 & -7.0482 & -1.0520 & -3.7452 & 0 \\ -2.2868 & 0.2347 & -2.8228 & -0.6451 & -0.8690 & 0 \\ 7.8847 & 5.9819 & 5.8447 & 3.8690 & 4.9591 & 0 \end{bmatrix}$$

Then, using the RR vectors, we can evaluate R as follows:

$R_1 = \{0, 5.9457, 10.1444, 13.7381, 5.9560, 11.5525\}$
$R_2 = \{6.8930, 0, 10.0685, 11.5841, 5.1504, 11.9537\}$
$R_3 = \{9.1371, 8.0376, 0, 12.7878, 10.6594, 6.6277\}$
$R_4 = \{14.3234, 11.9526, 7.3705, 0, 12.7073, 14.9105\}$
$R_5 = \{6.4377, 4.2224, 8.9959, 5.0325, 0, 12.2569\}$
$R_6 = \{12.2222, 10.2135, 12.5704, 12.5098, 8.3323, 0\}$

KH uses the average of the R vector's elements (d_s) as a threshold for the sensing distance. In this way, for every krill which meets this criterion α_n will be calculated as follows:

$$\alpha_n = \alpha_n - 2 \times \left(1 + r \times \left(\frac{Iter}{MI}\right)\right) \times \frac{K_n - K_i}{K_w\_K_{gb} \times R_n} \times \overrightarrow{RR_n} \qquad (18.5)$$

where $\overrightarrow{RR_n} = X_i - X_j$.

Therefore, in step 19 d_s was calculated as follows:

$d_s = \{1.5779, 1.4886, 1.6751, 2.1884, 1.5416, 1.9061\}$

In step 20 and 21, a counter called nn as well as the parameter α_n are initialized as zero. In the next step, the threshold for updating α_n has been checked and updated. In this study, the mentioned condition is not satisfied for any of the krill individuals. As a result, N_i for i-th krill individual is depicted accordingly.

$$N_i = \begin{bmatrix} -0.0042 & -0.0128 & 0.0135 & -0.0087 & 0 & 0.0163 \\ -0.0028 & -0.0182 & -0.0214 & -0.0249 & 0 & -0.0248 \\ -0.0125 & -0.0030 & -0.0101 & 0.0086 & 0 & -0.0151 \\ 0.0013 & -0.0052 & 0.0052 & 0.0251 & 0 & -0.0026 \\ -0.0005 & -0.0048 & -0.0023 & 0.0082 & 0 & 0.0148 \end{bmatrix}$$

The next movement is caused by foraging motion. Two effective factors in this motion are food attraction (β_f) and best position attraction (β_b). For the individuals with objective values more than K_f and K_{ib} the factors of β_f and β_b resulted from the Equations (18.4) and (18.5), respectively.

$$\beta_f = -2 \times \left(1 - \frac{Iter}{MI}\right) \times \frac{K_f - K_i}{K_w\_K_{gb} \times \sqrt{R_f \otimes R_f}} \times R_f \qquad (18.6)$$

$$\beta_b = -\frac{K_{ib} - K_i}{K_w\_K_{gb} \times \sqrt{R_{ib} \otimes R_{ib}}} \times R_{ib} \qquad (18.7)$$

In these relationships, R_f is defined as the difference between X_f and i-th krill and R_{ib} as the difference between X_{ib} and i-th krill. In step 23, R_f is computed as follows:

$R_{f1} = \{-2.0654, 0.8748, -4.3835, -0.0296, -0.5753\}$
$R_{f2} = \{-2.9979, -1.8734, 0.4476, -1.7426, -1.3843\}$
$R_{f3} = \{4.9770, -6.0782, -2.7770, 1.3678, -1.2348\}$
$R_{f4} = \{-3.0756, -6.1538, 3.9650, 7.7169, 2.3585\}$
$R_{f5} = \{-0.1980, 2.1127, 1.0990, -0.6125, -0.3372\}$
$R_{f6} = \{5.2785, -6.2122, -3.9592, -1.4815, 4.6219\}$

β_f values were calculated in step 24 using the R_f values. In this case of study β_f values for all the krill have been equal to zero. After that, the effect of β_b for evaluating F_i needed to be considered. Therefore, R_{ib} values as the difference between X_{ib} and the current location of each krill, are required in this step. During the first iteration of the algorithm as the values of X_{ib} are determined to be equal to X_i, the values for R_{ib} would be a zero vector. As a result, in this case, the values for β_b would be zero. Now, we can examine F_i using ω, V_f, β_f, and β_b. In step 27, F_i proved to be a zero vector in the first iteration.

In step 28, D_i is calculated using the following equation:

$$\beta_f = D_{max} \times \left(1 - \frac{t}{MI}\right) \times \left(rand + \frac{K_i - K_{gb}^t}{K_w\_K_{gb}}\right) \times (2 \times V_r - V_u) \qquad (18.8)$$

where t represents the current iteration, $rand$ is a random number, and K_{gb}^t is the t-th global best solution. D_{max} is equal to 0.005 as discussed earlier. Let dim be the number of design variables, so V_r is a vector of a random number, and V_u is a vector of unit elements both with the size of dim by one. Here, β_f is a zero vector with the size of dim by one.

In step 29, the movement interval (D_x) is computed by summing N_i, F_i, and D_i.

$$D_x = \begin{bmatrix} -0.02179 & -0.0654 & 0.0693 & -0.0444 & 0 & 0.0837 \\ -0.0144 & -0.0931 & -0.1096 & -0.1275 & 0 & -0.1272 \\ -0.0640 & -0.0152 & -0.0519 & 0.0442 & 0 & -0.0772 \\ 0.0068 & -0.0264 & 0.0265 & 0.1284 & 0 & -0.0133 \\ -0.0028 & -0.0244 & -0.0120 & 0.0416 & 0 & 0.0758 \end{bmatrix}$$

Before adding the term of D_x to the current positions of the herd of krill, a crossover operator is applied to the current solution (X^i) and one of the randomly selected krill (X^{random}). Therefore, in step 30, the rate of crossover is defined by the following expression:

$$C_{rate} = 0.8 + 0.2 \times \left(\frac{K_i - K_{gb}^t}{K_w\_K_{gb}} \right) \tag{18.9}$$

Then, a uniform random number called C_r will be produced for every variable of the problem. For every dimension with the random number greater than C_{rate} the design variable will be replaced by the following term:

$$X_{dim}^i = X_{dim}^{random} \times (1 - C_r) + X_{dim}^i \times C_r \tag{18.10}$$

In this study, the values for C_{rate} are summarized as follows:

$$C_{rate} = \{0.9063, 0.9919, 0.9586, 1, 0.8000, 0.9884\}$$

The values of C_r for every dimension of each krill are gathered in the following:

$C_{r1} = \{0.8687, 0.0046, 0.8173, 0.9619, 0.7749, 0.0496\}$
$C_{r2} = \{0.0844, 0.3998, 0.2599, 0.8001, 0.4314, 0.9027\}$
$C_{r3} = \{0.1818, 0.9106, 0.2638, 0.1455, 0.1361, 0.9448\}$
$C_{r4} = \{0.8693, 0.5797, 0.5499, 0.8530, 0.1449, 0.4909\}$
$C_{r5} = \{0.6220, 0.3509, 0.5132, 0.4018, 0.0760, 0.4892\}$
$C_{r6} = \{0.2399, 0.1233, 0.1839, 0.2399, 0.4173, 0.9991\}$

Randomly selected krill for the crossover are 5, 2, 4, 3, 3, and 2. The current population after applying the crossover operator changed into the following positions:

$K_1 = \{3.2228, -2.2682, 4.6814, 2.9922, 1.8302\}$
$K_2 = \{4.1553, 0.4801, -0.1497, 4.7052, 2.6392\}$
$K_3 = \{-3.8196, 4.6848, 3.0749, 1.5948, 2.4897\}$
$K_4 = \{4.2330, 4.7604, -3.6671, -4.7543, -1.1036\}$
$K_5 = \{1.3553, 4.5752, -2.7911, 3.5751, 1.5921\}$
$K_6 = \{-4.1212, 4.8189, 4.2571, 4.4441, -3.3670\}$

A mutation operator is applied to the output of the crossover function at the next step. To this end, the following equation is utilized:

$$X_{dim}^i = \begin{cases} K_{gb,dim}^t + \mu\,(X_{dim}^p - X_{dim}^q) & \text{if } rand_{dim}^i < Mu \\ X_{dim}^i & \text{otherwise} \end{cases} \qquad (18.11)$$

where p and q are two randomly selected krill. The results in our study are collected as below:

$rand^1 = \{0.16, 0.37, 0.35, 0.02, 0.29\}, p = 3, q = 6$
$K_1 = \{1.5061, -2.2682, 4.6814, 2.1504, 4.5204\}$
$rand^2 = \{0.24, 0.76, 0.76, 0.74, 0.74\}, p = 4, q = 4$
$K_2 = \{1.3554, 0.4801, -0.1497, 4.7052, 2.6392\}$
$rand^3 = \{0.10, 0.82, 0.17, 0.46, 0.66\}, p = 5, q = 3$
$K_3 = \{3.9429, 4.6849, -2.7392, 1.5948, 2.4897\}$
$rand^4 = \{0.95, 0.14, 0.28, 0.04, 0.11\}, p = 6, q = 4$
$K_4 = \{4.2330, -3.4768, 3.1609, 8.1743, 0.4603\}$
$rand^5 = \{0.95, 0.54, 0.04, 0.68, 0.81\}, p = 2, q = 3$
$K_5 = \{1.3554, 4.5752, 0.5119, 3.5751, 1.5921\}$
$rand^6 = \{0.95, 0.04, 0.54, 0.68, 0.81\}, p = 1, q = 6$
$K_6 = \{-4.1212, -7.0568, 4.2571, 4.4441, -3.3670\}$

Now, the term of D_x is added to the current solutions to update the positions of the krill herd as follows:

$K_1 = \{1.4843, -2.2826, 4.6174, 2.1572, 4.5177\}$
$K_2 = \{1.2900, 0.3870, -0.1650, 4.6788, 2.6148\}$
$K_3 = \{4.0121, 4.5752, -2.7911, 1.6213, 2.4777\}$
$K_4 = \{4.1886, -3.6043, 3.2051, 8.3027, 0.5019\}$
$K_5 = \{1.3553, 4.5752, 0.5119, 3.5751, 1.5921\}$
$K_6 = \{-4.0375, -7.1840, 4.1798, 4.4308, -3.2913\}$

After checking the boundary limitations, the objective values for every krill will be evaluated. In this example there is no boundary violation, therefore, the objective function values are calculated as follows:

$OF(K_1) = 2.2032$
$OF(K_2) = 1.6641$
$OF(K_3) = 16.0973$
$OF(K_4) = 17.5444$
$OF(K_5) = 1.8370$
$OF(K_6) = 16.3013$

Finally, the global best individual will be updated accordingly:

$$X_{gb} = \{1.2900, 0.3870, -0.1650, 4.6788, 2.6148\}$$
$$K_{gb} = 1.6641$$

After determination of the main loop of the algorithm, the global best solution found in the previous step will be proposed as the final result of the algorithm.

18.6 Conclusion

In this chapter, the strategy behind the KH algorithm to handle a given objective function has been explained. Therefore, all the steps for handling the objective function through a pseudo-code were provided. To better conveyance of the concept, we presented the source-codes for the original KH algorithm in both Matlab and C++ programming language were presented. In addition, a simple numerical example is tackled for detailed computation with the aim of better understanding the mechanism of the KH algorithm. The whole computational loads during the KH process were provided for this example. This chapter can be helpful to better understanding of the fundamentals of KH or desires to rewrite this algorithm in any other programming language.

References

1. A.H. Gandomi, A.H. Alavi. "Krill herd: a new bio-inspired optimization algorithm". *Communications in Nonlinear Science and Numerical Simulation*, vol. 17(12), pp. 4831-4845, 2012.

2. G.G Wang, L. Guo, A.H. Gandomi, G.S. Hao, H. Wang. "Chaotic krill herd algorithm". *Information Sciences*, vol. 274, pp. 17-34, 2014.

3. G.G Wang, A.H. Gandomi, A.H. Alavi. "Stud krill herd algorithm". *Neurocomputing*, vol. 128, pp. 363-370, 2014.

4. E. Fattahi, M. Bidar, H.R. Kanan. "Fuzzy krill herd optimization algorithm" in *IEEE First International Conference on Networks & Soft Computing (ICNSC)*, pp. 423-426, August 2014.

5. S. Sultana, P.K. Roy. "Oppositional krill herd algorithm for optimal location of capacitor with reconfiguration in radial distribution system". *International Journal of Electrical Power & Energy Systems*, vol. 74, pp. 78-90, 2016.

6. D. Rodrigues, L.A. Pereira, J.P. Papa, S.A. Weber. "A binary krill herd approach for feature selection" in *22nd IEEE International Conference on Pattern Recognition (ICPR)*, pp. 1407-1412, August 2014.

7. G.G. Wang, A.H. Gandomi, A.H. Alavi. "An effective krill herd algorithm with migration operator in biogeography-based optimization". *Applied Mathematical Modelling*, vol. 38(9-10), pp. 2454-2462, 2014.

8. J. Li, Y. Tang, C. Hua, X. Guan. "An improved krill herd algorithm: Krill herd with linear decreasing step". *Applied Mathematics and Computation*, vol. 234, pp. 356-367, 2014.

9. R.R. Bulatovic, G. Miodragovic, M.S. Boskovic. "Modified Krill Herd (MKH) algorithm and its application in dimensional synthesis of a four-bar linkage". *Mechanism and Machine Theory*, vol. 95, pp. 1-21, 2016.

19

Monarch Butterfly Optimization

Pushpendra Singh

Department of Electrical Engineering
Govt. Women Engineering College, Ajmer, India

Nand K. Meena

School of Engineering and Applied Science
Aston University, Birmingham, United Kingdom

Jin Yang

School of Engineering and Applied Science
Aston University, Birmingham, United Kingdom

Adam Slowik

Department of Electronics and Computer Science
Koszalin University of Technology, Koszalin, Poland

CONTENTS

19.1 Introduction

The natural inspired algorithms (NIAs) influenced by nature; these are found to be very effective and popular in the area of mathematics, computer science and decision making variables. Generally, the natural inspired algorithm can be divided into two categories, swarm based algorithm (SBAs) and evolutionary algorithm (EAs). The SBAs are a well-known model of NIA and are widely used techniques [1] in diversified areas of science and engineering. Many of such optimization techniques are inspired by the behaviour of cuckoos, chicken, honey bees, bats, etc. These methods are utilizing the swarm-intelligence for collective and interacting agents which are based on some set of defined rules [2]. The popularly known SBAs may include particle swarm optimization (PSO) [3], ant colony optimization (ACO) [4] etc. Similarly, the EAs are based on the prototype of biology and natural evolution; the popularly known EAs may include evolutionary programming (EP), genetic algorithm (GA), differential evolution (DE), and evolutionary strategy (ES) [5].

In the year 2015, Wang et al. [7] developed a new swarm-intelligence based optimization technique. The monarch butterfly, considered to be the most beautiful of butterflies, is known as the king of butterflies and is therefore named "monarch". The monarch butterfly is a species of insects generally found in the North American region. This creature has black and orange patterned wings and belongs to the milkweed butterfly part of the family of Nymphalidae [6, 8]. However, some white morphs of monarch are also found in the Hawaiian Islands. The male and female butterflies have different wing creation and therefore can be differentiated easily. These eastern North American butteries are well known for their migration from the USA and the southern part of Canada, to California and then Mexico, for winter. They fly thousands of miles over the Rocky Mountains, California to Mexico. When days begin to turn longer, they start their journey back to the North. They ley eggs somewhere during the journey and this process keeps repeating until they reach to north, e.g. for 4 to 5 generations.

A recent investigation [9], on preserved male and female monarch butterflies, carried out in 2015 revealed significant differences in wing sizes and body structure. The female usually has thicker wings to provide greater tensile strength which reduces the tendencies to get hurt during migration. On average, the male has wider wings and is heavier than a female.

During the journey to Mexico, females lay their eggs to produce offspring. In their life cycle, these butterflies pass through four major stages and a four-generation cycle in one year. The four stages may include egg, caterpillar, chrysalis and adult respectively. The monarch butterfly optimisation (MBO) is inspired from the migration behaviour of monarch butterflies from one continent to other, i.e., monarch butterfly optimization (MBO). Figure 19.1 shows the movement of the monarch butterfly from one region to another. They keep their reproduction cycle active during this cross county movement [10]. The migration behaviour is modelled into some set of mathematical formulation, presented in the following section.

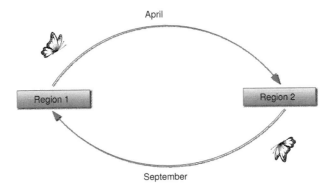

FIGURE 19.1
Migration behaviour of monarch butterflies.

19.2 Monarch butterfly optimization

In order to model the realistic migration behaviour of monarch butterflies, some set of pre-defined rules is formulated. These pre-defined rules are listed below [7].

- All the members of the monarch fleet should be located either in region 1 or 2. The complete flutter of that region is considered to be the total population in MBO.

- The new offspring of monarch butterflies in region 1 or 2 are produced by the migration operator which can be controlled by the migration ratio.

- In MBO, the total population of the monarch flutter remains unaltered. The offspring replace their respective parents, if they exhibit better fitness than their parent.

- The fittest butterfly of the flutter remains in the fleet and not affected by the migration operator.

19.2.1 Migration operator

As presented in Fig. 19.1, the monarch butterflies migrate from region 1 to 2 in the month of April every year and return in September of the same year. It is examined that the monarch butterflies flutter stays in region 1 from September to March and April to August in region 2. Suppose, the number of monarch butterflies staying in region 1 is considered as 'subpopulation 1' and represented by P_{n1}. Here, P_{n1} is determined as the nearest integer greater than or equal to $R \cdot (P_n \times P_{n1})$, where, R and P_n are the ratio of the butterfly flutter staying in region 1 and total number of monarch butterfly population respectively. Similarly, the butterflies found in region 2 are considered to be 'subpopulation 2', $P_{n2} = P_n - P_{n1}$.

This migration process of butterflies is mathematically expressed as

$$s_{ij}(t+1) = s_{r_1,j}(t) \qquad (19.1)$$

where, $s_{ij}(t+1)$, represents the jth position element/variable of the ith butterfly in $t+1$ generation, whereas, $s_{r_1,j}(t)$, represents the jth element of s_{r_1} which is the new position of monarch butterfly r_1 of current generation t. The butterfly r_1 is randomly selected from subpopulation 1 or region 1.

For decision making, a random number r is generated as follows

$$r = \psi \times rand \qquad (19.2)$$

where, $rand$ is a random number obtained from the uniform distribution; whereas, ψ represents the migration period considered to 1.2 for 12 months. If $r \leq R$, then the new element j of butterfly is produced by (19.1); otherwise, the new born monarch butterfly is produced as

$$s_{ij}(t+1) = s_{r_2,j}(t) \qquad (19.3)$$

where, r_2 is the randomly selected monarch butterfly from subpopulation 2 or region 2. It has been noted that by adjusting the value of R, the direction of migration can be balanced. For example, if the value of R is high then more members of monarch flutter P_{n1} will be expected. On the other hand, a low value of R increases the number of monarch butterflies P_{n2} in subpopulation 2 or region 2. The selection of R is very important in order to produce new monarch butterflies. In this chapter, the value of R is considered to be $5/12 = 0.4166$.

The important steps involved in the migration operator are listed below.

Migration operator

1. Calculate the value of subpopulation 1, i.e., P_{n1}, for region 1;

2. calculate r as suggested in (19.2). If $r \leq R$ then use (19.1) otherwise (19.3) for updating monarch butterfly of region 1;

3. a correction method may be applied to correct the infeasible positions of butterflies, if any;

4. repeat steps 1 to 3 until the positions of all individuals are updated.

19.2.2 Butterfly adjusting operator

Recently, it has been researched that monarch butterflies follow the Lévy flight movement pattern when moving from one place to the other [10]. Besides the migration operator, the positions of monarch butterflies are also updated by a 'butterfly adjusting operator'. Mathematically it is expressed as

$$s_{ij}(t+1) = s_{best,j}(t) \tag{19.4}$$

where, $s_{ij}(t+1)$, represents the jth variable or element of the ith monarch butterfly for generation $t+1$. Besides, $s_{best,j}(t)$ represents the jth element in the best monarch butterfly of region 1 and 2 for present generation t. Similar to the migration operator, a random number $rand$ is generated for decision making and then compared with R. If $rand \le R$ then the jth position of butterfly i is updated by (19.4); otherwise it is modified as follows

$$s_{ij}(t+1) = s_{r_3,j}(t) \tag{19.5}$$

where, $s_{r_3,j}(t)$ represents the jth element of $s_{r_3,j}$, selected randomly from region 2, $r_3 \in \{1, 2, 3, \ldots, P_{n2}\}$. If $rand>$BAR (Butterfly adjustment rate) then s is further updated under this condition as suggested below

$$s_{ij}(t+1) = s_{ij}(t+1) + \alpha \cdot (ds_j - 0.5) \tag{19.6}$$

where, ds_j is a small walk step of the ith butterfly which is calculated by Lévy flight movement as

$$ds_j = L[s_j(t)] \tag{19.7}$$

In (19.6), α is representing the weighting factor, determined as

$$\alpha = S_{max}/t^2 \tag{19.8}$$

where, S_{max} is the maximum value of the walk step taken by the monarch butterfly in a single move. It means that the larger the value of α, the higher will be the influence of ds on $s_{ij}(t+1)$. So, it encourages the exploration process of the method. Furthermore, the small size of α decreases the influence of ds on $s_{ij}(t+1)$, enhancing the exploitation process of MBO.

The algorithm for the butterfly adjusting operator is presented as

---

Butterfly adjusting operator

1. Calculate the value of subpopulation, P_{n2};

2. calculate the value of ds_j and weighing factor α, as suggested in (19.7) and (19.8) respectively for region 2;

3. generate a random number $rand$ following uniform distribution. If $rand \le R$ then update the jth element of the ith butterfly of region 2 by (19.4), otherwise by (19.5);

4. If $rand > R$ and $rand>$BAR then further update by (19.6);

5. apply a correction technique to correct infeasible individuals;

6. repeat steps 1 to 4 until all elements of the population are updated.

19.3   Algorithm of monarch butterfly optimization

In this section, the algorithm of MBO is presented for basic understanding. The essential steps of the algorithm involved in MBO are given below.

1. Set the value of required parameters and maximum number of iterations, e.g. $S_{Max.}$, p, BAR, ψ, t etc.;

2. initialize the random but feasible population P_n of the monarch flutter;

3. calculate the fitness of each butterfly i, based on suggested position elements;

4. sort all monarch butterfly individuals based on their fitness values and then divide into two populations P_{n1} and P_{n2} which correspond to region 1 and 2 respectively;

5. for P_{n1} of region 1, generate new subpopulation by migration operator and P_{n2} of region 2 by butterfly adjusting operator, as discussed in Sections 19.2.1 and 19.2.2 respectively;

6. repeat steps 3 to 5 until convergence criteria or maximum value of iteration is reached;

7. print the best solution along with its fitness value.

The flowchart of standard MBO, discussed in the previous section, is presented in Fig. 19.2.

19.4   Source-code of MBO algorithm in Matlab

In Listing 19.1, the source-code for the MBO algorithm is presented. Here, the $OF(.)$ is representing the address of the objective function.

```
1  clc
2  clear
3  LB=[0  0  0];  %set upper limits of each dimension
4  UB=[5  5  5];  %set upper limits of each dimension
5  %================MBO parameters================
6  Max_iter=50;        %maximum number of iterations.
7  Pn=30;              %population size.
8  D=length(UB);       %number of butterfly variables.
9  Stepmax=1.0;        %Maximum walk size
10 R=5/12;             %ratio of butterfly flutter...
11 Keep=2;             %populations keep in elitism...
12 phi=1.2;            %migration period
13 BAR=5/12;           %butterfly adjusting rate
14 Pn1= ceil(Pn*5/12); Pn2= Pn - Pn1;%butterflies in regions 1&2
15 Lnd1=zeros(Pn1,D); Lnd2=zeros(Pn2,D);%flutter of regions 1&2
16 s=[Lnd1;Lnd2];      %total flutter in both regions
17 fitness=zeros(Pn,1); %fitness values
```

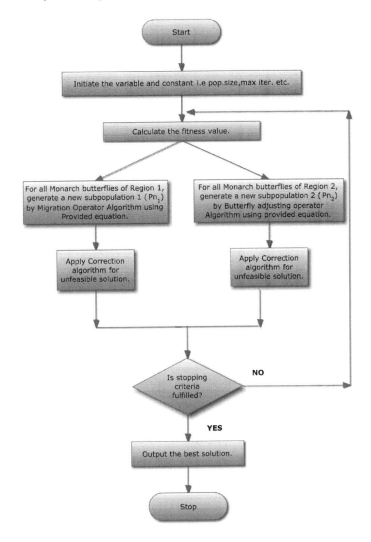

FIGURE 19.2
Flow chart of monarch butterfly optimization.

```
18  %=========================initialization=========================
19  for  i  =  1:Pn
20  for  j  =  1:D
21  s(i,j)  =  LB(j)  +  rand*(UB(j)−LB(j));  %random  populations
22  end
23  fitness(i)  =  OF(s(i,:));          %fitness  calculations...
24  end
25  best_fit  =  min(fitness);      %best  butterfly  fitness
26  nn_best  =  find(best_fit==fitness);  %position  of  best  butterfly
27  s_best  =  s(nn_best(1),:);      %best  butterfly
28  iter  =  0;                  %initialize  the  iterations.
29  while  iter  <  Max_iter        %iterations  start  here...
```

```
30  iter = iter + 1;              %update iteration
31  COMB_POP=[s, fitness];
32  s_srt=sortrows(COMB_POP, D+1);%sort the population
33  s_keep = s_srt(1:Keep,:);      %elitism...
34  s_srt((Pn-Keep+1):Pn,:)=s_keep;
35  s = s_srt(:,1:D);              %replace by sorted population
36  fitness = (s_srt(:,D+1));       %replace by sorted fitness
37  %===============Split population into two parts===========
38  pop1=s(1:Pn1,:);              %butterflies pop in region 1
39  pop2=s(Pn1+1:Pn,:);          %butterflies pop in region 2
40  %===============Migration operator================
41  for i =1:Pn1
42  for j = 1:D
43  r=phi*rand;        %decision making....
44  if r<=R
45  r1=randi([1 Pn1],1,1);
46  Lnd1(i,j)=pop1(r1,j);
47  else
48  r2=randi([1 Pn2],1,1);
49  Lnd1(i,j)=pop2(r2,j);
50  end
51  end
52  end
53  %===============Butterfly adjustment operator==========
54  for i=1:Pn2
55  alpha=Stepmax/(iter)^2;    %weighting factor
56  stepsize=ceil(exprnd(2*Max_iter,1,1)); %step size
57  ds=LevyFlight(stepsize,D);   %walk step
58  for j=1:D
59  if rand<=R
60  Lnd2(i,j)=s_best(1,j);
61  else
62  r3=randi([1 Pn2],1,1);
63  Lnd2(i,j)=pop2(r3,j);
64  if rand> BAR
65  Lnd2(i,j)=Lnd2(i,j)+ alpha*(ds(j)-0.5);
66  end
67  end
68  end
69  end
70  %combined updated butterflies population
71  s_new=[Lnd1;Lnd2];     %combined butterfly population
72  for i = 1:Pn
73  [fitness(i)]=OF(s_new(i,:)); %Fitness calculations
74  end
75  %===============Preserve the best solution=========
76  best_fit1 = min(fitness);%find best butterfly in new pop
77  nn_best = find(best_fit1==fitness); %location of best butterfly
78  s_best1 = s_new(nn_best(1),:);        %best butterfly of new pop
79  if best_fit1 < best_fit %replace if found better than this
80  best_fit = best_fit1;
81  s_best = s_best1;
82  end
83  s = s_new;          %replace the pop with new pop
84  end
85  disp(best_fit)
86  disp(s_best)
```

Listing 19.1
Source-code of MBO in Matlab.

19.5 Source-code of MBO algorithm in C++

```cpp
1  #include <iostream>
2  #include <math.h>
3  #include <time.h>
4  #include <algorithm>
5  using namespace std;
6  //---- function for swapping two values in the table
7  void swape(float *xp, float *yp)
8  {float temp=*xp; *xp=*yp; *yp=temp;}
9  //---- function for random generation of number in exp distribution
10 double ran_expo(float lambda)
11 {double u; u = rand()/(RAND_MAX + 1.0);
12 return -log(1-u)/lambda;}
13 //---- definition of objective function
14 float OF(float x[])
15 {float Fit=pow((x[0]-0.2),2)+pow((x[1]-1.7),2)+pow((x[2]-5.1),2);
16 return Fit;}
17 //---- function for generation of random values from the range [0, 1)
18 float ra(){return (float)(rand()%1000)/1000;}
19 //---- function for generation of Levy-Flight values
20 float LF(int StepSize)
21 {float suma=0;
22 for(int j=0; j<StepSize; j++){
23 float fx=tan(M_PI * ra());
24 suma=suma+fx;}
25 return suma;}
26 //---- main function
27 int main()
28 {    srand(time(NULL));
29 int Max_iter=50, Pn=30, D=3, Keep=2, Pn1=ceil(Pn*5/12);
30 int Pn2=Pn-Pn1, nr_fit[Pn];
31 float LB[3]={0, 0, 0}, UB[3]={5, 5, 5}, Stepmax=1;
32 float R=5/12, phi=1.2, BAR=5/12, s_best[D];
33 float Lnd1[Pn1][D], Lnd2[Pn2][D], s[Pn1+Pn2][D], fitness[Pn];
34 float s_srt[Pn1+Pn2][D];
35 //---- initialization
36 for(int i=0; i<Pn; i++){
37 for(int j=0; j<D; j++){s[i][j]=LB[j]+ra()*(UB[j]-LB[j]);}
38 fitness[i]=OF(s[i]); nr_fit[i]=i;}
39 float best_fit=*min_element(fitness,fitness+Pn);
40 int nn_best=min_element(fitness, fitness+Pn)-fitness;
41 for(int i=0; i<D; i++){s_best[i]=s[nn_best][i];}
42 int iter=0;
43 while(iter<Max_iter)
44 {    iter++;
45 for(int i=0; i<Pn; i++){
46 for(int j=0; j<Pn-i; j++){
47 if (fitness[j]>fitness[j+1]){
48 swape(&fitness[j],&fitness[j+1]);
49 int tmp=nr_fit[j];
50 nr_fit[j]=nr_fit[j+1];
51 nr_fit[j+1]=tmp;}}}
52 for(int i=0; i<Pn; i++){
53 for(int j=0; j<D; j++){
54 s_srt[i][j]=s[nr_fit[i]][j];}}
55 for(int i=0; i<Keep; i++){
56 for(int j=0; j<D; j++){
57 s_srt[Pn-1-i][j]=s_srt[i][j];
58 fitness[Pn-1-i]=fitness[i];}}
59 for(int i=0; i<Pn; i++){
60 for(int j=0; j<D; j++){
61 s[i][j]=s_srt[i][j];}}
62 //---- divide the whole population into two parts
```

```
63 float pop1[Pn1][D], pop2[Pn2][D];
64 for(int i=0; i<Pn1;i++){
65 for(int j=0; j<D; j++){pop1[i][j]=s[i][j];}}
66 for(int i=0; i<Pn2;i++){
67 for(int j=0; j<D; j++){pop2[i][j]=s[Pn1+i][j];}}
68 //---- migration operator
69 for(int i=0; i<Pn1; i++){
70 for(int j=0; j<D; j++){
71 float r=phi*ra();
72 if (r<=R){int r1=rand()%Pn1+1;
73 Lnd1[i][j]=pop1[r1][j];}
74 else {int r2=rand()%Pn2+1;
75 Lnd1[i][j]=pop2[r2][j];}}}
76 //---- butterfly adjustment operator
77 for(int i=0; i<Pn2; i++){
78 float alpha=Stepmax/(iter*iter);
79 float stepsize=ceil(ran_expo(2*Max_iter));
80 for (int j=0; j<D; j++){
81 if(ra()<=R){
82 Lnd2[i][j]=s_best[j];}
83 else {int r3=rand()%Pn2+1;
84 Lnd2[i][j]=pop2[r3][j];
85 if (ra()>BAR){
86 Lnd2[i][j]=Lnd2[i][j]+alpha*(LF(stepsize)-0.5);}}}}
87 //---- combined updated butterflies population
88 float s_new[Pn1+Pn2][D];
89 for(int i=0; i<Pn1; i++){
90 for(int j=0; j<D; j++){s_new[i][j]=Lnd1[i][j];}}
91 for(int i=0; i<Pn2; i++){
92 for(int j=0; j<D; j++){s_new[Pn1+i][j]=Lnd2[i][j];}}
93 for(int i=0; i<Pn; i++){
94 fitness[i]=OF(s_new[i]);}
95 //---- preserve the best solution
96 float best_fit1=*min_element(fitness,fitness+Pn);
97 int nn_best=min_element(fitness, fitness+Pn)-fitness;
98 if (best_fit1<best_fit){
99 best_fit=best_fit1;
100 for(int i=0; i<D; i++){s_best[i]=s_new[nn_best][i];}}
101 for(int i=0;i<Pn;i++){
102 for(int j=0; j<D; j++){s[i][j]=s_new[i][j];}}
103 //---- print the best result in each iteration
104 cout<<"Iteration: "<<iter<<" - The Best: "<<best_fit<<endl;}
105 getchar();
106 return 0;
107 }
```

Listing 19.2
Source-code of MBO algorithm in C++.

19.6   Step-by-step numerical example of MBO algorithm

Example 4 *Determine the global minima of function $f(x)$, where $x_i \epsilon [0,5] \; \forall \; i = 1,2,3$*

$$f(x) = (x_1 - 0.2)^2 + (x_2 - 1.7)^2 + (x_3 - 5.1)^2 \qquad (19.9)$$

Solution: In this problem, the optimal values of three variables x_1, x_2, x_3 have to be determined for the minima of $f(x)$. Listing 19.3 presents the Matlab source-code for function, $F(x)$, used in optimization.

```
1 function [Fit] = OF(x)
2 Fit=(x(1)-0.2)^2 + (x(2)-1.7)^2 +(x(3)-5.1)^2;
```

Listing 19.3
Definition of function $OF(.)$ in Matlab.

In this problem, x is a 3-dimensional column vector. In the first step, we set upper and lower limits of x for all dimensions as $UB = [5, 5, 5]$ and $LB = [0, 0, 0]$.

In the second step, we set the values of the algorithm parameters. In this example, we assume that maximum step size, $S_{max} = 1$; population size, $P_n = 6$; partition ratio $R = 5/12 = 0.4167$; subpopulation 1, $P_{n1} = ceil(Pn * R) = 3$; subpopulation 2, $P_{n2} = P_n - P_{n1} = 3$; butterfly adjusting rate, $BAR = 5/12 = 0.4167$; migration period, $\psi = 1.2$.

For the third step, we initialise a random but feasible population of P_n monarch butterflies as shown below.
$x_1 = \{0.1383\ 2.8725\ 3.0034\}$
$x_2 = \{0.3189\ 2.2329\ 3.0163\}$
$x_3 = \{2.5637\ 1.4946\ 1.3914\}$
$x_4 = \{1.8562\ 1.1483\ 3.0960\}$
$x_5 = \{3.5239\ 4.0837\ 4.3452\}$
$x_6 = \{3.4696\ 3.3391\ 4.7787\}$

In the fourth step, the fitness values of each butterfly is calculated by using $OF(.)$, as illustrated below.
$f_1 = OF(x_1) = 5.774$
$f_2 = OF(x_2) = 4.639$
$f_3 = OF(x_3) = 19.383$
$f_4 = OF(x_4) = 7.063$
$f_5 = OF(x_5) = 17.300$
$f_6 = OF(x_6) = 13.480$

In the fifth step, the best butterfly is identified based on best fitness value. We observed that butterfly x_2 is the best butterfly, as our goal is to minimise the function. Therefore,
$f_{best} = 4.639$
$x_{best} = \{0.3189, 2.2329, 3.0163\}$

In the sixth step, the main loop of the algorithm starts with iteration. We check whether the algorithm termination condition is fulfilled. If yes, we jump to step seventeen. If no, we move to step seven.

In step seven, we sort the population of butterflies according to their fitness values (best to worst fitnesses), as is done below.
$x_1 = \{0.3189, 2.2329, 3.016\}$ $f_1 = 4.639$
$x_2 = \{0.1383, 2.8725, 3.003\}$ $f_2 = 5.774$

$x_3 = \{1.8562, 1.1483, 3.096\}$ $f_3 = 7.063$
$x_4 = \{3.4696, 3.3391, 4.778\}$ $f_4 = 13.480$
$x_5 = \{3.5239, 4.0837, 4.345\}$ $f_5 = 17.300$
$x_6 = \{2.5637, 1.4946, 1.391\}$ $f_6 = 19.383$

In the eighth step, split the sorted butterflies population into two parts. The upper half will be known as subpopulation $1(P_{n1})$:
$Region1_1 = x_1 = \{0.3189, 2.2329, 3.0163\};$ $fRegion1_1 = f_1 = 4.639$
$Region1_2 = x_2 = \{0.1383, 2.8725, 3.0034\};$ $fRegion1_2 = f_2 = 5.774$
$Region1_3 = x_3 = \{1.8562, 1.1483, 3.0960\};$ $fRegion1_3 = f_3 = 7.063$
The second half will be known as subpopulation $2(P_{n2})$:
$Region2_1 = x_4 = \{3.4696, 3.3391, 4.7787\};$ $fRegion2_1 = f_4 = 13.480$
$Region2_2 = x_5 = \{3.5239, 4.0837, 4.3452\};$ $fRegion2_2 = f_5 = 17.300$
$Region2_3 = x_6 = \{2.5637, 1.4946, 1.3914\};$ $fRegion2_3 = f_6 = 19.383$

In the ninth step, we apply the migration operator to update positions of butterflies in subpopulation 1 (P_{n1}).
To do this, we generate a random number, e.g. $rand = 0.7750$, and then determine r as, $r = \psi \times rand = 1.2 \times 0.7750 = 0.9300$. Here, $r > R$ so $Region1_{1,1}$ is updated by (19.3), as illustrated below.
Now, choose a butterfly randomly from subpopulation 2 i.e., r_2. Let's assume $r_2 = 2$; then $Region1_{1,1}^{new}$ is updated as, $Region1_{1,1}^{new} = Region2(r_2, 1) = 3.5239$

For the second dimension, again we generate a random number, e.g. $rand = 0.1262$; then $r = 1.2 \times 0.1262 = 0.1514$. This time $r < R$ therefore $Region1_{1,2}$ will be updated as suggested in (19.1). Now, we choose a butterfly from subpopulation 1 randomly, i.e., r_1. We assume that $r_1 = 1$; then $Region1_{1,2}$ is updated as $Region1_{1,2}^{new} = Region1(r_1, 2) = 2.2329$.
For the third dimension, we repeat the process and assume that $rand = 0.2172$ and then $r = 1.2 \times 0.2172 = 0.2607$. Again, $r < R$ therefore, a butterfly is selected randomly from subpopulation 1. Let's assume that $r_1 = 3$; then $Region1_{1,3}$ is updated as, $Region1_{1,3}^{new} = Region1(r_1, 3) = 3.0960$.
Finally, $Region1_1^{new}$ can be expressed as $Region1_1^{new} = \{3.5239\ 2.2329\ 3.0960\}$

Similarly, other butterflies are updated and given below:
$Region1_2^{new} = \{3.4696\ 3.3391\ 3.0960\}$
$Region1_3^{new} = \{0.1383\ 2.8725\ 3.0034\}$

In the tenth step, we apply the butterfly adjustment operator to update butterflies in subpopulation $2(P_{n2})$.
For this, we determine $\alpha = Stepmax/t^2 = 1/(1^2) = 1$ and ds by using the Levy distribution. Let's assume that $ds = \{0.2857 - 7.2522 - 3.4262\}$.
Now, generate a random number, e.g., $rand = 0.3753$, as $rand < R$; then $Region2_{1,1}$ is updated by using (19.4). Therefore,
$Region2_{1,1}^{new} = x_{best}(1) = 0.3189$

For the second dimension $j = 2$, we again generate a random number, e.g. $rand = 0.6950$. This time, $rand > R$ thus $Region2_{1,2}$ will be updated by (19.5), as illustrated here.

We randomly select a butterfly r_3 from subpopulation 2. Let's assume $r_3 = 3$; then $Region2_{1,2}$ is updated as

$Region2_{1,2}^{new} = Region2_{r_3,2} = Region2_{3,2} = 1.4946$.

Now, we generate a random number, e.g. $rand = 0.9289$. If $rand > BAR$ then $Region2_{1,2}^{new}$ is further updated by (19.6), as illustrated here.

$Region2_{1,2}^{new} = Region2_{1,2}^{new} + \alpha(ds(2) - 0.5) = 1.4946 + 1.(-7.2522 - 0.5) = -6.2576$.

It is observed that this variable crosses the boundary limits $[0, 5]$ therefore, correction is applied as

$Region2_{1,2}^{new} = LB(2) + rand(UB(2) - LB(2)) = 0 + 0.7613(5 - 0) = 3.8065$.

For the third dimension $j = 3$, we repeat the process by generating a random number, e.g. $rand = 0.1903$ which is less than R therefore, $Region2_{1,3}$ is updated by (19.4) as: $Region2_{1,3}^{new} = x_{best}(3) = 3.0163$. Therefore $Region2_1^{new} = \{0.3189, 3.8065, 3.0163\}$.

Similarly, $Region2_2$ and $Region2_3$ are updated as
$Region2_2^{new} = \{0.3189\ 2.2329\ 3.0163\}$
$Region2_3^{new} = \{0.3117\ 0.1184\ 3.0163\}$

In the eleventh step, we combine both subpopulation 1 and 2, and then store in x^{new} as illustrated below.

$x_1^{new} = Region1_1^{new} = \{3.5239\ 2.2329\ 3.0960\}$
$x_2^{new} = Region1_2^{new} = \{3.4696\ 3.3391\ 3.0960\}$
$x_3^{new} = Region1_3^{new} = \{0.1383\ 2.8725\ 3.0034\}$
$x_4^{new} = Region2_1^{new} = \{0.3189\ 3.8065\ 3.0163\}$
$x_5^{new} = Region2_2^{new} = \{0.3189\ 2.2329\ 3.0163\}$
$x_6^{new} = Region2_3^{new} = \{0.3117\ 0.1184\ 3.0163\}$

In step twelve, we determine the fitness values of all butterflies at updated positions $x_{i,j}^{new}\ \forall i, j$ by using $OF(,)$, as

$f_1 = OF(x_1^{new}) = 15.348$
$f_2 = OF(x_2^{new}) = 17.393$
$f_3 = OF(x_3^{new}) = 5.774$
$f_4 = OF(x_4^{new}) = 8.793$
$f_5 = OF(x_5^{new}) = 4.640$
$f_6 = OF(x_6^{new}) = 6.856$

In the thirteenth step, the best butterfly is identified from the updated population, based on best fitness value. It is observed that x_5^{new} is the best butterfly. Therefore,

$f_{best}^{new} = 4.640$
$x_{best}^{new} = \{0.3189\ 2.2329\ 3.0163\}$

In the fourteenth step, the best fitness of the new population f_{best}^{new} is compared with f_{best}. If $f_{best}^{new} < f_{best}$ then we replace the x_{best} with x_{best}^{new} and f_{best} with f_{best}^{new}, otherwise proceed to step fifteen.

In this iteration $f_{best}^{new} = 4.640 > f_{best} = 4.639$; therefore, we move to step fifteen.

In the fifteenth step, we replace the $x_{i,j}$ with $x_{i,j}^{new}$ $\forall\ i,j$.

$$x_{i,j} = x_{i,j}^{new} \quad \forall\ i,j$$

In the sixteenth step, we return to step six.

In the seventeenth step, we print the optimal solution x_{best} with f_{best} and stop the algorithm.

19.7   Conclusion

A swarm based meta-heuristic optimization technique is presented in this chapter which is inspired by the migration behaviour of monarch butterflies and called MBO. The migration and butterfly adjusting operators are discussed and explained individually. To understand the application of this method, a simple step-by-step numerical example is demonstrated for one iteration of the algorithm. To help the new users, we have also presented the algorithm source-codes in Matlab and C++.

Acknowledgement

This work was supported by the Engineering and Physical Sciences Research Council (EPSRC) of United Kingdom (Reference Nos.: EP/R001456/1 and EP/S001778/1).

References

1. Cui Z, Gao X "Theory and applications of swarm intelligence" *Neural Comput Appl.*, 2012, pp. 205-206.

2. Fister Jr I, Yang XS, Fister I, Brest J, Fister D. "A brief review of nature-inspired algorithms for optimization" *arXiv preprint arXiv:1307.4186*, 2013.

3. Kennedy J, Eberhart R, "Particle swarm optimization" In paper presented at the Proceedings of *The IEEE International Conference on Neural Networks*, 2013.

4. Dorigo M, Maniezzo V, Colorni "A Ant system: optimization by a colony of cooperating agents" *IEEE Trans Syst Man Cybern B Cybern*, 1996, doi:10.1109/3477.484436

5. Back T, *"Evolutionary Algorithms in Theory and Practice: Evolution Strategies, Evolutionary Programming, Genetic Algorithms"* Oxford University Press, 1996.

6. Wang G-G, Gandomi AH, Alavi AH, Hao G-S, "Hybrid krill herd algorithm with differential evolution for global numerical optimization", *Neural Comput Appl*, 2014, pp. 297-308, doi:10.1007/s00521-013-1485-9.

7. Wang G-G, Deb S, Cui Z "Monarch butterfly optimization", *Neural Comput Appl.*, 2015. doi:10.1007/s00521-015-1923-y.

8. Garber SD, *"The Urban Naturalist"*, Dover Publications, Mineola, 1998.

9. A. K. Davis and M. T. Holden. "Measuring intraspecific variation in flight-related morphology of monarch butterflies (Danaus plexippus): which sex has the best flying gear?" *Journal of Insects*, vol. 2015, 6 pages, 2015. https://doi.org/10.1155/2015/591705. A. K. Davis and M. T. Holden. "Measuring intraspecific variation in flight-related morphology of monarch butterflies (Danaus plexippus): which sex has the best flying gear?". *Journal of Insects*, vol. 2015, 6 pages, 2015. https://doi.org/10.1155/2015/591705.

10. Breed GA, Severns PM, Edwards AM, "Apparent powerlaw distributions in animal movements can arise from intraspecific interactions" *J R Soc Interface*, 2015, doi:10.1098/rsif.2014.0927.

20

Particle Swarm Optimization

Adam Slowik
Department of Electronics and Computer Science
Koszalin University of Technology, Koszalin, Poland

CONTENTS

20.1   Introduction

In 1995, in the paper [1] by Kennedy et al., the first version of a particle swarm optimization (PSO) algorithm was proposed. The main inspiration for the PSO algorithm was the social behavior of such organisms as birds (a bird flock) or fish (a fish school). The PSO [1] and ACO [2] algorithms can be considered as the fathers of the new research area named swarm intelligence. The PSO algorithm is a global optimization algorithm which was originally developed for solving problems where the solutions are represented by the points in a multidimensional search space of parameters with continuous values (algorithm for real-valued optimization). Due to the PSO algorithm we can also obtain solutions for non-differentiable problems which may be irregular, dynamically changing with time, or noisy. Up until now many modifications of the PSO algorithm have been elaborated. Among them are, for example: the micro-PSO for a high dimensional optimization problem (it operates on a swarm consisting of a small number of particles) [5], discrete PSO (the combinatorial problem can be solved using this algorithm) [6], correlation-based binary PSO [7], heterogeneous strategy PSO (the proportion of particles adopts a fully informed strategy to enhance the converging speed while the rest is singly informed to maintain the diversity) [8], and multi-objective PSO [9]. The rest of this paper is organized as follows: in Section 20.2 the pseudo-code for the global version of the PSO algorithm is presented and discussed (also a short introduction to the local version of PSO algorithm is given), in Sections 20.3 and 20.4 the source-codes for the global version of the PSO algorithm are shown in Matlab and C++ programming language respectively, in Section 20.5 we present in detail the numerical example of the global version of the PSO algorithm, and in Section 20.6 some conclusions are provided.

20.2   Original PSO algorithm

20.2.1   Pseudo-code of global version of PSO algorithm

The pseudo-code for the global version of the PSO algorithm is presented in Algorithm 18.

Algorithm 18 Pseudo-code of the global version PSO.

1: determine the $D - th$ dimensional objective function $OF(.)$
2: determine the range of variability for each $j - th$ dimension $\left[P_{i,j}^{min}, P_{i,j}^{max}\right]$
3: determine the PSO algorithm parameter values such as N – number of particles in the swarm, c_1, c_2 – learning factor
4: randomly create swarm P which consists of N particles (each particle is a D-dimensional vector)
5: create the D-dimensional *Gbest* vector
6: **for** each i-th particle P_i from swarm P **do**
7:    create a *Pbest$_i$* particle equal to i-th particle P_i
8:    create a velocity V_i for each particle P_i
9:    evaluate the particle P_i using $OF(.)$ function
10: **end for**
11: assign the best particle P_i to the *Gbest*
12: **while** termination condition not met **do**
13:    **for** each i particle in the swarm P **do**
14:       **for** each j dimension **do**
15:          update the velocity $V_{i,j}$ using formula
16:          $V_{i,j} = V_{i,j} + c_1 \cdot r_1 \cdot (Pbest_{i,j} - P_{i,j}) + c_2 \cdot r_2 \cdot (Gbest_j - P_{i,j})$
17:          update the particle $P_{i,j}$ using formula
18:          $P_{i,j} = P_{i,j} + V_{i,j}$
19:          **if** $P_{i,j} < P_j^{min}$ **then** $P_{i,j} = P_j^{min}$
20:          **end if**
21:          **if** $P_{i,j} > P_j^{max}$ **then** $P_{i,j} = P_j^{max}$

22:           **end if**
23:         **end for**
24:         evaluate the particle P_i using $OF(.)$ function
25:         **if** $OF(P_i)$ is better than $OF(Pbest_i)$ **then**
26:             assign the particle P_i to the particle $Pbest_i$
27:         **end if**
28:       **end for**
29:       select the best particle from swarm P and assign it to particle T
30:       **if** $OF(T)$ is better than $OF(Gbest)$ **then**
31:           assign the particle T to the particle $Gbest$
32:       **end if**
33: **end while**
34: return the $Gbest$ as a result

20.2.2   Description of the global version of the PSO algorithm

In Algorithm 18 the pseudo-code of the global version of the PSO algorithm was presented. Now in this section we will discuss this algorithm in detail. Before we start our PSO algorithm we should have defined an objective function $OF(.)$ (step 1), and all the algorithm parameters such as learning coefficient c_1 and c_2, and number of particles N (step 2). In literature, we can find that the summation of c_1 and c_2 should be less than or equal to 4. Next, we randomly create the swarm P which consists of N particles (step 3). Each particle is represented by the D-dimensional vector, and its particular elements represent the decision variables from the $OF(.)$. Of course each j-th decision variable must be constrained by the lower P_j^{min} and upper P_j^{max} bound value. Therefore the particle P_i is represented by the following vector:

$$P_i = \{P_{i,1}, P_{i,2}, P_{i,3}, ..., P_{i,D-1}, P_{i,D}\}$$
where: $P_{i,1}, P_{i,2}$, and so on, are the values from the range $\left[P_j^{min}, P_j^{max}\right]$.

In step 4, the *Gbest* vector is created. In this vector the best solution found so far is stored. At the start, the *Gbest* is a D-dimensional zero vector. Next, the $Pbest_i$ particle is created for each corresponding P_i particle. At the start (step 6), the $Pbest_i$ particle is equal to the P_i particle. In the $Pbest_i$ particle the best particle P_i which was found so far will be stored. In the 7th step, the velocity vector V_i is created for particle P_i. The newly created vector V_i consists of only zero values. When the first swarm P is created, we can evaluate the quality of each particle P_i using the previously defined $OF(.)$ function (step 8). After swarm evaluation, we can select, from the swarm P, the best particle P_i which will be stored in the particle *Gbest* (step 9). By "the best" we mean the particle with the lowest value of $OF(.)$ function (for the minimization tasks) or with the highest value of $OF(.)$ function (for the maximization tasks). In the 11th step, the main algorithm loop is started until the termination criteria are fulfilled. As a termination criterion we can take into account, for example, the number of generations, the convergence of the algorithm, or the computational time. In the main algorithm loop, for all particles in swarm P, we update each j-th element value from the velocity vector V_i which is assigned to the particle P_i (step 15), and we update each j-th element value from the particle P_i using values from the previously computed velocity vector V_i (step 17). When the particle P_i is updated, we evaluate the quality of this particle using the $OF(.)$ function (step 19). Next, in step 20, if the value of $OF(P_i)$ is lower than the value of $OF(Pbest_i)$ (for the minimization tasks) or higher than the value of $OF(Pbest_i)$ then the particle P_i will be assigned to the corresponding particle $Pbest_i$ (step 21). When all particles from swarm P have been evaluated then the best particle is selected from the swarm and assigned to the temporary particle T (step 24). If the $OF(.)$ value for particle T is lower than the $OF(.)$ value for particle $Gbest$ (for minimization tasks) or higher than the $OF(.)$ value for particle $Gbest$ (for maximization tasks) (step 25) then the particle T will be assigned to the particle $Gbest$ (step 26). The algorithm (steps 11 to 28) is repeated until its termination criterion is satisfied. When the termination criterion is fulfilled, the particle $Gbest$ is returned as an algorithm result. In practical realization of the PSO algorithm the formula from the 16th step of the algorithm (see Algorithm 18) is very often replaced by the following expression:

$$V_{i,j} = \omega \cdot V_{i,j} + c_1 \cdot r_1 \cdot (Pbest_{i,j} - P_{i,j}) + c_2 \cdot r_2 \cdot (Gbest_j - P_{i,j}) \quad (20.1)$$

where ω is a factor called an inertia weight which significantly affects the convergence and exploration-exploitation trade-off in PSO, the typical value for inertia weight factor is from the range $[0.4, 0.9]$ [3].

20.2.3 Description of the local version of the PSO algorithm

In the global version of the PSO algorithm each particle goes toward the best particle in the whole swarm. In the local version of the PSO algorithm each particle goes toward the best particle in its neighborhood. Therefore the formula from the 16th step of the algorithm (see Algorithm 18) is replaced by the following expression (without inertia weight):

$$V_{i,j} = V_{i,j} + c_1 \cdot r_1 \cdot (Pbest_{i,j} - P_{i,j}) + c_2 \cdot r_2 \cdot (Lbest_{i,j} - P_{i,j}) \quad (20.2)$$

or by the expression with inertia weight factor:

$$V_{i,j} = \omega \cdot V_{i,j} + c_1 \cdot r_1 \cdot (Pbest_{i,j} - P_{i,j}) + c_2 \cdot r_2 \cdot (Lbest_{i,j} - P_{i,j}) \quad (20.3)$$

where in both expressions 20.2 and 20.3 the $Lbest_i$ is the best particle in the neighborhood of particle P_i. The neighborhood can be a geometrical one, or a social one. When the neighborhood is geometrical, we can compute the distances between all particles in the swarm (for example using Euclidean distance) and then for each particle P_i we select the best particle $Lbest_i$ among the m nearest neighborhoods of particle P_i. The social neighborhood is defined on the basis of a topology which depends only on the particle index i [4]. In practical application, the neighborhood of each particle is defined (at the start of the PSO algorithm) as a list of particles which does not change with the iterations. The topology which is more commonly used is a ring with a $k = 2$ (each particle has two neighbors – first on the left side, and second on the right side). In ring topology the left ($Left_i$) and right ($Right_i$) neighbor for the $i - th$ particle P_i can be determined using the formula:

$$Left_i = \begin{cases} i + 1 & \text{if } i < N \\ 1 & \text{if } i = N \end{cases} \tag{20.4}$$

$$Right_i = \begin{cases} i - 1 & \text{if } i > 1 \\ N & \text{if } i = 1 \end{cases} \tag{20.5}$$

Next, the best particle (from both the particles, $Left_i$ and $Right_i$) is selected as a $Lbest_i$ particle for particle P_i. The advantages of the local version of the PSO algorithm are that it may search a different area of the solutions space or explore different local optima.

20.3 Source-code of global version of PSO algorithm in Matlab

In Listing 20.1 the source-code for the objective function which will optimize by the PSO algorithm is shown. In the function $OF(.)$, the input parameter is a whole swarm P. The result of $OF(.)$ function is an N-dimensional column vector with the objective function values for each particle P_i from the swarm P. The objective function is given by formula 20.6. The PSO algorithm minimizes this objective function.

$$OF(P_i) = \sum_{j=1}^{D} P_{i,j}^2 \qquad \text{where } -5.12 \leqslant P_{i,j} \leqslant 5.12 \tag{20.6}$$

```matlab
function [out]=OF(P)
[x,y]=size(P);
out=zeros(x,1);
for i=1:x
    for j=1:y
    out(i,1)=out(i,1)+P(i,j)^2;
    end
end
```

Listing 20.1
Definition of objective function $OF(.)$ in Matlab.

```matlab
1   % declaration of the parameters of PSO algorithm
2   N=100; c1=2; c2=2; MAX=100; Iter=0; D=20;
3   % constraints definition for all decision variables
4   Pmin(1,1:D)=-5.12; Pmax(1,1:D)=5.12;
5   % the swarm P is created randomly
6   P=zeros(N,D)+(Pmax(1,:)-Pmin(1,:)).*rand(N,D)+Pmin(1,:);
7   % vector Gbest} is created
8   Gbest=zeros(1,D);
9   % all the particles from swarm P are remembered in swarm Pbest
10  Pbest=P;
11  % velocity vector is created for all particles in swarm P
12  V=zeros(N,D);
13  % the whole swarm P is evaluated using objective function OF(.)
14  Eval=OF(P);
15  % the swarm Pbest is evaluated using objective function OF(.)
16  EvalPbest(:,1)=Eval(:,1);
17  % the best particle is chosen from the swarm P
18  [Y,I]=min(Eval(:,1));
19  % the best particle from the swarm P is stored in Gbest
20  Gbest(1,:)=P(I,:);
21  % the OF(.) value of the best particle is stored in EvalGbest
22  EvalGbest=Y;
23  % main program loop starts
24  while(Iter<MAX)
25  % number of iteration is increased by one
26      Iter=Iter+1;
27  % the velocity vector for each particle is computed
28      V=V+c1*rand()*(Pbest-P)+c2*rand()*(Gbest-P);
29  % the new position for each particle is computed
30      P=P+V;
31  % the constraints for each decision variable are checked
32      for i=1:N
33          for j=1:D
34              if P(i,j)<Pmin(1,j)
35                  P(i,j)=Pmin(1,j);
36              end
37              if P(i,j)>Pmax(1,j)
38                  P(i,j)=Pmax(1,j);
39              end
40          end
41      end
42  % the whole swarm P is evaluated using objective function OF(.)
43      Eval=OF(P);
44  % the Pbest swarm is updated if needed
45      for i=1:N
46          if Eval(i,1)<EvalPbest(i,1)
47              Pbest(i,:)=P(i,:);
48          end
49      end
50  % the best particle is chosen from the swarm P
51      [Y,I]=min(Eval(:,1));
52  % the Gbest vector is updated if needed
53      if Eval(I,1)<EvalGbest
54          Gbest(1,:)=P(I,:);
55          EvalGbest=Y;
56      end
57  end
58  % the result of PSO algorithm is returned
59  disp(EvalGbest);
```

Listing 20.2
Source-code of the global version of the PSO in Matlab.

20.4 Source-code of global version of PSO algorithm in C++

```cpp
#include <iostream>
#include <time.h>
using namespace std;
// definition of the objective function OF(.)
float OF(float x[], int size_array)
{
    float t=0;
    for(int i=0; i<size_array; i++)
    {
        t=t+x[i]*x[i];
    }
    return t;
}
// generate pseudo random values from the range [0, 1)
float r()
{
    return (float)(rand()%1000)/1000;
}
// main program function
int main()
{
    srand(time(NULL));
// initialization of the parameters
    int N=100; int MAX=100; int Iter=0; int D=20;
    int i,j,k;
    int TheBest=0;
    float c1=2; float c2=2;
    float Pmin[D]; float Pmax[D];
// initialization of the constraints
    for (int j=0; j<D; j++)
    {
        Pmin[j]=-5.12; Pmax[j]=5.12;
    }
// initialization of all data structures which are needed by PSO
    float P[N][D]; float Pbest[N][D];
    float V[N][D]; float Eval[N]; float EvalPbest[N];
    float Gbest[D]; float EvalGbest;
// randomly create swarm P and Pbest; create velocity vectors V
    for(int i=0; i<N; i++)
    {
        for(int j=0; j<D; j++)
        {
        P[i][j]=(Pmax[j]-Pmin[j])*r()+Pmin[j];
        Pbest[i][j]=P[i][j];
        V[i][j]=0;
        }
// evaluate all the particles in the swarm P
        Eval[i]=OF(P[i],D);
        EvalPbest[i]=Eval[i];
// find the best particle in the swarm P
        if (Eval[i]<Eval[TheBest]) TheBest=i;
    }
// assign the best particle to Gbest
    for(j=0; j<D; j++) Gbest[j]=P[TheBest][j];
    EvalGbest=Eval[TheBest];
// main program loop
    while(Iter<MAX)
    {
        Iter++;
        TheBest=0;
```

```
61      for(int i=0; i<N; i++)
62      {
63          for(int j=0; j<D; j++)
64          {
65  // for each particle compute the velocity vector
66          V[i][j]=V[i][j]+c1*r()*(Pbest[i][j]-P[i][j])+c2*r()*(
            Gbest[j]-P[i][j]);
67  // update each particle using velocity vector
68          P[i][j]=P[i][j]+V[i][j];
69  // check the constraints for each dimension
70          if (P[i][j]<Pmin[j]) P[i][j]=Pmin[j];
71          if (P[i][j]>Pmax[j]) P[i][j]=Pmax[j];
72          }
73  // evaluate each particle from the swarm P
74          Eval[i]=OF(P[i],D);
75          if (Eval[i]<EvalPbest[i])
76          {
77  // for each particle update its Pbest vector
78          for(int j=0; j<D; j++) Pbest[i][j]=P[i][j];
79          }
80          if (Eval[i]<Eval[TheBest]) TheBest=i;
81      }
82  // update Gbest vector if better particle is found
83      if (Eval[TheBest]<EvalGbest)
84      {
85          for(int j=0; j<D; j++) Gbest[j]=P[TheBest][j];
86          EvalGbest=Eval[TheBest];
87      }}
88      cout << "EvalGbest = " << EvalGbest << endl;
89      getchar();
90      return 0;}
```

Listing 20.3
Source-code of the global version of the PSO in C++.

20.5   Step-by-step numerical example of global version of PSO algorithm

In the first step, let's assume that we want to minimize a mathematical function given by equation 20.6 where D is equal to 5. In the second step, we determine the PSO algorithm parameter values such as $N = 6$, $c_1 = 2$, and $c_2 = 2$. In the third step, the swarm P which consists of 6 particles (each particle is a 5-dimensional vector) is created.

$P_1 = \{-2.956, 2.622, -5.118, -1.737, 1.693\}$
$P_2 = \{1.314, 3.581, 1.902, 3.873, -4.420\}$
$P_3 = \{0.623, 1.662, 2.318, -3.087, 0.453\}$
$P_4 = \{-2.743, -2.752, -2.903, 3.926, 1.562\}$
$P_5 = \{-1.970, 4.433, -2.922, -1.918, -1.417\}$
$P_6 = \{-2.127, 0.680, -0.177, -1.718, 0.957\}$

In the fourth step, the 5-dimensional *Gbest* vector is created.
$Gbest = \{0, 0, 0, 0, 0\}$

In the fifth step, the $Pbest_i$ particle is created for each i-th particle P_i. At the start, the particles $Pbest_i$ are the same as particles P_i.

$Pbest_1 = \{-2.956, 2.622, -5.118, -1.737, 1.693\}$
$Pbest_2 = \{1.314, 3.581, 1.902, 3.873, -4.420\}$
$Pbest_3 = \{0.623, 1.662, 2.318, -3.087, 0.453\}$
$Pbest_4 = \{-2.743, -2.752, -2.903, 3.926, 1.562\}$
$Pbest_5 = \{-1.970, 4.433, -2.922, -1.918, -1.417\}$
$Pbest_6 = \{-2.127, 0.680, -0.177, -1.718, 0.957\}$

In the sixth step, for each particle P_i the 5-dimensional velocity vector V_i is created. At the start each V_i vector consists of zeros.
$V_1 = \{0, 0, 0, 0, 0\}$
$V_2 = \{0, 0, 0, 0, 0\}$
$V_3 = \{0, 0, 0, 0, 0\}$
$V_4 = \{0, 0, 0, 0, 0\}$
$V_5 = \{0, 0, 0, 0, 0\}$
$V_6 = \{0, 0, 0, 0, 0\}$

In the seventh step, we evaluate each particle P_i using the objective function $OF(.)$ given by equation 20.6. For particle $P_1 = \{-2.956, 2.622, -5.118, -1.737, 1.693\}$ we obtain:

$$OF(P_1) = \sum_{j=1}^{5} P_{1,j}^2 = (-2.956)^2 + (2.622)^2 + (-5.118)^2 + (-1.737)^2 + (1.693)^2 = 8.738 + 6.875 + 26.194 + 3.017 + 2.866 = 47.690$$

If we repeat the same computations for all particles in the swarm P, we obtain the $Eval\_P_i$ values which are the following:
$Eval\_P_1 = OF(P_1) = 47.690$
$Eval\_P_2 = OF(P_2) = 52.704$
$Eval\_P_3 = OF(P_3) = 18.258$
$Eval\_P_4 = OF(P_4) = 41.378$
$Eval\_P_5 = OF(P_5) = 37.757$
$Eval\_P_6 = OF(P_6) = 8.885$

In the eighth step, we take the best particle P_i (with the smallest value of objective function – our goal is $OF(.)$ minimization) and assign it to the *Gbest*. In our case, the best particle is P_6, therefore $Gbest = P_6 = \{-2.127, 0.680, -0.177, -1.718, 0.957\}$ and of course $OF(Gbest) = 8.885$.

In the ninth step, the main loop of the algorithm starts. We check whether the algorithm termination condition is fulfilled. If yes we jump to the sixteenth step. If no we go to the tenth step.

In the tenth step, the velocity vector V_i is updated for each particle P_i according to the formula:
$V_{i,j} = V_{i,j} + c_1 \cdot r_1 \cdot (Pbest_{i,j} - P_{i,j}) + c_2 \cdot r_2 \cdot (Gbest_j - P_{i,j})$

Now, we show in detail how the velocity vector V_1 is computed for single particle P_1.

$V_{1,1} = V_{1,1} + c_1 \cdot r_1 \cdot (Pbest_{1,1} - P_{1,1}) + c_2 \cdot r_2 \cdot (Gbest_1 - P_{1,1})$
Two numbers r_1 and r_2 from the range $[0; 1]$ are randomly chosen for computation of $V_{1,1}$. Let us assume that $r_1 = 0.89$ and $r_2 = 0.53$ have been randomly selected for $V_{1,1}$ computation.
$V_{1,1} = 0 + 2 \cdot 0.89 \cdot (-2.956 - (-2.956)) + 2 \cdot 0.53 \cdot (-2.127 - (-2.956))$
$V_{1,1} = 0 + 2 \cdot 0.89 \cdot 0 + 2 \cdot 0.53 \cdot 0.829 = 0.879$

$V_{1,2} = V_{1,2} + c_1 \cdot r_1 \cdot (Pbest_{1,2} - P_{1,2}) + c_2 \cdot r_2 \cdot (Gbest_2 - P_{1,2})$
Two numbers r_1 and r_2 from the range $[0; 1]$ are randomly chosen for computation of $V_{1,2}$. Let us assume that $r_1 = 0.21$ and $r_2 = 0.75$ have been randomly selected for $V_{1,2}$ computation.
$V_{1,2} = 0 + 2 \cdot 0.21 \cdot (2.622 - 2.622) + 2 \cdot 0.75 \cdot (0.680 - 2.622)$
$V_{1,2} = 0 + 2 \cdot 0.21 \cdot 0 + 2 \cdot 0.75 \cdot (-1.942) = 2.913$

$V_{1,3} = V_{1,3} + c_1 \cdot r_1 \cdot (Pbest_{1,3} - P_{1,3}) + c_2 \cdot r_2 \cdot (Gbest_3 - P_{1,3})$
Two numbers r_1 and r_2 from the range $[0; 1]$ are randomly chosen for computation of $V_{1,3}$. Let us assume that $r_1 = 0.33$ and $r_2 = 0.66$ have been randomly selected for $V_{1,3}$ computation.
$V_{1,3} = 0 + 2 \cdot 0.33 \cdot (-5.118 - (-5.118)) + 2 \cdot 0.66 \cdot (-0.177 - (-5.118))$
$V_{1,3} = 0 + 2 \cdot 0.33 \cdot 0 + 2 \cdot 0.66 \cdot 4.941 = 6.522$

$V_{1,4} = V_{1,4} + c_1 \cdot r_1 \cdot (Pbest_{1,4} - P_{1,4}) + c_2 \cdot r_2 \cdot (Gbest_4 - P_{1,4})$
Two numbers r_1 and r_2 from the range $[0; 1]$ are randomly chosen for computation of $V_{1,4}$. Let us assume that $r_1 = 0.63$ and $r_2 = 0.85$ have been randomly selected for $V_{1,4}$ computation.
$V_{1,4} = 0 + 2 \cdot 0.63 \cdot (-1.737 - (-1.737)) + 2 \cdot 0.85 \cdot (-1.718 - (-1.737))$
$V_{1,4} = 0 + 2 \cdot 0.63 \cdot 0 + 2 \cdot 0.85 \cdot 0.019 = 0.032$

$V_{1,5} = V_{1,5} + c_1 \cdot r_1 \cdot (Pbest_{1,5} - P_{1,5}) + c_2 \cdot r_2 \cdot (Gbest_5 - P_{1,5})$
Two numbers r_1 and r_2 from the range $[0; 1]$ are randomly chosen for computation of $V_{1,5}$. Let us assume that $r_1 = 0.68$ and $r_2 = 0.88$ have been randomly selected for $V_{1,5}$ computation.
$V_{1,5} = 0 + 2 \cdot 0.68 \cdot (1.693 - 1.693) + 2 \cdot 0.88 \cdot (0.957 - 1.693)$
$V_{1,5} = 0 + 2 \cdot 0.68 \cdot 0 + 2 \cdot 0.88 \cdot (-0.736) = -1.295$

After computations the velocity vector V_1 is as follows:
$V_1 = \{0.879, 2.913, 6.522, 0.032, -1.295\}$

If we do the same computations for the other velocity (V_2, V_3, V_4, V_5, and V_6) vectors we obtain the following results:
$V_2 = \{-3.832, -7.923, -2.693, -8.597, 7.839\}$
$V_3 = \{-3.498, -1.981, -4.954, 5.790, -0.170\}$

$V_4 = \{2.621, 6.047, 1.842, -7.112, -3.196\}$
$V_5 = \{1.472, -8.642, 4.252, 2.536, 4.918\}$
$V_6 = \{1.486, -1.254, 0.122, 2.522, -0.956\}$

In the eleventh step, we update each particle P_i using the velocity vector V_i corresponding to it ($P_{i,j} = P_{i,j} + V_{i,j}$).
Now, we show in detail how the particle vector P_1 is computed using velocity vector V_1.
$P_{1,1} = P_{1,1} + V_{1,1} = -2.956 + 0.879 = -2.077$
$P_{1,2} = P_{1,2} + V_{1,2} = 2.622 + 2.913 = 5.535$
$P_{1,3} = P_{1,3} + V_{1,3} = -5.118 + 6.522 = 1.404$
$P_{1,4} = P_{1,4} + V_{1,4} = -1.737 + 0.032 = -1.705$
$P_{1,5} = P_{1,5} + V_{1,5} = 1.693 + (-1.295) = 0.398$

So, the new vector for particle P_1 is as follows:
$P_1 = \{-2.077, 5.535, 1.404, -1.705, 0.398\}$

Now, we should check whether all constraints are satisfied for each decision variable from particle P_1. As we know, the constraint for each variable is $P_{i,j} = [P_j^{min}, P_j^{max}] = [-5.12, 5.12]$. We can see that the value for the second decision variable in particle P_1 does not satisfy the constraint ($P_{i,j} > P_j^{max}; 5.535 > 5.12$). Therefore the value of the second variable for particle P_i is set up to value $P_j^{max} = 5.12$, and the new vector for particle P_1 is as follows:
$P_1 = \{-2.077, 5.12, 1.404, -1.705, 0.398\}$

If we do the same computations for the other particles (P_2, P_3, P_4, P_5, and P_6) we obtain the following results:
$P_2 = \{-2.518, -4.342, -0.791, -4.724, 3.419\}$
$P_3 = \{-2.875, -0.319, -2.636, 2.703, 0.283\}$
$P_4 = \{-0.122, 3.295, -1.061, -3.186, -1.634\}$
$P_5 = \{-0.498, -4.209, 1.33, 0.618, 3.501\}$
$P_6 = \{-0.641, -0.574, -0.055, 0.804, 0.001\}$

In the twelfth step, we evaluate all the particles P_i from the swarm P using objective function $OF(.)$ given by the equation 20.6.
For particle $P_1 = \{-2.077, 5.12, 1.404, -1.705, 0.398\}$ we obtain:

$$OF(P_1) = \sum_{j=1}^{5} P_{1,j}^2 = (-2.077)^2 + (5.12)^2 + (1.404)^2 + (-1.705)^2 + (0.398)^2 = 4.314 + 26.214 + 1.971 + 2.907 + 0.158 = 35.564$$

If we repeat the same computations for all particles in the swarm P, we obtain the $Eval\_P_i$ values which are the following:
$Eval\_P_1 = OF(P_1) = 35.564$
$Eval\_P_2 = OF(P_2) = 59.825$
$Eval\_P_3 = OF(P_3) = 22.702$

$Eval\_P_4 = OF(P_4) = 24.818$
$Eval\_P_5 = OF(P_5) = 32.371$
$Eval\_P_6 = OF(P_6) = 1.390$

In the thirteenth step, we compare the value of the objective function for particle P_i with the value of the objective function for particle $Pbest_i$ corresponding to it. When the value of the objective function for particle P_i is smaller than the value of the objective function for particle $Pbest_i$ then the particle P_i is written down to the particle $Pbest_i$.

$OF(P_1) = 35.564 < OF(Pbest_1) = 47.690$ then $Pbest_1$ is updated by P_1
$OF(P_2) = 59.825 > OF(Pbest_2) = 52.704$ then $Pbest_2$ is not updated
$OF(P_3) = 22.702 > OF(Pbest_3) = 18.258$ then $Pbest_3$ is not updated
$OF(P_4) = 24.818 < OF(Pbest_4) = 41.378$ then $Pbest_4$ is updated by P_4
$OF(P_5) = 32.371 < OF(Pbest_5) = 37.757$ then $Pbest_5$ is updated by P_5
$OF(P_6) = 1.390 < OF(Pbest_6) = 8.885$ then $Pbest_6$ is updated by P_6

Therefore after this operation, we have a new set of *Pbest* particles which is as follows:
$Pbest_1 = \{-2.077, 5.12, 1.404, -1.705, 0.398\}$
$Pbest_2 = \{1.314, 3.581, 1.902, 3.873, -4.420\}$
$Pbest_3 = \{0.623, 1.662, 2.318, -3.087, 0.453\}$
$Pbest_4 = \{-0.122, 3.295, -1.061, -3.186, -1.634\}$
$Pbest_5 = \{-0.498, -4.209, 1.33, 0.618, 3.501\}$
$Pbest_6 = \{-0.641, -0.574, -0.055, 0.804, 0.001\}$

In the fourteenth step, we select the best particle from swarm P. We can see that the best particle is particle P_6 (objective function $OF(P_6) = 1.390$). After this, we check whether the $OF(P_6)$ is better than $OF(Gbest)$.
$OF(P_6) = 1.390 < OF(Gbest) = 8.885$ then *Gbest* is updated by P_6.

After this operation the new vector *Gbest* is the following:
$Gbest = \{-0.641, -0.574, -0.055, 0.804, 0.001\}$

In the fifteenth step we return to the ninth step.

In the sixteenth step, we return the particle *Gbest* as a result of the algorithm operation, and we stop the algorithm.

20.6   Conclusions

In this chapter we have aimed to show the main principles of the PSO algorithm. We have shown how this algorithm works (in both the global and local version). In addition, we have provided source-codes in Matlab and in C++ programming language which could help with others' implementation of this algorithm. Finally, the step-by-step numerical example of the PSO algorithm has been presented in detail for a better understanding of the particular operations which occur in the PSO algorithm. We believe that this chapter will make the implementation of one's own PSO algorithm in any programming language easier for anybody to achieve.

References

1. J. Kennedy, R.C. Eberhart. "Particle swarm optimization" in *Proc. of IEEE International Conference on Neural Networks*, 1995, pp. 1942-1948.

2. A. Colorni, M. Dorigo, V. Maniezzo. "Distributed optimization by ant colonies" in *Proc. of the European Conf. on Artificial Life*, 1991, pp. 134-142.

3. J. Xin, G. Chen, Y. Hai. "A particle swarm optimizer with multi-stage linearly-decreasing inertia weight" in Proc. of International Joint Conference on Computational Sciences and Optimization, 2009, pp. 505-508.

4. F. Marini, B. Walczak. "Particle swarm optimization (PSO). A tutorial". Chemometrics and Intelligent Laboratory Systems, vol. 149, part B, pp. 153-165, 2015.

5. W.H. Han. "A new simple micro-PSO for high dimensional optimization problem". Applied Mechanics and Materials, vol. 236-237, pp. 1195-1200, 2012.

6. Z. Wen-Liang, G.Z. Jun, C. Wei-Neng. "A novel discrete particle swarm optimization to solve traveling salesman problem" in *IEEE Congress on Evolutionary Computation*, 2007, pp. 3283-3287.

7. H.J. Wang, R.M. Ke, J.H. Li, Y. An, K. Wang, L. Yu. "A correlation-based binary particle swarm optimization method for feature selection in human activity recognition". *International Journal of Distributed Sensor Networks*, vol. 14(4), 2018

8. W.B. Du, W. Ying, G. Yan, Y.B. Zhu, X.B. Cao. "Heterogeneous strategy particle swarm optimization". *IEEE Transactions on Circuits and Systems II - Express Briefs*, vol. 64(4), pp. 467-471, 2017.

9. K. Sethanan, W. Neungmatcha. "Multi-objective particle swarm optimization for mechanical harvester route planning of sugarcane fields operations". *European Journal of Operational Research*, vol. 252(3), pp. 969-984, 2016.

21

Salp Swarm Algorithm: Tutorial

Essam H. Houssein
Faculty of Computers and Information
Minia University, Minya, Egypt

Ibrahim E. Mohamed
Faculty of Computers and Information
South Valley University, Luxor, Egypt

Aboul Ella Hassanien
Faculty of Computers and Information
Cairo University, Cairo, Egypt

CONTENTS

21.1   Introduction

Meta-heuristics are techniques for generating, finding or selecting heuristic partial search algorithms in computer science and optimization. This can provide a sustainable solution to a problem. The optimal solution found depends on a set of generated random parameters [1]. Meta-heuristics can indeed find suitable solutions with less mathematical potential and optimization techniques than simple heuristics or iterative methodologies by having to search for a wide range of possible solutions [2]. Swarm Intelligence (SI) algorithms are a category of meta-heuristic algorithms that mimic the collective behaviour of natural or artificial decentralized self-organized systems such as plants, animals, fish, birds, ants and other elements in our ecosystem that use the intuitive intelligence of the entire swarm, and provide solutions for a set of fundamental problems that could not be solved if the agent does not work collectively [3, 4, 5].

In [6, 7], meta-heuristics can be divided into three main classes: evolution-based, physics-based, and swarm-based methods. Swarm Intelligence (SI) meta-heuristic algorithms mimic the self-organized and collective behaviors of nature's systems. Swarm-inspired algorithms mimic the social behavior of groups of animals, birds, plants and humans. These algorithms include Pity Beetle Algorithm (PBA) [8], Emperor Penguin Optimizer (EPO) [9], Grasshopper optimisation algorithm [10], Artificial Flora (AF) [11], Grey Wolf Optimizer (GWO) [12, 13], Elephant Herding Optimization (EHO) [14] and Whale Optimization Algorithm [15, 16].

Recently, various meta-heuristic optimization algorithms have been developed to solve a wide variety of real life problems. All these algorithms are nature inspired and simulate some principle of biology, physics, ethology or swarm intelligence [17]. Surprisingly, some of them such as Genetic Algorithm (GA) [18], and Particle Swarm Optimization (PSO) [19] are fairly well-known among not only computer scientists but also scientists from different fields.

The work presented in [20] proposed a new meta-heuristic algorithm, salp swarm algorithm (SSA), influenced heavily by deep-sea swarming action salps (as shown in Figure 21.1). SSA seeks to establish a new population-based optimizer by trying to mimic the swarming behaviour of salps in the natural habitat. Several applications have been solved by SSA [21, 22]. In this chapter the modification and the mathematical model of the SSA is presented.

The remainder of this paper is structured as follows: in Section 21.2 the pseudo-code for the standard version of the SSA algorithm is presented and discussed (including a brief introduction to the traditional version of the SSA algorithm), in Sections 21.3 and 21.4 the source codes for the SSA algorithm are shown in the Matlab and C++ programming language respectively, and in Section 21.5 step by step example is presented.

21.2   Salp swarm algorithm (SSA)

21.2.1   Pseudo-code of SSA algorithm

The pseudo-code for the global version of the SSA algorithm is presented in algorithm 19.

Algorithm 19 Pseudo-code of the SSA.

1: Initialize the salp population $x_i (i = 1, 2, \ldots, n)$ considering ub and lb
2: **while** (end condition is not satisfied) **do**
3:    Calculate the fitness of each search agent (salp)
4:    F = the best search agent
5:    Update c_1
6:    **for** each salp in (X_i) **do**
7:       **if** $(i == 1)$ **then**
8:          Update the position of the leading salp
9:       **else**
10:          Update the position of the follower salp.
11:       **end if**
12:    **end for**
13:    Amend the salps based on the upper and lower bounds of variables
14: **end while**
15: return F

21.2.2 Description of SSA algorithm

The pseudo-code and flowchart of SSA were presented in Algorithm 19 and Fig. 21.2. In the next section, the SSA algorithm will be discussed extensively. Before starting to discuss the SSA algorithm, the objective function $OFun(.)$ (Step 1), and all parameters of the algorithm such as c_1, c_2, c_3, number of iterations, size of population and swarm population should have initialized.

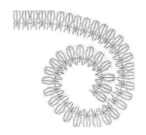

FIGURE 21.1
Demonstration of Salp's series.

Mathematically, The series Salp consists of two groups: leaders and followers. The first salp in the series of salps is called the leader salp, while the residual salps are considered followers. The first swarm (leader) directs the remaining swarm's movements [20]. Let D be the number of variables for a given problem; the positions of the Salps are denoted in a search space of D-dimension. Thus, the Salps X population consists of N swarms with a D

dimension. Hence, a matrix of $N \times D$ could be described as outlined in the equation below:

$$X_i = \begin{bmatrix} x_1^1 & x_2^1 & \cdots & x_d^1 \\ x_1^2 & x_2^2 & \cdots & x_d^2 \\ \vdots & \vdots & \ddots & \vdots \\ x_1^N & x_2^N & \cdots & x_D^N \end{bmatrix} \tag{21.1}$$

A food source F is also thought to be the target of the swarm. The position of the leader is updated by the next equation:

$$X_j^1 = \begin{cases} F_j + c_1((ub_j - lb_j)c_2 + lb_j) & c_3 \geq 0 \\ F_j - c_1((ub_j - lb_j)c_2 + lb_j) & c_3 < 0 \end{cases} \tag{21.2}$$

where X_j^1 and F_j respectively represent leadership positions and food source positions in the jth dimension. The ub_j and lb_j indicate j^{th} dimension in the upper and lower boundaries. c_2 and c_3 are two random numbers. In fact, usually, in addition to defining the step size, they govern the next position in the jth dimension towards the $+\infty$ or $-\infty$. Equation 21.2 shows that the leader only updates its food source position. The coefficient c_1, the most important parameter in SSA, progressively decreases over the next iterations to balance exploration and exploitation, and is defined as follows:

$$c_1 = 2e^{-(\frac{4l}{L})^2} \tag{21.3}$$

Respectively, l and L represent the current iteration and maximum number of iterations. To update the position of the followers, the next equation is used (Newton's motion law):

$$X_j^i = \frac{X_j^i + X_j^{i-1}}{2} \tag{21.4}$$

where $i \geq 2$ and X_j^i is the location of the ith follower at the jth dimension.

21.3   Source code of SSA algorithm in Matlab

In Listing 21.1 the source-code for the objective function for De Jong's function 1 which will optimize by the SSA algorithm is shown. In the function $OFun(X)$, the input parameter is the D-dimensional row vector for the positions of swarm elements. The result of the $OFun(X)$ function is the minimum value. The objective function equation was formulated in equation 21.5

$$OFun(x) = \sum_{i=1}^{n} ix_i^2 \quad -5.12 \leq x_i \leq 5.12 \tag{21.5}$$

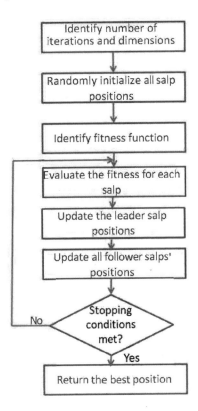

FIGURE 21.2

Flowchart of SSA.

```
1  %% De Jong's function 1
2  function [objective]=OFun(x)
3  [w,h]=size(x);
4  objective = zeros(w,1);
5  for i =1: w
6      for j =1:h
7          objective(i,1)= objective(i,1)+ (i*input(i,j)^2);
8      end
9  end
```

Listing 21.1

Definition of objective function $OFun(x)$ in Matlab.

```
1  clear all
2  clc
3  SearchAgents_no=60; % Number of search agents
4  Max_iteration=1000; % Maximum number of iterations
5  % Load details of the selected benchmark function
6  lb=[-5.12 -5.12 -5.12 -5.12];
7  ub=[5.12 5.12 5.12 5.12];
8  dim=4;
9  [Best_score,Best_pos,SSA_cg_curve]=SSA(SearchAgents_no,Max_iteration,
       lb,ub,dim,OFun);
```

```
10  display (['The best solution obtained by SSA is   ', num2str(Best_pos)])
    ;
11  display (['The best optimal value of the objective function found by
        SSA is ', num2str(Best_score)]);
12  bbest=min(SSA_cg_curve);
13  mbest=mean(SSA_cg_curve);
14  wbest=max(SSA_cg_curve);
15  stdbest=std(SSA_cg_curve);
16  fprintf('\n best=%f',bbest);
17  fprintf('\n mean=%f',mbest);
18  fprintf('\n worst=%f',wbest);
19  fprintf('\n std. dev.=%f',stdbest);
20  %This function randomly initializes the position of agents in the
        search space.
21  function [Positions]=initialization(SearchAgents_no,dim,ub,lb)
22      Boundary_no= size(ub,2); % number of boundaries
23  % If the boundaries of all variables are equal and user enters a
        single
24  % number for both ub and lb
25  if Boundary_no==1
26      Positions=rand(SearchAgents_no,dim).*(ub-lb)+lb;
27  end
28  % If each variable has a different lb and ub
29  if Boundary_no>1
30      for i=1:dim
31          ub_i=ub(i);
32          lb_i=lb(i);
33          Positions(:,i)=rand(SearchAgents_no,1).*(ub_i-lb_i)+lb_i;
34      end
35  end
36  function [FoodFitness,FoodPosition,Convergence_curve]=SSA(N,Max_iter,
        lb,ub,dim,fobj)
37  if size(ub,1)==1
38      ub=ones(dim,1)*ub;
39      lb=ones(dim,1)*lb;
40  end
41  Convergence_curve = zeros(1,Max_iter);
42  %Initialize the positions of salps
43  SalpPositions=initialization(N,dim,ub,lb);
44  FoodPosition=zeros(1,dim);
45  FoodFitness=inf;
46  %calculate the fitness of initial salps
47  for i=1:size(SalpPositions,1)
48      SalpFitness(1,i)=OFun(SalpPositions(i,:));
49  end
50  [sorted_salps_fitness,sorted_indexes]=sort(SalpFitness);
51  for newindex=1:N
52      Sorted_salps(newindex,:)=SalpPositions(sorted_indexes(newindex),:)
        ;
53  end
54  FoodPosition=Sorted_salps(1,:);
55  FoodFitness=sorted_salps_fitness(1);
56  %Main loop
57  l=2; % start from the second iteration since the first iteration was
        dedicated to calculating the fitness of salps
58  while l<Max_iter+1
59      c1 = 2*exp(-(4*l/Max_iter)^2);
60      for i=1:size(SalpPositions,1)
61          SalpPositions= SalpPositions';
62          if i<=N/2
63              for j=1:1:dim
64                  c2=rand();
65                  c3=rand();
66                  if c3<0.5
67                      SalpPositions(j,i)=FoodPosition(j)+c1*((ub(j)-lb(j
        ))*c2+lb(j));
68                  else
```

```
69          SalpPositions(j,i)=FoodPosition(j)-c1*((ub(j)-lb(j
   ))*c2+lb(j));
70              end
71          end
72      elseif i>N/2 && i<N+1
73          point1=SalpPositions(:,i-1);
74          point2=SalpPositions(:,i);
75          SalpPositions(:,i)=(point2+point1)/2;
76      end
77      SalpPositions= SalpPositions';
78  end
79  for i=1:size(SalpPositions,1)
80      Tp=SalpPositions(i,:)>ub';Tm=SalpPositions(i,:)<lb';
   SalpPositions(i,:)=(SalpPositions(i,:).*(~(Tp+Tm)))+ub'.*Tp+lb'.*
   Tm;
81      SalpFitness(1,i)=OFun(SalpPositions(i,:));
82      if SalpFitness(1,i)<FoodFitness
83          FoodPosition=SalpPositions(i,:);
84          FoodFitness=SalpFitness(1,i);
85      end
86  end
87  Convergence_curve(l)=FoodFitness;
88  l = l + 1;
89 end
```

Listing 21.2
Source-code of the SSA in Matlab.

21.4 Source-code of SSA algorithm in C++

```cpp
 1 #include <iostream>
 2 #include <math.h>
 3 #include<ctime>
 4 using namespace std;
 5 float OFun(float x[], int size)
 6 {
 7   float sum = 0;
 8   for (int i = 1; i <= size; ++i)
 9   {
10     sum = sum + i * pow(x[i], 2);
11   }
12   return sum;
13 }
14 //initialization population randomly between upper and lower
        boundaries
15 float ** initialization(int SearchAgents_no, int dim, float ub[],
        float lb[])
16 {
17   int Boundary_no;
18   float ** Positions = new float *[SearchAgents_no];
19   for (int i = 0; i < SearchAgents_no; ++i)
20     Positions[i] = new float[dim];
21   for (int i = 0; i < SearchAgents_no; i++)
22   {
23     for (int j = 0; j < dim; j++)
24     {
25       Positions[i][j] = rand() / (float(ub[j] - lb[j])) + lb[j];
26     }
27   }
```

```
28    return Positions;
29  }
30  void SSA(int N, int Max_iter, float lb[], float ub[], int dim)
31  {
32    //initialization of all used data structures in SS Algorithm
33    float  * Convergence_curve = new float[Max_iter]; // equal float
          Convergence_curve[N] where N is constant
34    float * FoodPosition = new float[dim];
35    float  FoodFitness = 0;
36    float * SalpFitness = new float[N];
37    // initialization  Salp Positions randomly
38    float ** SalpPositions = initialization(N, dim, ub, lb);// equal
          float SalpPositions[M][N] where M, N are constant
                              //calculate fitness values and select best position
          as first iteration;
39    for (int i = 0; i <N; i++)
40    {
41      SalpFitness[i] = OFun(SalpPositions[i], dim);
42      if (SalpFitness[i] < FoodFitness || i == 0)
43      {
44        FoodFitness = SalpFitness[i];
45        FoodPosition = SalpPositions[i];
46      }
47    }
48    //calculate Convergence_curve for first iteration
49    Convergence_curve[0] = FoodFitness;
50    int l = 1;
51    float c1, c2, c3;
52    // Main loop
53    //start from the second iteration since the first iteration was
          dedicated to calculating the fitness of salps
54    while (l < Max_iter + 1)
55    {
56      c1 = 2 * exp(-pow((4 * l / Max_iter), 2));
57      for (int i = 0; i < N; i++)
58      {
59        for (int j = 0; j < dim; j++)
60        {
61          //if we consider that we have N/2 leaders in the chain
62          if (i <= N / 2)
63          {
64            srand(time(0));
65            c2 = (float)(rand() % 10000) / 10000;//generate number
          between [0,1)
66            c3 = (float)(rand() % 10000) / 10000;
67            cout << c3;
68            if (c3 < 0.5)
69               SalpPositions[i][j] = FoodPosition[j] + c1 * ((ub[j] - lb[
          j])*c2 + lb[j]);
70            else
71               SalpPositions[i][j] = FoodPosition[j] - c1 * ((ub[j] - lb[
          j])*c2 + lb[j]);
72          }
73          else if (i > N / 2 && i < N + 1)
74          {
75            SalpPositions[i][j] = (SalpPositions[i][j] + SalpPositions[i
          - 1][j]) / 2;
76          }
77        }
78        //evaluate current agent (swarm) and compare with best fitness
          value
79      }
80      for(int k=0; k < N;k++)
81      {
82        for(int cc=0;cc < dim;cc++)
83        {
84          if (SalpPositions[k][cc] < lb[cc])
```

```
85        SalpPositions[k][cc] = lb[cc];
86      if (SalpPositions[k][cc] > ub[cc])
87        SalpPositions[k][cc] = ub[jcc];
88      }
89      SalpFitness[k] = OFun(SalpPositions[k],dim);
90      if (SalpFitness[k] < FoodFitness)
91      {
92        FoodPosition = SalpPositions[k];
93        FoodFitness = SalpFitness[k];
94      }
95    }
96    Convergence_curve[l] = FoodFitness;
97    l = l + 1;
98    }
99    cout << "The best solution obtained by SSA is ";
100   for (int i = 0; i < dim; i++)
101   {
102     cout << " " << FoodPosition[i];
103   }
104   cout << endl << "The best optimal value of the objective function
          found by SSA is " << FoodFitness << endl;
105 }
106 int main()
107 {
108   int SearchAgents_no = 60; //Number of search agents
109   int Max_iteration = 1000; //Maximum number of iterations
110            //Load details of the selected objective function
111   float  lb[] = { -5.12, -5.12, -5.12, -5.12 };
112   float  ub[] = { 5.12, 5.12, 5.12, 5.12 };
113   int dim = 4;
114   SSA(SearchAgents_no, Max_iteration, lb, ub, dim);
115   return 0;
116 }
```

Listing 21.3
Source-code of SSA algorithm in C++.

21.5 Step-by-step numerical example of SSA algorithm

To demonstrate the details of the SSA algorithm, an objective function in equation 21.5 is considered. With five search agents, Let x_j^1 is the position of the leader salp and $(x_j^2, x_j^3, x_j^4$ and $x_j^5)$ be positions of follower salps.

The initial positions are randomly generated within the boundaries of the design parameters and the value of objective function are shown in Table 21.1. From the first iteration in Table 21.1 it can be seen that the position of the 2nd agent has the minimum value for the objective function. So it is identified as the food source $\{4.1553, -2.2682, 4.8189, -0.8012\}$.

Now for the second iteration, the value of c1(0.9076) is calcuated using equation 21.3 and the values of c2 and c3 are selected randomly. For the first agent(x_j^1), the new values of x_1^1, x_2^1, x_3^1 and x_4^1 are updated according to the equation 21.2; then the new positions of leader salp are placed on Table 21.2, and the calculation is done as shown below:

$$X_j^1 \text{ positions} = \left\{ X_1^1,\ X_2^1,\ X_3^1,\ X_4^1 \right\}$$

TABLE 21.1

Initial Positions.

NO	x_1	x_2	x_3	x_4	$F(X)$	status
1	3.2228	-4.1212	-3.5060	-3.6671	5.311026e+01	
2	4.1553	-2.2682	4.8189	-0.8012	4.627460e+01	food source
3	-3.8197	0.4801	4.6814	4.2571	5.485881e+01	
4	4.2330	4.6849	-0.1498	2.9922	4.884174e+01	
5	1.3554	4.7605	3.0749	4.7052	5.609273e+01	

To calculate X_1^1 position let $c1 = 0.90758$, $c2 = 0.6557$ and $c3 = 0.0357$.

$X_1^1 = 4.1553 + 0.90758 * ((5.12 - (-5.12)) * 0.65574 + (-5.12))) = 5.6027$
where $c3 \leq 0.5$

To calculate X_2^1 position let $c1 = 0.90758$, $c2 = 0.8491$ and $c3 = 0.9340$.

$X_2^1 = (-2.2682) - 0.90758 * ((5.12 - (-5.12)) * 0.84913 + (-5.12))) = -5.5128$
where $c3 \geq 0.5$

To calculate X_3^1 position let $c1 = 0.9076$, $c2 = 0.6787$ and $c3 = 0.7577$.

$X_3^1 = 4.8189 - 0.90758 * ((5.12 - (-5.12)) * 0.67874 + (-5.12))) = 3.1578$
where $c3 \geq 0.5$

To calculate X_1^1 position let $c1 = 0.90758$, $c2 = 0.7431$ and $c3 = 0.3922$.

$X_1^4 = (-0.80116) + 0.90758 * ((5.12 - (-5.12)) * 0.74313 + (-5.12))) = 1.4584$
where $c3 \leq 0.5$.

For the second agent(follower : x_j^2), the new values of $\{X_1^2, X_2^2, X_3^2, X_4^2\}$) are updated according to equation 21.4 and placed on Table 21.2 as shown below:

$x_j^2 = (x_j^2 + x_j^1)/2$

$x_1^2 = (x_1^2 + x_1^1)/2 = (5.6027 + 4.1553)/2 = 4.8790$

$x_2^2 = (x_2^2 + x_2^1)/2 = (-5.5128 + -2.2682)/2 = -3.8905$

$x_3^2 = (x_3^2 + x_3^1)/2 = (3.1578 + 4.8189)/2 = 3.9883$

$x_4^2 = (x_4^2 + x_4^1)/2 = (1.4584 + -0.8012)/2 = 0.3286$

TABLE 21.2

After iteration 2.

NO	x_1	x_2	x_3	x_4	$F(X)$	*status*
1	5.1200	-5.1200	3.1578	1.4584	6.452732e+01	
2	4.8790	-3.8905	3.9883	0.3286	5.495549e+01	
3	0.5297	-1.7052	4.3349	2.2929	2.723661e+01	
4	2.3813	1.4898	2.0926	2.6425	1.925206e+01	food source
5	1.8683	3.1251	2.5837	3.6739	3.343010e+01	

Similarly, the new values of $x_j^3 = \{x_1^3, x_2^3, x_3^3, x_4^3\}$ based on values on Table 21.1 and Table 21.2

$$x_1^3 = (x_1^3 + x_1^2)/2 = (4.8790 + (-3.8197))/2 = 0.5297$$

$$x_2^3 = (x_2^3 + x_2^2)/2 = (-3.8905 + 0.4801)/2 = -1.7052$$

$$x_3^3 = (x_3^3 + x_3^2)/2 = (3.9883 + 4.6814)/2 = 4.3349$$

$$x_4^3 = (x_4^3 + x_4^2)/2 = (0.3286 + 4.2571)/2 = 2.2929$$

Repeat the same steps until each search agent's position is updated. Before the end of the current iteration, the new values for the position of salps should not be beyond the boundary of design variables for the welded problem. If there is a case where positions of salps exceed the limits of design variables, then the new values must be updated to the boundary of the problem.

Before update:

$$\begin{bmatrix} x_1 & x_2 & x_3 & x_4 \\ 5.6027 & -5.5128 & 3.1578 & 1.4584 \\ 4.8790 & -3.8905 & 3.9883 & 0.3286 \\ 0.5297 & -1.7052 & 4.3349 & 2.2929 \\ 2.3813 & 1.4898 & 2.0926 & 2.6425 \\ 1.8683 & 3.1251 & 2.5837 & 3.6739 \end{bmatrix}$$

After update:

$$\begin{bmatrix} x_1 & x_2 & x_3 & x_4 \\ 5.1200 & -5.1200 & 3.1578 & 1.4584 \\ 4.8790 & -3.8905 & 3.9883 & 0.3286 \\ 0.5297 & -1.7052 & 4.3349 & 2.2929 \\ 2.3813 & 1.4898 & 2.0926 & 2.6425 \\ 1.8683 & 3.1251 & 2.5837 & 3.6739 \end{bmatrix}$$

At the end of the current iteration, the fitness values for new positions are computed. Now, the values of f(x) of Table 21.2 in addition to the value of the 2nd agent in Table 21.1 (last food source) are compared and the best value of f(x) is considered the new food source.

The next iterations follow the same steps until the maximum number of iterations is achieved. In the last step in the last iteration, the best solution obtained by SSA is returned as a result of the algorithm operation, and the algorithm is stopped.

21.6   Conclusion

The paper aimed to show the fundamental principles of the SSA algorithm. The algorithm works have been demonstrated. In addition, source codes were introduced in two programming languages, Matlab and C++, that could assist with the implementation of this algorithm by researchers. At last, for a clearer picture of the particular operations that take place in the SSA algorithm, the step-by-step mathematical illustration of the SSA algorithm has been described in detail. It is assumed that this section of the book will make it easier for anyone to complete the development of his modification of the SSA algorithm.

References

1. M.N. Ab Wahab, S. Nefti-Meziani, A. Atyabi. "A comprehensive review of swarm optimization algorithms". PLoS One, vol. 10(5), pp. 1-36, 2015.

2. S. Russell, P. Norvig. "Artificial Intelligence: A Modern Approach". Prentice Hall, 1995.

3. V. Pandiri, A. Singh. "Swarm intelligence approaches for multidepot salesmen problems with load balancing". Applied Intelligence, vol. 44(4), pp. 849-861, 2016.

4. A.A. Ewees, M.A. Elaziz, E.H. Houssein. "Improved grasshopper optimization algorithm using opposition-based learning". Expert Systems with Applications, vol. 112, pp. 156-172, 2018.

5. A.G. Hussien, E.H. Houssein, A.E. Hassanien. "A binary whale optimization algorithm with hyperbolic tangent fitness function for feature selection" in *Proc. of Eighth International Conference on Intelligent Computing and Information Systems (ICICIS)*, pp. 166-172, 2017.

6. S. Mirjalili, A. Lewis. "The whale optimization algorithm". Advances in Engineering Software, vol. 95, pp. 51-67, 2016.

7. A.G. Hussien, A.E. Hassanien, E.H. Houssein, S. Bhattacharyya, M. Amin. "S-shaped binary whale optimization algorithm for feature selection" in Recent Trends in Signal and Image Processing, Advances in Intelligent Systems and Computing, vol. 727, Springer, pp. 79-87, 2019.

8. N.A. Kallioras, N.D. Lagaros, D.N. Avtzis. "Pity beetle algorithm-A new metaheuristic inspired by the behavior of bark beetles". Advances in Engineering Software, vol. 121, pp. 147-166, 2018.

9. G. Dhiman, V. Kumar. "Emperor penguin optimizer: A bio-inspired algorithm for engineering problems". Knowledge-Based Systems, vol. 159, pp. 20-50, 2018.

10. A. Tharwat, E.H. Houssein, M.M. Ahmed, A.E. Hassanien, T. Gabel. "MOGOA algorithm for constrained and unconstrained multi-objective optimization problems". Applied Intelligence, vol. 48(8), pp. 2268-2283, 2018.

11. L. Cheng, X.-H. Wu, Y. Wang. "Artificial Flora (AF) optimization algorithm". Applied Sciences, vol. 8(3), pp. 329-351, 2018.

12. S. Mirjalili, S.M. Mirjalili, A. Lewis. "Grey wolf optimizer". Advances in Engineering Software, vol. 69, pp. 46-61, 2014.

13. A. Hamad, E.H. Houssein, A.E. Hassanien, A.A. Fahmy. "A hybrid EEG signals classification approach based on grey wolf optimizer enhanced SVMs for epileptic detection" in *Proc. of International Conference on Advanced Intelligent Systems and Informatics*, pp. 108-117, 2017.

14. A.A. Ismaeel, I.A. Elshaarawy, E.H. Houssein, F.H. Ismail, A.E. Hassanien. "Enhanced Elephant Herding Optimization for global optimization". IEEE Access, vol. 7, pp. 34738-34752, 2019.

15. M.M. Ahmed, E.H. Houssein, A.E. Hassanien, A. Taha, E. Hassanien. "Maximizing lifetime of large-scale wireless sensor networks using multi-objective whale optimization algorithm". Telecommunication Systems, vol. 72, pp. 243-259, 2019.

16. E.H. Houssein, A. Hamad, A.E. Hassanien, A.A. Fahmy. "Epileptic detection based on whale optimization enhanced support vector machine". Journal of Information and Optimization Sciences, vol. 40(3), pp. 699-723, 2019.

17. A. LaTorre, S. Muelas, Santiago, J.-M. Peña. "A comprehensive comparison of large scale global optimizers". Information Sciences, vol. 316, pp. 517-549, 2015.

18. E. Bonabeau, G. Theraulaz, M. Dorigo. "Swarm Intelligence: From Natural to Artificial Systems". Oxford University Press, 1999.

19. E. Russell, J. Kennedy. "A new optimizer using particle swarm theory" in *Proc. of Sixth International Symposium on Micro Machine and Human Science*, pp. 39-43, 1995.

20. S. Mirjalili, A.H. Gandomi, S.Z. Mirjalili, S. Saremi, H. Faris, S.M. Mirjalili. "Salp Swarm Algorithm: A bio-inspired optimizer for engineering design problems". Advances in Engineering Software, vol. 114, pp. 163-191, 2017.

21. H.M. Kanoosh, E.H. Houssein, M.M. Selim. "Salp swarm algorithm for node localization in wireless sensor networks". Journal of Computer Networks and Communications, vol. 2019, Article ID 1028723, 12 pages, 2019.

22. A.G. Hussien, A.E. Hassanien, E.H. Houssein. "Swarming behaviour of salps algorithm for predicting chemical compound activities" in *Proc. of Eighth International Conference on Intelligent Computing and Information Systems (ICICIS)*, pp. 315-320, 2017.

22

Social Spider Optimization

Ahmed F. Ali
Department of Computer Science
Suez Canal University, Ismaillia, Egypt

Mohamed A. Tawhid
Department of Mathematics and Statistics
Faculty of Science, Thompson Rivers University, Kamloops, Canada

CONTENTS

22.1 Introduction

Social spider optimization algorithm (SSO) is a recent population based swarm intelligence algorithm proposed by Cuevas et al. [4]. The SSO algorithm was inspired by the social behavior of the social spider colony that consists of social members and a communal web. The SSO has been applied in many applications due to its efficiency especially when it is applied to solve global optimization problems [5]. The authors in [8] have combined the SSO algorithm and Nelder-Mead search method to solve minimax and integer programming problems. The authors in [3] have applied the SSO algorithm for text document clustering, while the authors in [9] have proposed a new hybrid SSO and genetic algorithm to minimize the potential energy function and large scale optimization problems. The rest of the chapter is organized as follows. In Section 22.2, we present the standard SSO algorithm. In Sections 22.3 and 22.4, we give Matlab and C++ source codes of the SSO algorithm, respectively. In Section 22.5, we illustrate a step by step numerical example of the SSO algorithm. Finally, we outline the conclusion of this chapter in Section 22.6.

22.2 Original SSO algorithm

In the following subsections, we describe the main concepts of the SSO algorithm and how it works.

22.2.1 Social behavior and inspiration

The SSO algorithm mimics the social behavior of the social spider colony that consists of social members and a communal web. The social members are divided into males and females. The number of female and male spiders reaches to 70% and 30% of the total colony members [1], [2], respectively. Each member in the colony is responsible for a specific task such as building and maintaining the communal web, prey capturing, and mating [6]. The female spiders produce an attraction or dislike over others. The vibrations in the communal web are based on the weight and distance of the members which are the main features of the attraction or dislike of a particular spider [7]. Male spiders have two categories, dominant and non-dominant [10]. The dominant male spiders have better fitness characteristics than the non-dominant spiders. The dominant male can mate with one or all females in the colony to exchange information among members and produce offspring. In the social spider optimization algorithm (SSO), each solution represents a spider position, while the communal web represents the search space. The value of each solution is represented by calculating its fitness function which represents the weight of each spider.

22.2.2 Population initialization

In the SSO algorithm, the population consists of N solutions (spiders) and can be divided into females f_i and males m_i. The number of females N_f is selected within the range of $65\% - 90\%$ and it can be calculated by the following equation:

$$N_f = floor[(0.9 - rand(0, 1) \cdot 0.25) \cdot N] \tag{22.1}$$

where rand is a random number between $(0,1)$ and floor$(.)$ converts a real number to an integer number. The number of male spiders N_m can be calculated as follows.

$$N_m = N - N_f \qquad (22.2)$$

The female spider position f_i is randomly generated within the lower p_j^{low} and the upper p_j^{high} initial parameter bounds as follows.

$$f_{i,j}^0 = p_j^{low} + rand(0,1).(p_j^{high} - p_j^{low}) \qquad (22.3)$$
$$i = 1, 2, \ldots, N_f; j = 1, 2, \ldots, n$$

The male spider position m_i is randomly generated as follows.

$$m_{k,j}^0 = p_j^{low} + rand(0,1).(p_j^{high} - p_j^{low}) \qquad (22.4)$$
$$k = 1, 2, \ldots, N_m; j = 1, 2, \ldots, n$$

The zero signals represent the initial population and j, i and k are the parameter and individual indices, respectively. The value of $f_{i,j}$ is the jth parameter of the ith female spider position.

22.2.3 Evaluation of the solution quality

In the SSO algorithm, the solution quality is represented by the weight of each spider. Each solution i is evaluated by calculating its fitness function value as follows.

$$w_i = \frac{J(s_i) - worst_s}{best_s - worst_s} \qquad (22.5)$$

where $J(s_i)$ is the fitness value of the spider position s_i with regard to the substituted objective function $J(\cdot)$. The value $worst_s$ represents the maximum solution's value while the $best_s$ represents the minimum value of the solution in the population. These values are defined by considering the following minimization problem as follows.

$$best_s = \min_{k \in \{1,\ldots,N\}} J(s_k) \quad \text{and} \quad worst_s = \max_{k \in \{1,\ldots,N\}} J(s_k) \qquad (22.6)$$

22.2.4 Modeling of the vibrations through the communal web

The colony members share and transmit their information through the communal web by encoding it as small vibrations. These vibrations are critical for the collective coordination of all members in the population. The weight and the distance of the spider are responsible for generating these vibrations. The transmitted information (vibrations) perceived by the solution i from solution j is modeled as follows.

$$Vib_{i,j} = w_j \cdot e^{-d_{i,j}^2} \tag{22.7}$$

where the $d_{i,j}$ is the Euclidian distance between the spiders i and j.

The vibrations between any pair of individuals can be defined as follows.

- **Vibrations** $Vibc_i$. The vibrations (transmitted information) between the solution i and the nearest solution to it (which is solution c (s_c) that has a higher weight) can be defined as follows.

$$Vibc_i = w_c \cdot e^{-d_{i,c}^2} \tag{22.8}$$

- **Vibrations** $Vibb_i$. The vibrations (transmitted information) between the solution i and the best solution b (s_b) in the population can be defined as follows.

$$Vibb_i = w_b \cdot e^{-d_{i,b}^2} \tag{22.9}$$

- **Vibrations** $Vibf_i$. Finally, the vibrations (transmitted information) between the solution i and the nearest female solution $f(s_f)$ can be defined as

$$Vibf_i = w_f \cdot e^{-d_{i,f}^2} \tag{22.10}$$

22.2.5 Female cooperative operator

The female spiders attract or dislike other males. The movement of attraction or repulsion based on several random phenomena. A uniform random number r_m is generated within the range [0,1]. The attraction movement is generated if r_m is smaller than a threshold PF, otherwise, a repulsion movement is produced as follows.

$$f_i^{t+1} = \begin{cases} f_i^t + \alpha \cdot Vibc_i \cdot (s_c - f_i^t) + \beta \cdot Vibb_i \cdot (s_b - f_i^t) \\ \qquad\qquad\qquad\qquad +\delta \cdot (rand - 0.5) \text{ at } PF \\ f_i^t - \alpha \cdot Vibc_i \cdot (s_c - f_i^t) - \beta \cdot Vibb_i \cdot (s_b - f_i^t) \\ \qquad\qquad\qquad\qquad +\delta \cdot (rand - 0.5) \text{ at } 1 - PF \end{cases} \tag{22.11}$$

where α, β, δ and rand are random numbers in [0,1], whereas t is the number of iterations.

22.2.6 Male cooperative operator

The dominant male spider D is the spider with a weight value above the median value of the other males in the population, while the other males with weights under the median are called non-dominant ND. The median weight is indexed by $N_f + m$. The position of the male spider can be defined as the following.

$$m_i^{t+1} = \begin{cases} m_i^t + \alpha \cdot Vibf_i \cdot (s_f - m_i^t) + \delta \cdot (rand - 0.5) & \text{if } w_{N_f+i} > w_{N_f+m} \\ m_i^t + \alpha \cdot \left(\frac{\sum_{h=1}^{N_m} m_h^t \cdot w_{N_f+h}}{\sum_{h=1}^{N_m} w_{N_f+h}} - m_i^t \right) \end{cases}$$

$$(22.12)$$

where the solution s_f represents the nearest female solution to the male solution i.

22.2.7 Mating operator

The dominant male is responsible for mating a set E^g of female members when it locates them within a specific range r (range of mating), which can be calculated as follows.

$$r = \frac{\sum_{j=1}^{n} (p_j^{high} - p_j^{low})}{2.n} \qquad (22.13)$$

The spider with a heavier weight has a big chance to influence the new product. The influence probability Ps_i of each solution is assigned by the roulette wheel selection method as follows.

$$Ps_i = \frac{w_i}{\sum_{j \in T^t} w_j} \qquad (22.14)$$

22.2.8 Pseudo-code of SSO algorithm

In this subsection, we present the pesudo-code of the SSO algorithm as shown in Algorithm 20.

Algorithm 20 Social spider optimization algorithm.

1: Set the initial value of total number of solutions N in the population size S, threshold PF, and maximum number of iterations Max_{itr}
2: Set the number of female spiders N_f and number of males spiders N_m as in (22.1) and (22.2)
3: Set $t := 0$ ▷ **Counter initialization**
4: **for** $(i = 1; i < N_f + 1; i++)$ **do**
5:     **for** $(j = 1; j < n + 1; j++)$ **do**
6:         $f_{i,j}^t = p_j^{low} + rand(0, 1) \cdot (P_j^{high} - p_j^{low})$
7:     **end for**
8: **end for**   ▷ **Initialize randomly the female spider**
9: **for** $(k = 1; k < N_m + 1; k++)$ **do**
10:     **for** $(j = 1; j < n + 1; j++)$ **do**
11:         $m_{k,j}^t = p_j^{low} + rand(0, 1) \cdot (P_j^{high} - p_j^{low})$
12:     **end for**
13: **end for**   ▷ **Initialize randomly the male spider**
14: **repeat**

15:    **for** $(i = 1; i < N + 1; i + +)$ **do**

16:       $w_i = \frac{J(s_i) - worst_s}{best_s - worst_s}$

17:    **end for**   ▷ **Evaluate the weight (fitness function) of each spider**

18:    **for** $(i = 1; i < N_f + 1; i + +)$ **do**

19:       Calculate the vibrations of the best local and best global solutions $Vibc_i$ and $Vibb_i$ as in (22.8) and (22.9)

20:       **if** $(r_m < PF)$ **then**

21:          $f_i^{t+1} = f_i^t + \alpha \cdot Vibc_i \cdot (s_c - f_i^t) + \beta \cdot Vibb_i \cdot (s_b - f_i^t) + \delta \cdot (rand - 0.5)$

22:       **else**

23:          $f_i^{t+1} = f_i^t - \alpha \cdot Vibc_i \cdot (s_c - f_i^t) - \beta \cdot Vibb_i \cdot (s_b - f_i^t) + \delta \cdot (rand - 0.5)$

24:       **end if**

25:    **end for**

26:    Find the median male individual $(w_{N_f + m})$ from M

27:    **for** $(i = 1; i < N_m + 1; i + +)$ **do**

28:       Calculate $Vibf_i$ as in (22.10)

29:       **if** $(w_{N_f i} > w_{N_f + m})$ **then**

30:          $m_i^{t+1} = m_i^t + \alpha \cdot Vibf_i \cdot (s_f - m_i^t) + \delta \cdot (rand - 0.5)$

31:       **else**

32:          $m_i^{t+1} = m_i^t + \alpha \cdot \left(\frac{\sum_{h=1}^{N_m} m_h^t \cdot w_{N_f + h}}{\sum_{h=1}^{N_m} w_{N_f + h}} - m_i^t \right)$

33:       **end if**

34:    **end for**

35:    Calculate the radius of mating r, where $r = \frac{\sum_{j=1}^{n} (p_i^{high} - p_j^{low})}{2 \cdot n}$ ▷ **Perform the mating operation**

36:    **for** $(i = 1; i < N_m + 1; i + +)$ **do**

37:       **if** $(m_i \in D)$ **then**

38:          Find E^i

39:          **if** E^i is not empty **then**

40:             Form s_{new} using the roulette method

41:             **if** $w_{new} > w_{wo}$ **then**

42:                Set $s_{wo} = s_{new}$

43:             **end if**

44:          **end if**

45:       **end if**

46:    **end for**

47:    $t = t + 1$ ▷ **Iteration counter is increasing**

48: **until** $(t \geq Max_{itr})$ ▷ **Termination criteria are satisfied**

49: Produce the best solution

22.2.9   Description of the SSO algorithm

In this subsection, we give a description of the SSO algorithm. The algorithm initializes the values of the number of solutions N in the population size P, threshold PF and the maximum number of iterations Max_{itr}. The number of female and male solutions are assigned as shown in (22.1) and (22.2). The counter of the initial iteration is initialized and the initial population is randomly generated which contains the female and the male solutions. The following steps are repeated until termination criteria are satisfied.

- The fitness function of each solution is calculated to determine its weight as shown in (22.5).

- The female spiders are moving according to their cooperative operator by calculating the vibrations of the local and global best spiders as shown in (22.8) and (22.9).

- The male spiders are moving according to their cooperative operator by calculating the median male solution w_{N_f+m} from all male spiders.

- The mating operation is applied by calculating the radius of mating as shown in (22.13).

- The number of iterations is increased.

The overall process is repeated until termination criteria are satisfied. Finally, the best obtained solution is presented as the optimal or near optimal solution.

22.3 Source-code of SSO algorithm in Matlab

In this section, we present the source code of the fitness function, which we need to minimize by using SSO algorithm as shown in Listing 22.1. The fitness function is shown in Equation 22.15. The function takes the whole population X and outputs the objective function of each solution in the population. Also, we present the source codes of the main SSO algorithm and its functions in Matlab [4] as shown in Listing 22.2.

$$f(X_i) = \sum_{j=1}^{D} X_{i,j}^2 \qquad \text{where } -10 \leqslant X_{i,j} \leqslant 10 \tag{22.15}$$

```
1  function [out]=fun(X)
2  [x,y]=size(X);
3  out=zeros(x,1);
4  for i=1:x
5      for j=1:y
6          out(i,1)=out(i,1)+X(i,j)^2;
7      end
8  end
```

Listing 22.1
Definition of objective function $fun(.)$ in Matlab.

```
1  function [bfit,befit,spbest,spbesth] = SSO(spidn,L,U,fun,dim,itern)
2
3      f=fun;
4      xd=L;
5      xu=U;
6      dims = dim;
7      for i=1:dims
8          lb(i,:)=xd;
9          ub(i,:)=xu;
10     end
11     rand(state,0);   % Reset the random generator
```

```
12      % Define the population of females and males
13      fpl = 0.65;      % Lower Female Percent
14      fpu = 0.9;       % Upper Female Percent
15      fp = fpl+(fpu-fpl)\cdot rand; % Aleatory Percent
16      fn = round(spidn\cdot fp);     % Number of females
17      mn = spidn-fn;                % Number of males
18   %Probabilities of attraction or repulsion
19   % Proper tuning for better results
20      pm = exp(-(0.1:(3-0.1)/(itern-1):3));
21      % Initialization of vectors
22      fsp = zeros(fn,dims);     % Initialize females
23      msp = zeros(mn,dims);     % Initialize males
24      fefit = zeros(fn,1);      % Initialize fitness females
25      mafit = zeros(mn,1);      % Initialize fitness males
26      spwei = zeros(spidn,1);   % Initialize weigth spiders
27      fewei = zeros(fn,1);  % Initialize weigth spiders
28      mawei = zeros(mn,1);  % Initialize weigth spiders
29   %% Population Initialization
30   % Generate Females
31      for i=1:fn
32          fsp(i,1:dims)=lb(1)+rand(1,dims)* (ub(1)-lb(1));
33          fsp=round(fsp);
34      end
35   % Generate Males
36      for i=1:mn
37          msp(i,1:dims)=lb(1)+rand(1,dims).* (ub(1)-lb(1));
38          msp=round(msp);
39      end
40   %% **** Evaluations *****
41   % Evaluation of function for females
42      for i=1:fn
43          fefit(i)=f(fsp(i,:),dims);
44      end
45   % Evaluation of function for males
46      for i=1:mn
47          mafit(i)=f(msp(i,:),dims);
48      end
49   %% ***** Assign weight or sort ***********
50   % Obtain weight for every spider
51      spfit = [fefit mafit];    % Mix Females and Males
52      bfitw = min(spfit);       % best fitness
53      wfit = max(spfit);        % worst fitness
54      for i=1:spidn
55          spwei(i) = 0.001+((spfit(i)-wfit)/(bfitw-wfit));
56      end
57      fewei = spwei(1:fn);       % Separate the female mass
58      mawei = spwei(fn+1:spidn);% Separate the male mass
59   %% Memory of the best
60   % Check the best position
61      [~,Ibe] = max(spwei);
62   % Check if female or male
63      if Ibe > fn
64          % Is Male
65          spbest=msp(Ibe-fn,:);    % Assign best position to spbest
66          bfit = mafit(Ibe-fn);      % Get best fitness for memory
67      else
68          % Is Female
69          spbest=fsp(Ibe,:);       % Assign best position to spbest
70          bfit = fefit(Ibe);       % Get best fitness for memory
71      end
72   %% Start the iterations
73      for i=1:itern
74      %% ***** Movement of spiders *****
75      % Move Females
76      [fsp] = FeMove(spidn,fn,fsp,msp,spbest,Ibe,spwei,dims,lb,ub,pm(i)
         );
77      % Move Males
```

```
78        |msp| = MaMove(fn ,mn, fsp ,msp, fewei ,mawei ,dims ,lb ,ub ,pm( i ));
79        %% **** Evaluations *****
80        % Evaluation of function for females
81        for j=1:fn
82            fefit(j)=f(fsp(j ,:) ,dims);
83        end
84        % Evaluation of function for males
85        for j=1:mn
86            mafit(j)=f(msp(j ,:) ,dims);
87        end
88
89 end
```

Listing 22.2
The main code for the social spider optimization algorithm $SSO(.)$ in Matlab.

The rest of the Matlab code is presented in
https://www.mathworks.com/matlabcentral/fileexchange/ 46942-a-swarm-
-optimization-algorithm-inspired-in-the-behavior-of-the-social-spider

22.4 Source-code of SSO algorithm in C++

In this section, we highlight the C++ code of SSO algorithm as follows.

```cpp
1  #include "SSA.h"
2
3  using namespace std;
4
5  class fun : public Problem {
6  public:
7      fun(unsigned int dimension) : Problem(dimension) { }
8
9      double eval(const std::vector<double>& solution) {
10         double sum = 0.0;
11         for (int i = 0; i < solution.size(); ++i) {
12             sum += solution[i] * solution[i];
13         }
14         return sum;
15     }
```

Listing 22.3
Definition of objective function $fun(.)$ and the main file in C++.

```cpp
1  #ifndef SSA_SSA_H
2  #define SSA_SSA_H
3
4  #include <chrono>
5  #include <iostream>
6  #include <math.h>
7  #include <stdio.h>
8  #include <stdlib.h>
9  #include <vector>
10
11 class Problem {
12 public:
13     unsigned int dimension;
14
15     Problem(unsigned int dimension) : dimension(dimension) { }
```

```
16      virtual double eval(const std::vector<double>& solution) = 0;
17  };
18
19  class Position {
20  public:
21      double fitness;
22      std::vector<double> solution;
23
24      Position() { };
25      Position(const std::vector<double>& solution, double fitness) :
26              solution(solution), fitness(fitness) { }
27
28      friend bool operator==(const Position& p1, const Position& p2) {
29          for (int i = 0; i < p1.solution.size(); ++i) {
30              if (p1.solution[i] != p2.solution[i]) {
31                  return false;
32              }
33          }
34          return true;
35      }
36
37      friend double operator-(const Position& p1, const Position& p2) {
38          double distance = 0.0;
39          for (int i = 0; i < p1.solution.size(); ++i) {
40              distance += fabs(p1.solution[i] - p2.solution[i]);
41          }
42          return distance;
43      }
44
45      static Position init_position(Problem* problem);
46  };
47
48  class Vibration {
49  public:
50      double intensity;
51      Position position;
52      static double C;
53
54      Vibration() { }
55      Vibration(const Position& position);
56      Vibration(double intensity, const Position& position);
57
58      double intensity_attenuation(double attenuation_factor, double
          distance) const;
59      static double fitness_to_intensity(double fitness);
60  };
```

Listing 22.4
SSA header file in C++.

The rest of the social spider optimization C++ code is represented in
https://github.com/James-Yu/SocialSpiderAlgorithm.

22.5   Step-by-step numerical example of SSO algorithm

In this section, we apply the steps of the SSO algorithm in Algorithm 20 to
minimize the function in Equation 22.15. We set the initial population size
$N = 6$ and the maximum number of iterations Max_{itr} in step 1. In step 2, we

set the number of female spiders N_f to 5 and the number of male spiders N_m to 1. We initialize the iteration counter $t = 0$ in step 3. The initial population of female population fsp as shown in Equation 22.3 is created in steps 4–8 as follows.

$N1 = \{-5.3772, 2.1369, -0.2804, 7.8260, 5.2419\}$
$N2 = \{-0.8706, -9.6299, 6.4281, -1.1059, 2.3086\}$
$N3 = \{5.8387, 8.4363, 4.7641, -6.4747, -1.8859\}$
$N4 = \{8.7094, 8.3381, -1.7946, 7.8730, -8.8422\}$
$N5 = \{-2.9426, 6.2633, -9.8028, -7.2222, -5.9447\}$

while the population of male msp as shown in Equation 22.4 is generated in steps 9–13 as follows. $N6 = \{-6.0256, 2.0758, -4.5562, -6.0237, -9.6945\}$

Each solution in the population (females and males) is evaluated by calculating its weight (fitness function) as shown in Equation 22.5 in steps 15–17 as follows.

In step 16, the values of the objective function for each solution in the populations $J(.)$ are $\{j = 122.2832, 141.3675, 173.4364, 288.7659, 231.4821, 191.6448\}$. The values of the overall worst and best solutions in the population are $worsts = 288.7659$, $bests = 122.2832$, respectively. The quality for each solution is assigned by calculating the weight of each solution in the population as follows $w_i = \{1.0010, 0.8864, 0.6937, 0.0010, 0.3451, 0.5844\}$. In steps 18–19, we calculate the vibrations of the best local and global solutions $Vibc_i$ and $Vibb_i$, where $Vibc_i = 1.5753$ and $Vibb_i = 1.7244$. In step 20, the random number $rm = 0.1934$ is less than the threshold $PF = 0.9048$ so the female spider is moving as shown in Equation 22.11 as follows:

α, β are random vectors, where $\alpha = \{0.6822, 0.3028, 0.5417, 0.1509, 0.6979\}$, $\beta = \{0.3784, 0.8600, 0.8537, 0.5936, 0.4966\}$ and $\delta = 2 \times PF$, where $\delta = 1.8097$. The new female solutions $f1 = \{-5.3772, 2.1369, -0.2804, 7.8260, 5.2419\}$

$f2 = \{-0.8706, -9.6299, 6.4281, -1.1059, 2.3086\}$
$f3 = \{5.8387, 8.4363, 4.7641, -6.4747, -1.8859\}$
$f4 = \{8.7094, 8.3381, -1.7946, 7.8730, -8.8422\}$
$f5 = \{-2.9426, 6.2633, -9.8028, -7.2222, -5.9447\}$.

Based on the previous values, the new female spiders' values are shown below

$f1 = \{-5.5953, 2.7373, -0.2753, 8.2051, 5.1133\}$
$f2 = \{7.8709, -32.3970, 22.2858, -11.8858, -2.9965\}$
$f3 = \{20.3480, 20.9845, 11.6847, -32.3357, -13.7400\}$
$f4 = \{23.3709, 16.5485, -3.0839, 7.9691, -12.0286\}$
$f5 = \{-1.3881, 11.5068, -24.9425, -21.5623, -11.1657\}$.

In step 26, the median male equals 0.5844. In steps 27–34, the male solutions are updated according to the male weight $w_{N_f i}$ value and the median weight $w_{N_f + m}$, where $w_{N_f i} = 0.3451$ and $w_{N_f + m} = 0.5844$. The random vector is α where $\alpha = \{0.9891, 0.0359, 0.3798, 1.3668, 0.8667\}$. The vibration of the male spider $Vibf_i = 1.0172$.

The nearest female solution is $s_f - m_i^t$, where

$$s_f - m_i^t = \{-1.6925, -9.6998, 5.7772, -5.7207, -3.5993\}.$$

According to the previous values, the new male spider m_i^{t+1} is $m_i^{t+1} = \{-2.1862, 2.2598, -0.3307, -6.2653, -4.8513\}$. In step 35, the radius of mating in Equation 22.13 is calculated and set to $r = 3.4991$. The surviving operator in lines 36-46 is starting to generate a new male solution as follows $spbest = \{-1.9968, 2.7438, -0.6281, 1.4694, 2.3207\}$. In step 47 the iteration counter is increased and the termination criteria is checked in step 48. Finally the best solution is produced, which is $\{-2.1862, 2.2598, -0.3307, -6.2653, -4.8513\}$ and its value $= 72.7846$. The overall processes are repeated until termination criteria are satisfied.

22.6 Conclusion

In this chapter, we show the main steps of the SSO algorithm and how it works. We demonstrate the source code in Matlab and C++ language in order to help the user to implement it on various applications. In order to give a more thorough understanding of the SSO algorithm, we present a step by step numerical example and show how it can solve a global optimization problem.

References

1. L. Aviles. "Sex-Ratio bias and possible group selection in the social spider snelosimus eximius". The American Naturalist, vol. 128, no. 1, pp. 1–12, 1986.

2. L. Avilés. "Causes and consequences of cooperation and permanent-sociality in spiders". In B. C. Choe, The Evolution of Social Behavior in Insects and Arachnids pp. 476-498, Cambridge, Massachusetts.: Cambridge University Press, 1997.

3. T.R. Chandran, A. V. Reddy, and B. Janet. "An effective implementation of Social Spider Optimization for text document clustering using single cluster approach". In 2018 Second International Conference on Inventive Communication and Computational Technologies (ICICCT) (pp. 508-511). IEEE, 2018.

4. E. Cuevas, M. Cienfuegos, D. Zaldívar and M. Pérez-Cisneros. "A swarm optimization algorithm inspired in the behavior of the social-spider". Expert Systems with Applications, vol. 40, no. 16, pp. 6374–6384, 2013.

5. J.Q. James and V.O. Li. "A social spider algorithm for global optimization". Applied Soft Computing, vol. 30, pp. 614–627, 2015.

6. C. Eric and K. S. Yip. "Cooperative capture of large prey solves scaling challenge faced by spider societies". Proceedings of the National Academy of Sciences of the United States of America, vol. 105, no. 33, pp. 11818–11822, 2008.

7. S. Maxence. "Social organization of the colonial spider Leucauge sp. in the Neotropics: vertical stratification within colonies". The Journal of Arachnology, vol. 38, pp. 446–451, 2010.

8. M.A. Tawhid, A. F. Ali. "A simplex social spider algorithm for solving integer programming and minimax problems". Memetic Computing, vol. 8, no. 3, pp. 169–188, 2016.

9. M.A. Tawhid, A. F. Ali. "A hybrid social spider optimization and genetic algorithm for minimizing molecular potential energy function". Soft Computing, vol. 21, no. 21, pp. 6499–6514, 2017.

10. A. Pasquet. "Cooperation and prey capture efficiency in a social spider, Anelosimus eximius (Araneae, Theridiidae)". Ethology, vol. 90, pp. 121–133, 1991.

23

Stochastic Diffusion Search: A Tutorial

Mohammad Majid al-Rifaie

School of Computing and Mathematical Sciences
University of Greenwich, Old Royal Naval College, London, United Kingdom

J. Mark Bishop

Department of Computing
Goldsmiths, University of London, United Kingdom

CONTENTS

23.1 Introduction

Noisy environments and incomplete data are often at the heart of hard, real-world search and optimisation-related problems, generating input that established search heuristics (e.g. tabu search [1], simulated annealing [2], etc.) sometimes have difficulty dealing with [3]. Conversely, ever since their inception, researchers have been attracted to the complex emergent behaviour, robustness and easy-to-understand architecture of nature-inspired swarm intelligence algorithms, and, particularly in challenging search environments. These algorithms have often proved more useful than conventional approaches [4]. SDS has been applied to various fields, including but not limited to: computational creativity [5, 6, 7, 8, 9] digital arts [10, 11, 12, 13] optimisation [14, 15, 16, 17] medical imaging [18, 19, 20, 21] machine learning [22, 23, 24, 25, 26] and other theoretical or practical domains [27, 28, 29, 30, 31, 32, 33].

This chapter explains the principles of Stochastic Diffusion Search, a multi-agent global search and optimisation swarm intelligence algorithm based upon simple iterated interactions between agents. First a high-level description of the algorithm is presented in the form of a search metaphor driven by social interactions. This is then followed by an example of a trivial 'text search' application to illustrate the core algorithmic processes by which standard SDS operates.

23.2   Stochastic Diffusion Search

Stochastic Diffusion Search (SDS) [34] has introduced a new probabilistic approach for solving best-fit pattern recognition and matching problems. SDS, as a multi-agent population-based global search and optimisation algorithm, is a distributed mode of computation utilising interaction between simple agents [36]. Its computational roots stem from Geoff Hinton's interest in 3D object classification and mapping (see [37, 38] for Hinton's work and [34, 35] for the connection between Hinton mapping and SDS).

Unlike many nature inspired search algorithms, SDS has a strong mathematical framework, which describes the behaviour of the algorithm by investigating its resource allocation [39], convergence to global optimum [40], robustness and minimal convergence criteria [41] and linear time complexity [42]. In order to introduce SDS, a social metaphor, *the Mining Game*, is introduced.

23.2.1   The mining game

The mining game provides a simple metaphor outlining the high-level behaviour of agents in SDS:

A group of friends (miners) learn that there is gold to be found on the hills of a mountain range but have no information regarding its distribution. On their maps, the mountain range is divided into a set of discrete hills and each hill contains a discrete set of seams to mine. Over time, on any day the probability of finding gold at a seam is proportional to its net wealth.

To maximise their collective wealth, the miners need to identify the hill with the richest seams of gold so that the maximum number of miners can dig there (this information is not available a-priori). In order to solve this problem, the miners decide to employ a simple Stochastic Diffusion Search.

- At the start of the mining process each miner is randomly allocated a hill to mine (his hill hypothesis, h).

- Every day, each miner is allocated a randomly selected seam on his hill to mine.

- At the end of each day, the probability that a miner is happy is proportional to the amount of gold he has found.

- At the end of the day, the miners congregate and over the evening each miner who is unhappy selects another miner at random to talk to. If the chosen miner is happy, he happily tells his colleague the identity of the hill he is mining (that is, he communicates his hill hypothesis, h, which thus both now maintain). Conversely, if the chosen miner is unhappy he says nothing and the original miner is once more reduced to selecting a new hypothesis – identifying the hill he is to mine the next day – at random.

In the context of SDS, agents take the role of miners; active agents being 'happy miners', inactive agents being 'unhappy miners' and the agent's hypothesis being the miner's 'hill-hypothesis'. It can be shown that this process is isomorphic to SDS, and thus that the miners will naturally self-organise and rapidly congregate over hill(s) on the mountain range with a high concentration of gold.

Algorithm 21 The Mining Game.

```
 1   Initialisation phase
 2   Allocate each miner (agent) to a random
 3     hill (hypothesis) to pick a region randomly
 4
 5   Until (all miners congregate over the highest concentration of gold)
 6
 7   Test phase
 8     - Each miner evaluates the amount of gold they have mined (hypotheses
       evaluation)
 9     - Miners are classified into happy (active) and unhappy (inactive)
       groups
10
11   Diffusion phase
12     - Unhappy miners consider a new hill by either communicating with
       another miner;
13     - or,if the selected miner is also unhappy, there will be no
       information flow between the miners; instead the selecting miner must
       consider another hill (new hypothesis) at random
14   End
15
```

23.2.2 Refinements in the metaphor

There are some refinements in the miners analogy, which will elaborate more on the correlation between the metaphor and different implementations of the algorithm. Whether an agent is active or not can be measured probabilistically

or gold may be considered as a resource of discrete units. In both cases, the agents are either active or inactive at the end of each iteration[1]; this is isomorphic to standard SDS. The Mining Game can be further refined through either of the following two assumptions at each location:

1. Finite resources: the amount of gold is reduced each time a miner mines the area

2. Infinite resources: a conceptual situation with potentially infinite amount of gold

In the case of having finite resources, the analogy can be related to a real world experiment of robots looking for food to return to a notional nest site [43]. Hence the amount of food (or gold, in the mining analogy) is reduced after each discovery. In this case, the goal is identifying the location of the resources throughout the search space. This type of search is similar to conducting a search in a dynamically, agent-initiated changing environment where agents change their congregation from one area to another.

The second assumption has similarities with discrete function optimisation where values at certain points are evaluated. However further re-evaluation of the same points does not change their values and they remain constant.

23.3   SDS architecture

The SDS algorithm commences a search or optimisation by initialising its population (e.g. miners, in the mining game metaphor). In any SDS search, each agent maintains a hypothesis, h, defining a possible problem solution. In the mining game analogy, agent hypothesis identifies a hill. After initialisation, two phases are followed (see Algorithm 21 for these phases in the mining game; for high-level SDS description see Algorithm 22):

• Test Phase (e.g. testing gold availability)

• Diffusion Phase (e.g. congregation and exchanging of information)

In the test phase, SDS checks whether the agent hypothesis is successful or not by performing a partial hypothesis evaluation and returning a domain independent boolean value. Later in the iteration, contingent on the strategy employed, successful hypotheses diffuse across the population and in this

---

[1]Whether an agent is active or not is defined using the following two methods:

• probabilistically: a function f takes a probability p as input and returns either true or false, $f(p) \implies Active|Inactive$

• discretely: if there is gold, the agent will be active, otherwise it will be inactive.

way information on potentially good solutions spreads throughout the entire population of agents.

In the Test phase, each agent performs *partial function evaluation, pFE*, which is some function of the agent's hypothesis; $pFE = f(h)$. In the mining game the partial function evaluation entails mining a random selected region on the hill, which is defined by the agent's hypothesis (instead of mining all regions on that hill).

In the Diffusion phase, each agent recruits another agent for interaction and potential communication of the hypothesis. In the mining game metaphor, diffusion is performed by communicating a hill hypothesis.

Algorithm 22 SDS Algorithm.

```
1   Initialising agents()
2   While (stopping condition is not met)
3       Testing hypotheses()
4           Determining agents' activities (active/inactive)
5       Diffusing hypotheses()
6           Exchanging of information
7   End While
```

23.4   Step by step example: text search

In order to demonstrate the process through which SDS functions, an example is presented which shows how to find a set of letters within a larger string of letters. The goal is to find a 3-letter model (Table 23.1) in a 16-letter search space (Table 23.2). In this example, there are four agents. For simplicity of exposition, a perfect match of the model exists in the Search Space (SS).

In this example, a hypothesis, which is a potential problem solution, identifies three adjacent letters in the search space (e.g. hypothesis '1' refers to Z-A-V, hypothesis '10' refers to G-O-L).

TABLE 23.1

Model.

Index:	0	1	2
Model:	*S*	*I*	*B*

TABLE 23.2

Search Space.

Index:	0	1	2	3	4	5	6	7	8	9	10	11	12	13	14	15
Search Space	X	Z	A	V	M	Z	S	I	B	V	G	O	L	B	E	H

In the first step, each agent initially randomly picks a hypothesis from the search space (see Table 23.3). Assume that:

- the first agent points to the 12th entry of the search space and in order to partially evaluate this entry, it randomly picks one of the letters (e.g. the first one, L):

| **L** | B | E |

- the second agent points to the 9th entry and randomly picks the second letter (G):

| V | **G** | O |

- the third agent refers to the 2nd entry in the search space and randomly picks the first letter (A):

| **A** | V | M |

- the fourth agent goes the 3rd entry and randomly picks the third letter (Z):

| V | M | **Z** |

TABLE 23.3

Initialisation and Iteration 1.

Agent No:	1	2	3	4
Hypothesis position:	12	9	2	3
	L-B-E	**V-G-O**	**A-V-M**	**V-M-Z**
Letter picked:	1st	2nd	1st	3rd
Status:	×	×	×	×

The letters picked are compared to the corresponding letters in the model, which is S-I-B (see Table 23.1).

In this case:

- The 1st letter from the first agent (L) is compared against the 1st letter from the model (S) and because they are not the same, the agent is set inactive.

- For the 2nd agent, the second letter (G) is compared with the second letter from the model (I) and again because they are not the same, the agent is set inactive.

- For the third and fourth agents, letters 'A' and 'Z' are compared against 'S' and 'B' from the model. Since none of the letters correspond to the letters in the model, the statuses of the agents are set inactive.

In the next step, as in the mining game, each inactive agent chooses another agent and adopts the same hypothesis if the selected agent is active. If the selected agent is inactive, the selecting agent generates a random hypothesis.

Assume that the first agent chooses the second one; since the second agent is inactive, the first agent must choose a new random hypothesis from the search space (e.g. 6). See Figure 23.1 for the communication between agents.

FIGURE 23.1
Agents Communication 1.

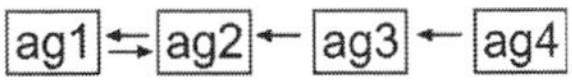

ag1 ⇌ ag2 ← ag3 ← ag4

The process is repeated for the other three agents. As the agents are inactive, they all choose new random hypotheses (see Table 23.4).

TABLE 23.4
Iteration 2.

Agent No:	1	2	3	4
Hypothesis position:	6	10	0	5
	S-I-B	G-O-L	X-Z-A	Z-S-I
Letter picked:	2nd	3rd	1st	1st
Status:	√	×	×	×

In Table 23.4, the second, third and fourth agents do not refer to their corresponding letter in the model, therefore they become inactive. The first agent, with hypothesis '6', chooses the 2nd letter (I) and compares it with the 2nd letter of the model (I). Since the letters are the same, the agent becomes active.

At this stage, consider the following communication between the agents: (see Figure 23.2)

• the fourth agent chooses the second one

• the third agent chooses the second one

• the second agent chooses the first one

FIGURE 23.2
Agents Communication 2.

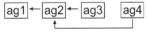

In this case, the third and fourth agents, which chose an inactive agent (the second agent), have to choose other random hypotheses each from the search space (e.g. agent three chooses hypothesis '1' which points to Z-A-V and agent four chooses hypothesis 4 which points to M-Z-S), but the second agent adopts the hypothesis of the first agent, which is active. As shown in Table 23.5:

- The first agent, with hypothesis '6', chooses the 3rd letter (B) and compares it with the 3rd letter of the model (B). Since the letters are the same, the agent remains active.

- The second agent, with hypothesis '6', chooses the 1st letter (S) and compares it with the 1st letter of the model (S). Since the letters are the same, the agent stays active.

- the third and fourth agents do not refer to their corresponding letter in the model, therefore they are set inactive.

TABLE 23.5

Iteration 3.

Agent No:	1	2	3	4
Hypothesis position:	6	6	1	4
	S-I-**B**	**S**-I-B	Z-**A**-V	M-Z-**S**
Letter picked:	3rd	1st	2nd	3rd
Status:	✓	✓	✗	✗

Because the third and fourth agents are inactive, they try to contact other agents randomly. For instance (see Figure 23.3):

- agent three chooses agent two

- agent four chooses agent one

FIGURE 23.3

Agents Communication 3.

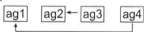

Since agent three chose an active agent, it adopts its hypothesis (6). As for agent four, because it chose agent one, which is active too, it adopts its hypothesis (6). Table 23.6 shows:

- The first agent, with hypothesis '6', chooses the 1st letter (S) and compares it with the 1st letter of the model (S). Since the letters are the same, the agent remains active.

- The second agent, with hypothesis '6', chooses the 2nd letter (I) and compares it with the 2nd letter of the model (I). Since the letters are the same, the agent stays active.

- The third agent, with hypothesis '6', chooses the 3rd letter (B) and compares it with the 3rd letter of the model (B). Since the letters are the same, the agent becomes active.

- The fourth agent, with hypothesis '6', chooses the 1st letter (S) and compares it with the 1st letter of the model (S). Since the letters are the same, the agent is set active.

TABLE 23.6

Iteration 4.

Agent No:	1	2	3	4
Hypothesis position:	6	6	6	6
	S-I-B	**S-I-B**	**S-I-B**	**S-I-B**
Letter picked:	1st	2nd	3rd	1st
Status:	√	√	√	√

At this stage, the entire agent populations are active pointing to the location of the model inside the search space.

23.5 Source code

The source code in this section provides the standard implementation of SDS in three programming languages, Matlab (listings 23.1), C++ (listing 23.2) and Python (listing 23.3).

23.5.1 Matlab

```matlab
clear
% search space (ss); target text (model); population size (N)
ss = 'try to find sds in this sentence';
model = 'sds';
N = 10; maxIter = 30;

% INITIALISE AGENTS
hypo = randi([1 length(ss)-length(model)],1,N) ;
status = false(1,N);

for itr=1:maxIter
  activeAgents = 0;
  % TEST PHASE
  for i=1:N
    % PICK A MICROFEATURE TO PARTIALLY EVALUATE HYPOTHSIS
    microFeature = randi([1 length(model)]);
    if ss( hypo(i)+microFeature ) == model(microFeature)
      status(i) = true;
      activeAgents = activeAgents+1;
    else
      status(i) = false;
    end
  end

  % DIFFUSION PHASE
  for i=1:N
```

```matlab
27    if status(i) == false % INACTIVE AGENT
28      rand = randi([1 N]); % PICK RANDOM AGENT TO COMMUNICATE
29      if status(rand) == true % SHARE HYPOTHESIS
30        hypo(i) = hypo(rand);
31      else % PICK A RANDOM HYPOTHSIS
32        hypo(i) = randi([1 length(ss)-length(model)]);
33      end
34    else % ACTIVE AGENT
35      microFeature = randi([1 length(model)]); % PICK MICROFEATURE
36      if ss( hypo(i)+microFeature ) == model(microFeature)
37        status(i) = true;
38      else
39        status(i) = false;
40      end
41    end
42  end
43  activityPercentage = activeAgents * 100 / N;
44  % DISPLAYING ACTIVITY PERCENTAGE AND THE FIRST AGENT'S HYPOTHESIS
45  disp(['Active agents: ' num2str(activityPercentage) '% ... found: '
      ss(hypo(1):length(model))])
46 end
```

Listing 23.1
SDS code in Matlab.

23.5.2 C++

```cpp
1 #include <iostream>
2 #include <string>
3 #include <stdlib.h>
4 #include <ctime>
5 using namespace std;
6
7 float r() { // GENERATE RANDOM NUMBER IN RANGE [0, 1)
8   return (float) rand() / (RAND_MAX);
9 }
10
11 int main() {
12   srand(time(NULL)); // TO GENERATE DIFFERENT RANDOM NUMBERS
13   string ss = "try to find sds in this sentence"; // SEARCH SPACE
14   string model = "sds"; // TARGET TEXT
15   int N = 10; int maxIter = 30; int hypo[N]; bool status[N];
16
17   // INITIALISE AGENTS
18   for (int i=0; i<N; i++) {
19     hypo[i] = r()*(ss.length() - model.length());
20     status[i] = false;
21   }
22
23   // MAIN LOOP
24   for (int itr=0; itr < maxIter; itr++) {
25     int activeAgents = 0;
26     // TEST PHASE
27     for (int i=0; i<N; i++) {
28       // PICK A MICROFEATURE TO PARTIALLY EVALUATE HYPOTHESIS
29       int microFeature = r()*model.length();
30       if (ss[ hypo[i]+microFeature ] == model[microFeature] ) {
31         status[i] = true;
32         activeAgents++;
33       }
34       else
35         status[i] = false;
36     }
37     // DIFFUSION PHASE
38     for (int i=0; i<N; i++) {
39       if (status[i] == false){ // INACTIVE AGENT
```

```
40    int rand = r()*N; // PICK RANDOM AGENT TO COMMUNICATE
41    if (status[rand] == true) // SHARE HYPOTHESIS
42      hypo[i] = hypo[rand];
43    else // PICK A RANDOM HYPOTHESIS
44      hypo[i] = r()*(ss.length() - model.length());
45    }
46    else { // ACTIVE AGENT
47      int microFeature = r()*model.length(); // PICK MICROFEATURE
48      if (ss[ hypo[i]+microFeature ] == model[microFeature] )
49        status[i] = true;
50      else
51        status[i] = false;
52    }
53    }
54    int activityPercentage = activeAgents * 100 / N;
55    // DISPLAYING ACTIVITY PERCENTAGE AND THE FIRST AGENT'S HYPOTHESIS
56    cout << "Active agents:" << activityPercentage << "%\t   ... found:
      " << ss.substr(hypo[0],model.length()) << endl;
57    }
58    return 0;
59 }
```

Listing 23.2
SDS code in C++.

23.5.3 Python

```
1  import numpy as np
2  # search space (ss); target text (model); population size (N)
3  ss = 'try to find sds in this sentence'
4  model = 'sds'
5  N = 10; maxIter = 30
6
7  # INITIALISE AGENTS
8  hypo = np.random.randint(len(ss)-len(model), size=(N));
9  status = np.zeros((N), dtype=bool)
10
11 for itr in range(maxIter):
12   activeAgents = 0;
13   # TEST PHASE
14   for i in range(N):
15     # PICK A MICROFEATURE TO PARTIALLY EVALUATE HYPOTHESIS
16     microFeature = np.random.randint(len(model))
17     if ss[ hypo[i]+microFeature ] == model[microFeature]:
18       status[i] = True
19       activeAgents += 1
20     else:
21       status[i] = False
22
23   # DIFFUSION PHASE
24   for i in range(N):
25     if status[i] == False: # INACTIVE AGENT
26       rand = np.random.randint(N) # PICK RANDOM AGENT TO COMMUNICATE
27       if status[rand] == True: # SHARE HYPOTHESIS
28         hypo[i] = hypo[rand];
29       else: # PICK A RANDOM HYPOTHESIS
30         hypo[i] = np.random.randint(len(ss)-len(model))
31     else: # ACTIVE AGENT
32       microFeature = np.random.randint(len(model)) # PICK MICROFEATURE
33       if ss[ hypo[i]+microFeature ] == model[microFeature]:
34         status[i] = True;
35       else:
36         status[i] = False;
37
38   activityPercentage = activeAgents * 100 / N;
39   # DISPLAYING ACTIVITY PERCENTAGE AND THE FIRST AGENT'S HYPOTHESIS
```

```
40    print('Active agents:', activityPercentage, '%  ...  found:  ', ss[
          hypo[0]:hypo[0]+len(model)  ]  )
```

Listing 23.3
SDS code in Python.

___

23.6   Conclusion

This chapter aimed at providing an introduction to the main principles of
Stochastic Diffusion Search (SDS). After providing a simple social metaphor,
outlining the high-level behaviour of agents in SDS, the architecture of the
algorithm is presented where the two main phases of SDS (test and diffusion) are detailed. This is then followed by a step by step example, which
is demonstrated through simple text search using SDS. A complete code of
SDS is then provided in Matlab, C++ and Python. Therefore, researchers
and students alike are able to expand their understanding of SDS and apply
the algorithms to the problems of their choice, especially where the concept
of 'partial function evaluation' can be applied.

___

References

1. F. Glover et al. "Tabu search-part I". ORSA Journal on Computing,
   pp. 190-206, 1989.

2. S. Kirkpatric, C.D. Gelatt, M.P. Vecchi. "Optimization by simulated
   annealing". Science, vol. 220(4598), pp. 671-680, 1983.

3. Y. Jin, J. Branke. "Evolutionary optimization in uncertain
   environments-a survey". IEEE Transactions on Evolutionary Computation, vol. 9(3), pp. 303-317, 2005.

4. J.F. Kennedy, R.C. Eberhart, Y. Shi. "Swarm Intelligence". Morgan
   Kaufmann Publishers, 2001.

5. M.M. al-Rifaie, J.M. Bishop, S. Caines. "Creativity and Autonomy
   in Swarm Intelligence Systems". Cognitive Computation, vol. 4(3),
   pp. 320-331, 2012.

6. M.M. al-Rifaie, J.M. Bishop. "Weak and Strong Computational Cre-
   ativity" in Computational Creativity Research: Towards Creative
   Machines. Atlantis Thinking Machines, vol. 7, pp. 37-49, Atlantis
   Press, 2015.

7. M.M. al-Rifaie, F.F. Leymarie, W. Latham, M. Bishop. "Swarmic
   autopoiesis and computational creativity". Connection Science, pp.
   1-19, 2017.

8. J.M. Bishop, M.M. al-Rifaie. "Autopoiesis, creativity and dance". Connection Science, vol. 29(1), pp. 21-35, 2017.

9. M.M. al-Rifaie, A. Cropley, D. Cropley, M. Bishop. "On evil and computational creativity". Connection Science, vol. 28(1), pp. 171-193, 2016.

10. M.M. al-Rifaie, A. Aber, M. Bishop. "Cooperation of Nature and Physiologically Inspired Mechanisms in Visualisation" in Biologically-Inspired Computing for the Arts: Scientific Data through Graphics, IGI Global, United States, 2012.

11. M.M. al-Rifaie, M. Bishop. "Swarmic Paintings and Colour Attention". Lecture Notes in Computer Science, vol. 7834, pp. 97-108, Springer, 2013.

12. M.M. al-Rifaie, M. Bishop. "Swarmic Sketches and Attention Mechanism". Lecture Notes in Computer Science, vol. 7834, pp. 85-96, Springer, 2013.

13. A.M. al-Rifaie, M.M. al-Rifaie. "Generative Music with Stochastic Diffusion Search". Lecture Notes in Computer Science, vol. 9027, pp. 1-14, Springer, 2015.

14. M.M. al-Rifaie, M.J. Bishop, T. Blackwell. "An investigation into the merger of stochastic diffusion search and particle swarm optimisation" in *Proc. of the 13th Annual Conference on Genetic and Evolutionary Computation, GECCO 2011*, pp. 37-44, 2011.

15. M.M. al-Rifaie, J.M. Bishop, T. Blackwell. "Information sharing impact of stochastic diffusion search on differential evolution algorithm". Memetic Computing, vol. 4(4), pp. 327-338, 2012.

16. M.M. al-Rifaie, M. Bishop, T. Blackwell. "Resource Allocation and Dispensation Impact of Stochastic Diffusion Search on Differential Evolution Algorithm" in Nature Inspired Cooperative Strategies for Optimization (NICSO 2011). Studies in Computational Intelligence, vol. 387, pp. 21-40, Springer, 2012.

17. M.G.H. Omran, A. Salman. "Probabilistic stochastic diffusion search" in Proc. of International Conference on Swarm Intelligence (ANTS 2012), Lecture Notes in Computer Science, vol. 7461, pp. 300-307, Springer, 2012.

18. M.M. al-Rifaie, A. Aber, D.J. Hemanth. "Deploying swarm intelligence in medical imaging identifying metastasis, micro-calcifications and brain image segmentation". Systems Biology, IET, vol. 9(6), pp. 234-244, 2015.

19. M.M. al-Rifaie, A. Aber. "Identifying Metastasis in Bone Scans with Stochastic Diffusion Search" in Proc. of International Symposium on Information Technologies in Medicine and Education, pp. 519-523, 2012.

20. M.M. al-Rifaie, A. Aber, A.M. Oudah. "Utilising Stochastic Diffusion Search to identify metastasis in bone scans and microcalcications on mammographs" in Proc. of IEEE International Conference on Bioinformatics and Biomedicine Workshops, pp. 280-287, 2012.

21. M.M. al-Rifaie, A. Aber, A.M. Oudah. "Ants intelligence framework; identifying traces of cancer" in The House of Commons, UK Parliment. SET for BRITAIN 2013. Poster exhibitions in Biological and Biomedical Science, 2013.

22. M.M. al-Rifaie, M. Y.-K. Matthew and M. d'Inverno. "Investigating swarm intelligence for performance prediction" in *Proc. of the 9th International Conference on Educational Data Mining*, pp. 264-269, 2016.

23. H. Alhakbani, M.M. al-Rifaie. "Feature selection using stochastic diffusion search" in *Proc. of the Genetic and Evolutionary Computation Conference*, pp. 385-392, 2017.

24. M.M. al-Rifaie, D. Joyce, S. Shergill, M. Bishop. "Investigating stochastic diffusion search in data clustering" in SAI Intelligent Systems Conference (IntelliSys), pp. 187-194, 2015.

25. H.A. Alhakbani, M.M. al-Rifaie. "Exploring Feature-Level Duplications on Imbalanced Data Using Stochastic Diffusion Search" in Multi-Agent Systems and Agreement Technologies, pp. 305-313, Springer, 2016.

26. I. Aleksander, T.J. Stonham. "Computers and Digital Techniques 2(1)" in Lecture Notes in Artificial Intelligence, vol. 1562, pp. 29-40, Springer, 1979.

27. M.A.J. Javid, M.M. al-Rifaie, R. Zimmer. "Detecting symmetry in cellular automata generated patterns using swarm intelligence" in *Proc. of 3rd International Conference on the Theory and Practice of Natural Computing (TPNC 2014)*, pp. 83-94, 2014.

28. F.M. al-Rifaie, M.M. al-Rifaie. "Investigating stochastic diffusion search in DNA sequence assembly problem" in *Proc. of SAI Intelligent Systems Conference (IntelliSys)*, pp. 625-631, 2015.

29. M.A.J. Javid, W. Alghamdi, A. Ursyn, R. Zimmer, M.M. al-Rifaie. "Swarmic approach for symmetry detection of cellular automata behaviour". Soft Computing, vol. 21(19), pp. 5585-5599, 2017.

30. E. Grech-Cini. "Locating Facial Features". PhD Thesis, University of Reading, Reading, UK, 1995.

31. P.D. Beattie, J.M. Bishop. "Self-localisation in the SENARIO autonomous wheelchair". Journal of Intellingent and Robotic Systems, vol. 22, pp. 255-267, 1998.

32. A.K. Nircan. "Stochastic Diffusion Search and Voting Methods". PhD Thesis, Bogaziki University, 2006.

33. K. de-Meyer, M. Bishop, S. Nasuto. "Small World Effects in Lattice Stochastic Diffusion Search". Lecture Notes in Computer Science, vol. 2415, pp. 147-152, 2002.

34. J.M. Bishop. "Stochastic searching networks" in *Proc. of 1st IEE Conf. on Artificial Neural Networks*, pp. 329-331, 1989.

35. J.M. Bishop, P. Torr. "The Stochastic Search Network" in Neural Networks for Images, Speech and Natural Language, pp. 370-387, Chapman & Hall, New York, 1992.

36. K. de-Meyer, J.M. Bishop, S.J. Nasuto. "Stochastic diffusion: Using recruitment for search" in Proc. of Symposium on Evolvability and Interaction, pp. 60-65, The University of London, UK, 2003.

37. G.F. Hinton. "A parallel computation that assigns canonical object-based frames of reference" in *Proc. of the 7th International Joint Conference on Artificial Intelligence*, vol. 2, pp. 683-685, 1981.

38. J.L. McClelland, D.E. Rumelhart, et al. "Parallel Distributed Processing, Volume 2: Explorations in the Microstructure of Cognition: Psychological and Biological Models", A Bradford Book, MIT Press, 1986.

39. S.J. Nasuto. "Resource Allocation Analysis of the Stochastic Diffusion Search". PhD Thesis, University of Reading, Reading, UK.

40. S.J. Nasuto, J.M. Bishop. "Convergence analysis of Stochastic Diffusion Search". Parallel Algorithms and Applications, vol. 14(2), pp. 89-107, 1999.

41. D.R. Myatt, J.M. Bishop, S.J. Nasuto. "Minimum stable convergence criteria for Stochastic Diffusion Search". Electronics Letters, vol. 40(2), pp. 112-113, 2004.

42. S.J. Nasuto, J.M. Bishop, S. Lauria. "Time Complexity Analysis of Stochastic Diffusion Search" in Proc. of International ICSC IFAC Symposium on Neural Computation, pp. 260-266, 1998.

43. M.J. Krieger, J.B. Billeter, L. Keller. "Ant-like task allocation and recruitment in cooperative robots". Nature, vol. 406, pp. 992-995, 2000.

24

Whale Optimization Algorithm

Ali R. Kashani
Department of Civil Engineering
University of Memphis, Memphis, United States

Charles V. Camp
Department of Civil Engineering
University of Memphis, Memphis, United States

Moein Armanfar
Department of Civil Engineering
Arak University, Arak, Iran

Adam Slowik
Department of Electronics and Computer Science
Koszalin University of Technology, Koszalin, Poland

CONTENTS

24.1   Introduction

In 2016, Mirjalili and Lewis [1] developed a swarm-based optimization algorithm called the whale optimization algorithm (WOA). This algorithm numerically models the social behavior and hunting strategies of humpback whales. A WOA explores the solution search space using a bubble-net feeding method. Humpback whales tend to entrap school of krill or small fishes through a circular or '9'-shaped bubble-made cage. To accomplish this task whales use two different tactics: upward-spirals or double-loops. More detail about those tactics can be found in [1, 2]. In general, a WOA uses three main steps: encircling prey; bubble-net attacking method; and search for prey. The encircling the prey method leads all the search agents towards the best-found solution (leader). Next the bubble-net attacking phase (exploitation) simulates the path which whales use to get close to their prey. Based on this strategy, whales move on both circular and spiral-shape paths simultaneously. Finally, in the searching for prey (exploration) phase, randomly selected agents modify the position of the i-th search agent. Successful applications of the WOA for different engineering problems have attracted many in the research community [3-5]. From this research, several variations and improvements on the WOA have been developed such as: a Lévy flight trajectory-based WOA which prevents premature convergence and helps to avoid local optimum solutions [6], an adaptive autoregressive WOA (used for handling a traffic-aware routing in VANET) [7], an improved WOA that uses a dynamic strategy for updating control parameters and applies a quadratic interpolation to the leader that enhanced its ability to handle large scale optimization problem [8], a chaos WOA (applies chaos theory tried to optimize the Elman neural network) [9], and a multi-objective WOA that considers a multi-level threshold for image segmentation [10]. The remainder of this chapter is organized accordingly. In Section 24.2 a step-by-step pseudo-code and the description for the original WOA algorithm are provided, in Sections 24.3 and 24.4 source-codes of the WOA algorithm in both Matlab and C++ are presented, respectively, in Section 24.5 a detailed numerical example is solved by WOA, and finally a conclusion is presented in Section 24.6.

24.2   Original WOA

24.2.1   Pseudo-code of the WOA

Algorithm 23 presents the pseudo-code for the global version of the WOA.

Algorithm 23 Pseudo-code of the original WOA.

1: define the $D - th$ dimensional objective function $OF(.)$
2: define the range of variability for each $j - th$ dimension $\left[X_{i,j}^{min}, X_{i,j}^{max}\right]$
3: determine the WOA algorithm parameter values such as $SearchAgents_no$
   – the number of search agents, MI – maximum iteration
4: randomly create positions X_i for $SearchAgents\_no$ number of search agents (each agent is a D-dimensional vector)
5: find the best search agent and call it leader
6: $Iter = 0$
7: **while** termination condition not met (here is reaching MI) **do**

8:      **for** each i-th search agent **do**

9:          update α value to decrease from 2 to 0 using formula:

10:          $\alpha = 2 - Iter \times \left(\frac{2}{MI}\right)$

11:          $A = 2\alpha \times rand - \alpha$

12:          $C = 2 \times rand$

13:          determine p as a random number between 0 and 1

14:          **if** p<0.5 **then**

15:             **if** $\mid A \mid$<1 **then**

16:                update the position of i-th agent using:

17:                $X\left(Iter\right) = X^*\left(Iter - 1\right) - A \cdot \mid C \cdot X^*\left(Iter - 1\right) - X\left(Iter - 1\right) \mid$

18:             **else**

19:                randomly select one of the search agents as a leader

20:                update the position of the i-th agent using:

21:                $X\left(Iter\right) = X_{rand} - A \cdot \mid C \cdot X_{rand} - X\left(Iter - 1\right) \mid$

22:             **end if**

23:          **else**

24:             update the position of i-th agent using:

25:             $X\left(Iter\right) = D' \cdot exp\left(bl\right) \cdot \cos\left(2\pi l\right) + X^*\left(Iter - 1\right)$

26:          **end if**

27:      **end for**

28:      calculate $OF(.)$

29:      update best-found solution

30: **end while**

31: post-processing the results

24.2.2   Description of the WOA

A step-by-step procedure describing the fundamentals of the WOA is listed in Algorithm 23. In the first step, the optimization problem is defined in the form of a D-dimensional objective function $OF(.)$. Next, boundary constraints are defined for the i-th agent of the j-th variable as $\left[X_{i,j}^{min}, X_{i,j}^{max}\right]$ where $j = \{1, ..., number\,of\,decision\,variables\}$. The necessary parameters of WOA algorithm have been set in the next step such as: number of search agents (SA_no) and maximum number of iterations (MI). In the fourth step, a randomly generated population of SA_{no} search agent has been initialized. A leader has been selected considering two different attitudes: the best-found solution to provide exploitation, and a randomly selected solution to guarantee the exploration. Then, the search agents explore the solution space around the leader agent. In step 5, the first attitude, using the best-found solution as a leader, has been tackled. Then, the current iteration counter is reset to zero to start the main loop of WOA algorithm. For each i-th search agent an iterative loop updates its position. In step 10, the value of α has been updated using a reducing trend. Next, two coefficient vectors, A and C, are updated in steps 11 and 12, respectively. Humpback whales get close to their prey following a shrinking circle (step 16 and 19) and spiral-shaped path (step 22), simultaneously. WOA will select one of those paths using a probability of 50%. In this way, a randomly generated p with value of lower than 50% leads the algorithm toward using a shrinking circle (as shown in step 14). Now, we check the absolute value of previously determined A in step 15. If this condition is met with values of less than a unit, the best found solution will be utilized as a leader to update the position via a shrinking circle. Otherwise, an agent is selected randomly and the position of the i-th agent is updated in steps 18 and 19, respectively. In step 22, the position of the i-th agent is updated following a spiral-shaped path in case the probability p is more than or equal to 50%. In step 25, we estimate the $OF(.)$ for all the search agents. Finally, the global best solution is updated. Like any other algorithm the best found solution and its relevant vector are proposed as the final result when the termination criteria are satisfied.

24.3   Source-code of the WOA in Matlab

Listing 24.1 shows the source-code for the objective function defined in the WOA. In the objective function $OF(.)$, the input parameters are all search agents (SA). The result of $OF(.)$ function is an N-dimensional column vector with the objective function values for each search agent SA_i for the whole SA population. Equation (24.1) gives the objective function for this example. Listing 24.2 shows the Matlab source-code for the WOA.

$$OF(SA_i) = \sum_{j=1}^{D} SA_{i,j}^2; where - 5.12 < SA_{i,j} < 5.12 \qquad (24.1)$$

```
1   function [output]=OF(SA)
2   [x,y]=size(SA);
3   output=zeros(x,1);
4   for i=1:x
5       for j=1:y
6         output(i,1)=output(i,1)+SA(i,j)^2;
7       end
8   end
```

Listing 24.1
Definition of objective function $OF(.)$ in Matlab.

```
1  % parameter setting for WOA algorithm
2  SearchAgents_no=6; Max_iter=1; dim=5;
3  % Bound constraint definition
4  lb=-5.12*ones(1,dim); ub=5.12*ones(1,dim);
5  % initialize position vector and score for the leader
6  Leader_pos=zeros(1,dim); Leader_score=inf;
7  %Initialize the positions of search agents
```

```matlab
 8  for i=1:dim
 9      Positions(:,i)=rand(SearchAgents_no,1).*(ub(i)-lb(i))+lb(i);
10  end
11  t=0;% Loop counter
12  % Main loop
13  while t<Max_iter
14      for i=1:size(Positions,1)
15  % Return back the search agents that go beyond the boundaries of the
        search space
16          Flag4ub=Positions(i,:)>ub; Flag4lb=Positions(i,:)<lb;
            Positions(i,:)=(Positions(i,:).*(~(Flag4ub+Flag4lb)))+ub.*Flag4ub+
            lb.*Flag4lb;
17  % Calculate objective function for each search agent
18          fitness=fobj(Positions(i,:));
19  % Update the leader
20          if fitness<Leader_score
21              Leader_score=fitness;
22              Leader_pos=Positions(i,:);
23          end
24      end
25  % a decreases linearly from 2 to 0 (Step 10 in Algorithm 1)
26      a=2-t*((2)/Max_iter);
27  % a2 linearly decreases from -1 to - b2 to calculate l in step 22
28      a2=-1+t*((-1)/Max_iter);
29  % Update the Position of search agents
30      for i=1:size(Positions,1)
31  % Step 11 in Algorithm 1
32          A=2*a*rand()-a;
33  % Step 12 in Algorithm 1
34          C=2*rand();
35          b=1;              %  parameter used in step 22 in Algorithm 1
36          l=(a2-1)*rand+1;% parameter used in step 22 in Algorithm 1
37          p = rand();       %  define p as a random number between 0 and 1
        (step 13 in Algorithm 1)
38          for j=1:size(Positions,2)
39  % Shrinking encircling mechanism
40              if p<0.5
41                  if abs(A)>=1
42  % select one of the search agents as a leader randomly
43                      rand_leader_index=floor(SearchAgents_no*rand()+1);
44                      X_rand=Positions(rand_leader_index,:);
45  % Update i-th agent based on step 19 in Algorithm 1
46                      D_X_rand=abs(C*X_rand(j)-Positions(i,j));
47                      Positions(i,j)=X_rand(j)-A*D_X_rand;
48                  elseif abs(A)<1
49  % Update i-th agent based on the best solution as a leader (step 16 in
        Algorithm 1)
50                      D_Leader=abs(C*Leader_pos(j)-Positions(i,j));
51                      Positions(i,j)=Leader_pos(j)-A*D_Leader;
52                  end
53              elseif p>=0.5
54  % Spiral updating position(step 22 in Algorithm 1)
55                  distance2Leader=abs(Leader_pos(j)-Positions(i,j)); %
        distance between the whale and prey
56                  Positions(i,j)=distance2Leader*exp(b.*l).*cos(l.*2*pi)
        +Leader_pos(j);
57              end
58          end
59      end
60      t=t+1;
61      Convergence_curve(t)=Leader_score;
62      [t Leader_score];
63  end
```

Listing 24.2
Source-code of the WOA in Matlab.

24.4   Source-code of the WOA in C++

The C++ source-code for the objective function and the WOA are presented
in Listing 24.3 and Listing 24.4, respectively.

```cpp
#include <iostream>
using namespace std;
/* Function Definitions */
double OF(double SA[], int size_array)
{
    double output;
    int j;
    output = 0.0;
    for (j = 0; j < size_array; j++) {
        output += SA[j] * SA[j];
    }
    return output;
}
```

Listing 24.3

Definition of objective function $OF(.)$ in C++.

```cpp
/* Include files */
#include <iostream>
#include <stdlib.h>
#include <math.h>
using namespace std;
/* Function Definitions */
WOA(int SearchAgents_no, int Max_iter, int dim, double lb[], double ub
    [])
{
double leader_pos[dim]; double leader_score; double fitness;
double Positions[SearchAgents_no][dim]; double fitness[1][
    SearchAgents_no]; double convergence[1][SearchAgents_no];
for (int i = 0; i < SearchAgents_no; i++) {
    leader_pos[i] = { 0 }; leader_score = {1.79769e+308 };    /* change
    this to -inf for maximization problems */
    srand(time(0));
    for (int j = 0; j < dim; j++) {
        double r = (rand() % 10000) / 10000;
        Positions[i][j] = (ub[j] - lb[j])*r + lb[j];
    }
}
double t = 0; /* Loop counter */
double Convergence_curve[1][Max_iter];
/* Main loop */
while (t < Max_iter) {
    for (int i = 0; i < SearchAgents_no; i++) {
        /* Return back the search agents that go beyond the boundaries
        of the search space */
        for (int j = 0; j < dim; j++) {
            if (Positions[i][j] > ub[j]) Positions[i][j] = ub[j];
            if (Positions[i][j] < lb[j]) Positions[i][j] = lb[j];
        }
        fitness = OF(Positions[i], dim);
        if (fitness < leader_score) {
            leader_score = fitness;
            for (int k = 0; k < dim; k++) leader_pos[k] = fitness[k];
        }

    }
    a = 2.0 - t*((2.0) / Max_iter);  /* a decreases linearly fron 2 to
    0 */
```

```
37    a2 = -1.0 + t*((-1.0) / Max_iter); /* a2 linearly decreases from
      -1 to -2 */
38    /* Update the Position of search agents */
39    for (i = 0; i < (int)SearchAgents_no; i++) {
40        srand(time(0));
41        double r1 = (rand() % 10000) / 10000;
42        double r2 = (rand() % 10000) / 10000;
43        double A = 2.0 * a * r1 - a;
44        double C = 2.0 * r2;
45        double l = (a2 - 1.0) * ((rand() % 10000) / 10000) + 1.0;
46        double p = (rand() % 10000) / 10000;
47        for (int j = 0; j < dim; j++) {
48            if (p < 0.5) {
49                if (abs(A) >= 1) {
50                    int rand_leader_index = floor(searchAgents_no*((rand() %
      10000) / 10000) + 1);
51                    for (int k = 0; K < dim; k++) double X_rand = Positions[
      rand_leader_index][k];
52                    double D_X_rand = abs(C*X_rand[j] - Positions[i][j]);
53                    Positions[i][j] = X_rand[j] - A*D_X_rand;
54                }
55                else if (abs(A) < 1) {
56                    double D_leader = abs(C*leader_pos[j] - Positions[i][j]);
57                    Positions[i][j] = leader_pos[j] - A*D_leader;
58                }
59            }
60            else if (p >= 0.5) {
61                double distance2LEader = abs(leader_pos[j] - Positions[i][j
      ]);
62                Positions[i][j] = distance2leader*exp(b*l)*cos(2 * l* M_PI)
      + leader_pos[j];
63            }
64        }
65    }
66    t++;
67    Convergence_curve[1][t] = leader_score;}
```

Listing 24.4
Source-code of the WOA in C++.

24.5 A step-by-step numerical example of WOA

In this section, a detailed computational process is provided for the WOA using the objective function defined in Equation 24.1. In this optimization example, there are 5 design variables and 6 search agents. In the first step, the position of the leader is set to be zero and its objective value is set to an initial value (infinity for a minimization problem). The leader search agent is updated by the best solution found in the population. In the second step, an initial population is developed by randomly generating values for the design variables within the permitted solution domain. In this example, the initial population is:

$SA_1 = \{-0.0597, 2.0352, 4.1444, -3.2469, 4.9017\}$
$SA_2 = \{2.8575, -3.0944, 1.1250, -2.6631, 2.1780\}$

$SA_3 = \{2.2020, -4.8073, 1.2049, 3.9579, 0.0048\}$
$SA_4 = \{4.1341, 2.4993, 3.6807, -4.8264, -0.2961\}$
$SA_5 = \{4.0030, 0.0002, 3.1282, -0.1034, -4.5095\}$
$SA_6 = \{-1.6982, -0.2056, 0.7856, -3.4004, 1.8634\}$

In the third step, the main loop of the algorithm starts and continues until termination criteria are satisfied. Within this loop, the objective values for the population of search agents are evaluated as:

$OF(SA_1) = 55.8899$
$OF(SA_2) = 30.8422$
$OF(SA_3) = 45.0751$
$OF(SA_4) = 60.2664$
$OF(SA_5) = 46.1564$
$OF(SA_6) = 18.5784$

In the next step, the position and objective function value of the leader search agent is updated based on the best-found solution. In this example, the new leader is SA_6:

$Leader\_position = \{-1.6982, -0.2056, 0.7856, -3.4004, 1.8634\}$
$Leader\_score = 18.5784$

Now, values of α (step 10 in Algorithm 23) and α_2 (for calculating l in line 36 of Listing 24.4) are updated on a linearly decreasing pattern from 2 to 0 and from -1 to -2, respectively.

$\alpha = 2$ (line 25 in Listing 24.4); α_2=-1 (line 28 in Listing 24.4).

In the next step, three possible movements can be considered for all the search agents via the $'for'$ loop (line 38 in Listing 24.4). Within this loop, in each iteration two different random numbers are produced between 0 and 1 to calculate A and C as follows:

$r_1 = 0.0424 \rightarrow A = 2\alpha \cdot r_1 - \alpha$
$A = \{-1.8303, 0.0744, -0.1848, -1.4673, 1.2435, -0.3328\}$

$r_1 = 0.0714 \rightarrow A = 2 \cdot r_2$
$C = \{0.1429, 1.9459, 0.8648, 0.3468, 0.1209, 1.3137\}$

Next, determine the spiral updating position l, a random number in $[-1, 1]$.

$l = (\alpha_2 - 1) \times rand + 1$
$l = \{-0.0433, -0.2980, -0.6506, 0.2181, 0.2015, -0.2559\}$

An important fact about humpback whale behaviors is the way of they approach their prey either by moving in a shrinking circle or spiral path. The WOA decides between these motions using a threshold probability of 50%. To this end, a random number p is generated between 0 and 1 (step 13 in Algorithm 23).

$$p = \{0.0967, 0.8003, 0.0835, 0.8314, 0.5269, 0.2920\}$$

In this example, it can be seen that for the first, third, and sixth search agents the shrinking encircling mechanism is used and the others are updated by the spiral updating position method. For the first search agent, the p value is less than 0.5 and $|A| > 1$. Hence, the shrinking circle method based on the randomly selected leader is used to update its position. To this end, $rand\_leader\_index$ in line 43 of Listing 24.4 proposes the 5-th search agent as a leader. In this case, the position of the search agent is updated using:

$$X(t+1) = X_{leader} - A \cdot D \text{ where } D = |C \cdot X_{leader} - X(t)|$$

In this example, these values are:

$$D = \{0.6317, 2.0351, 3.6974, 2.7829, 4.9440\}$$
$$SA_1 = \{5.1592, 3.7251, 9.8954, 1.8466, 8.7528\}$$

For the second search agent, the value of p is 0.8003 and is greater than 0.5. Therefore, the spiral updating position strategy is selected for the particle's movement. For this agent, the new position is obtained using:

$$X() = D' \cdot e^{bl} \cdot \cos(2\pi l) + X^*(t)$$

The updated position is computed as:

$$D' = \{4.5557, 2.8888, 0.3394, 0.7373, 0.3146\}$$
$$SA_2 = \{-2.7023, -0.8424, 0.7108, -3.5629, 1.7940\}$$

For the third search agent, $p = 0.0835$ and $|A| < 1$, so the shrinking encircling mechanism is utilized for updating the position. As a result, the third search agent uses the following information for updating its position:

$$D = \{3.6705, 4.6295, 0.5255, 6.8985, 1.6066\}$$
$$SA_3 = \{-1.0198, 0.6500, 0.8827, -2.1255, 2.1603\}$$

The p value for the fourth search agent is more than 0.5, therefore, a spiral updating strategy is uses with the following result:

$$D' = \{5.8323, 2.7049, 2.8951, 1.4260, 2.1594\}$$
$$SA_4 = \{-0.2551, 0.4637, 1.5020, -3.0476, 2.3977\}$$

For the fifth search agent, $p = 0.5269$ and the spiral updating position strategy is used with the following results:

$$D' = \{5.7012, 0.2058, 2.3426, 3.2970, 6.3729\}$$
$$SA_5 = \{0.3949, -0.1300, 1.6457, -2.1900, 4.2031\}$$

For the last search agent, $p = 0.2920$ and $|A| < 1$, so the shrinking encircling approach is used by the best solution leader to update the individual's position:

$$D = \{0.5327, 0.0645, 0.2465, 1.0668, 0.5846\}$$
$$SA_6 = \{-1.5209, -0.1841, 0.8677, -3.0454, 2.0579\}$$

Next, the boundary conditions are checked and every violated individual is pushed back into the permitted solution domain.

$$Valid\_SA_1 = \{5.1200, 3.7251, 5.1200, 1.8466, 5.1200\}$$
$$Valid\_SA_2 = \{-2.7023, -0.8424, 0.7108, -3.5629, 1.7940\}$$
$$Valid\_SA_3 = \{-1.0198, 0.6500, 0.8827, -2.1255, 2.1603\}$$
$$Valid\_SA_4 = \{-0.2551, 0.4637, 1.5020, -3.0476, 2.3977\}$$
$$Valid\_SA_5 = \{0.3949, -0.1300, 1.6457, -2.1900, 4.2031\}$$
$$Valid\_SA_6 = \{-1.5209, -0.1841, 0.8677, -3.0454, 2.0579\}$$

Next, the objective function values are computed for each search agent as:

$$OF(Valid\_SA_1) = 95.9295$$
$$OF(Valid\_SA_2) = 24.4300$$
$$OF(Valid\_SA_3) = 11.4263$$
$$OF(Valid\_SA_4) = 17.5729$$
$$OF(Valid\_SA_5) = 25.3433$$
$$OF(Valid\_SA_6) = 25.3433$$

In the final step, the leader's objective function value and position are updated based on the best-found solution.

$$Leader\_position = \{-1.0198, 0.6500, 0.8827, -2.1255, 2.1603\}$$
$$Leader\_score = 11.4263$$

24.6   Conclusions

In this chapter, the fundamentals of a WOA are described and examined. In support of this effort, pseudo-code for a WOA is presented to help to demonstrate how the algorithm works. Source-codes for a WOA are provided in both Matlab and C++ to help readers comprehend the fundamentals of the WOA. Finally, a step-by-step numerical example analysis is presented to help explain the mechanism of the WOA.

References

1. S. Mirjalili, A. Lewis, A. "The whale optimization algorithm". *Advances in Engineering Software*, vol. 95, pp. 51-67, 2016.

2. J.A. Goldbogen, A.S. Friedlaender, J. Calambokidis, M.F. Mckenna, M. Simon, D.P. Nowacek. "Integrative approaches to the study of baleen whale diving behavior, feeding performance, and foraging ecology". *BioScience*, vol. 63(2), pp. 90-100, 2013.

3. H.M. Hasanien. "Performance improvement of photovoltaic power systems using an optimal control strategy based on whale optimization algorithm". *Electric Power Systems Research*, vol. 157, pp. 168-176, 2018.

4. M. Mafarja, S. Mirjalili. "Whale optimization approaches for wrapper feature selection". *Applied Soft Computing*, vol. 62, pp. 441-453, 2018.

5. K. ben oualid Medani, S. Sayah, A. Bekrar. "Whale optimization algorithm based optimal reactive power dispatch: A case study of the Algerian power system". *Electric Power Systems Research*, vol. 163, pp. 696-705, 2018.

6. Y. Ling, Y. Zhou, Q. Luo. "Lévy Flight Trajectory-Based Whale Optimization Algorithm for Global Optimization". *IEEE Access*, vol. 5(99), pp. 6168-6186, 2017.

7. D. Rewadkar, D. Doye. "Adaptive-ARW: Adaptive autoregressive whale optimization algorithm for traffic-aware routing in urban VANET". *International Journal of Computer Sciences and Engineering*, vol. 6(2), pp. 303-312, 2018.

8. Y. Sun, X. Wang, Y. Chen, Z. Liu. "A modified whale optimization algorithm for large-scale global optimization problems". *Expert Systems with Applications*, vol. 114, pp. 563-577, 2018.

9. W.Z. Sun, J.S. Wang. "Elman Neural Network Soft-Sensor Model of Conversion Velocity in Polymerization Process Optimized by Chaos Whale Optimization Algorithm". *IEEE Access*, vol. 5, pp. 13062-13076, 2017.

10. M.A. El Aziz, A.A. Ewees, A.E. Hassanien, M. Mudhsh, S. Xiong. "Multi-objective Whale Optimization Algorithm for Multi-level Thresholding Segmentation". *Advances in Soft Computing and Machine Learning in Image Processing*, Springer, pp. 23-39, 2018.

Index

Printed and bound by CPI Group (UK) Ltd, Croydon, CR0 4YY

21/10/2024

01777049-0016